Recent Developments in Clustering and Data Analysis

Proceedings of the Japanese-French Scientific Seminar
March 24–26, 1987

Recent Developments in Clustering and Data Analysis

Proceedings of the Japanese-French Scientific Seminar March 24–26, 1987

Edited by

Edwin Diday
INRIA
Domaine de Voluceau
Le Chesnay Cedex
France

Michel Jambu
Centre National d'Etudes des Télécommunications
Issy Les Moulineaux
France

Chikio Hayashi
University of the Air
Wakaba, Chiba
Japan

Noboru Ohsumi
The Institute of Statistical Mathematics
Minato-ku, Tokyo
Japan

ACADEMIC PRESS, INC.
Harcourt Brace Jovanovich, Publishers
Boston San Diego New York
Berkeley London Sydney
Tokyo Toronto

ACADEMIC PRESS, INC.
1250 Sixth Avenue, San Diego, CA 92101

United Kingdom Edition published by
ACADEMIC PRESS INC. (LONDON) LTD.
24-28 Oval Road, London NW1 7DX

Library of Congress Cataloging-in-Publication Data

Japanese-French Scientific Seminar (1987 : Institute of Statistical Mathematics)
Recent developments in clustering and data analysis : proceedings of the Japanese-French Scientific Seminar / edited by Edwin Diday . . . [et al.].
p. cm.
Seminar sponsored by JSPS (Japan Society for the Promotion of Science and CNRS (Centre national de la recherche scientifique) held 24–26 March 1987, the Institute of Statistical Mathematics, Tokyo.
Bibliography: p.
ISBN 0-12-215485-1
1. Cluster analysis—Congresses. 2. Multivariate analysis—Congresses. I. Diday, E. II. Nihon Gakujutsu Shinkōkai. III. Centre national de la recherche scientifique (France). IV. Title.
QA278.J37 1987
519.5′3—dc19 88-6167
CIP

88 89 90 91 9 8 7 6 5 4 3 2 1
Printed in the United States of America

Contents

Contributors

Numbers in parentheses refer to the pages on which the authors' contributions begin.

Hideki Asoh (97), *Electrotechnical Laboratory, 1-1-4 Umezono, Sakura-mura, Niihari-gun, Ibaraki, 305 Japan*

Yasumasa Baba (87), *Institute of Statistical Mathematics, 4-6-7, Minami-Azabu, Minato-ku, Tokyo, Japan*

Edwin Diday (119), *INRIA, Domaine de Voluceau, Rocquencourt, B.P. 105, 78153 Le Chesnay Cedex, France and Universite Paris IX-Dauphine, Place de Lattre de Tassigny, 75775 Paris Cedex 16, France*

Brigitte Escofier (307), *INSA, 20 avenue des Buttes de Coesmes, 35043 Rennes Cedex and IRISA, Capus de Beaulieu, 35042 Rennes Cedex, France*

Yves Escoufier (231), *ENSA-INRA-USTL, Unite de Biometrie, 9 place Pierre Viala, 34060 Montpellier Cedex, France*

Masashi Goto (49), *SHIONOGI Kaiseki Center, 1-22-41, Izumicho, Suita City, Osaka, Japan*

Alice Gueguen (413), *CNRS/INSERM U.88, 91 Boulevard de l'Hopital, 75634, Paris Cedex 13, France*

Toshiro Haga (295), *Faculty of Engineering, Science University of Tokyo, 1-3, Kagurazaka, Shinjuku-ku, Tokyo 162, Japan*

Chikio Hayashi (3), *University of the Air, 11 Wakaba-2, Chiba 206, Japan*

Tadashi Imaizumi (171), *St. Paul's (Rikkyo) University, 3 Nishi-Ikebukuro, Toshima-ku, Tokyo Japan*

Tomio Ito (373), *Nihon University, College of Industrial Technology, 1-2-1 Izumicho, Narashino, Chiba 275, Japan*

Shuichi Iwatsubo (267), *The National Center for University Entrance Examination, 19-23 Komaba 2-chome, Meguro-ku, Tokyo, Japan*

Michel Jambu (17), *Centre National d'Etudes des Telecommunications, PAA/TIM/ATL/AD, 38 rue du General Leclerc, 92131 Issy Les Moulineaux, France*

Koya Kishida (373), *Takasaki City University of Economics, 1300 Kaminamie, Takasaki, Gunma 370, Japan*

Tsutomu Komazawa (401), *Institute of Statistical Mathematics, 4-6-7, Minami-Azabu, Minato-ku, Tokyo, Japan*

Takio Kurita (97), *Electrotechnical Laboratory, 1-1-4 Umezono, Sakura-mura, Niihari-gun, Ibaraki, 305 Japan*

Ludovic Lebart (331), *Centre de Recherche pour l'Etude et l'Observation des Conditions de Vie (CREDOC), 140 rue du Chevaleret, 75013, Paris, France*

I.C. Lerman (187), *IRISA, Campus de Beaulieu, 35042 Rennes Cedex, France*

Guy der Megreditchian (381), *Direction de la Météorologie Nationale, 2 avenue Rapp, 75007, Paris, France*

Akihiko Miyake (425), *Nihon Medical School, Division of Mathematics, 2-297-2, Kosugi, Nakahara-ku, Kawasaki, Japan*

Alain Morineau (67), *CEPREMAP, 142 rue du Chevaleret, 75013 Paris, France*

Jean-Pierre Nakache (413), *CNRS/INSERM U.88, 91 Boulevard de l'Hopital, 75634, Paris Cedex 13, France*

Satoki P. Ninomiya (437), *College of Science and Engineering, Aoyama Gakuin University, 6-16-1, Chitosedai, Setagaya-ku, Tokyo 157, Japan*

Yasuo Ohashi (137), *University Hospital Computer Center, University of Tokyo Hospital, 7-3-1, Hongo, Bunkyo-ku, Tokyo, Japan*

Noboru Ohsumi (201), *The Institute of Statistical Mathematics, 4-6-7, Minami-Azabu, Minato-ku, Tokyo, Japan*

Nobuyuki Otsu (97), *Electrotechnical Laboratory, 1-1-4 Umezono, Sakura-mura, Niihari-gun, Ibaraki, 305 Japan*

M. Roux (151), *Centre d'Etudes Phytosociologiques et Ecologiques (CNRS), B.P. 5051, 34033 Montpellier Cedex, France*

Shidou Sai (103), *Okayama College of Commerce, 2-10-1, Tsushimakyou-machi, Okayama 700, Japan*

Seiroku Sakai (179), *Institute of Biology and Life Sciences, Daito Bunka University, No. 560, Iwadono-Nagasaka, Higashi-Matsuyama, Saitama, Japan 355*

Yves Schektman (37), *Greco-CNRS 59, Universite Paul Sabatier, 118 route de Narbonne, 31062 Toulouse Cedex, France*

Yoshio Sugasawa (373), *Nihon University, College of Industrial Technology, 1-2-1 Izumicho, Narashino, Chiba 275, Japan*

Meiko Sugiyama (343), *NHK Broadcasting Culture Research Institute, 2-1-1, Atago, Minato-ku, Tokyo 105, Japan*

Shigeru Suzuki (223), *National Institute of Agrobiological Resources, Yatabe, Tsukuba, Ibaraki-ken, 305 Japan*

Masaaki Taguri (103), *Chiba University, Faculty of Science, 1-33, Yayoi-cho, Chiba 260, Japan*

Yutaka Tanaka (281), *Okayama University, 2-1-1, Tsushima-naka, Okayama 700, Japan*
Toshiro Tango (109), *Division of Clinical Epidemiology, The Tokyo Metropolitan Institute of Medical Science, 3-18-22, Honkomagome, Bunkyo-ku, Tokyo 113, Japan*
Tomoyuki Tarumi (281), *Okayama University, 2-1-1, Tsushima-naka, Okayama 700, Japan*
Takenobu Tasaki (49), *SHIONOGI Kaiseki Center, 1-22-41, Izumicho, Suita City, Osaka, Japan*
Shoichi Ueda (323), *Ryukoku University, Fushimi-ku, Kyoto, Japan*
Kazumasa Wakimoto (79), *Okayama University, 2-1-1, Tsushima-Naka, Okayama 700, Japan*
Keiji Yajima (445), *Institute of Japanese Union of Scientists and Engineers, 4-3-30, Sendagaya, Shibuya-ku, Tokyo 151, Japan*
Haruo Yanai (259), *The National Center for University Entrance Examination, 19-23 Komaba 2-chome, Meguro-ku, Tokyo, Japan*
Tadashi Yoshizawa (241), *Yamanashi University, 4-3-11, Takeda, Kofu, Japan*

Preface

That data analysis, particularly multidimensional data analysis, is a useful tool has become commonly accepted. However, the term "data analysis" has also come to have a number of meanings. For example, the corresponding term in French, *analyse de données,* has a completely different origin from the English term, data analysis. Even in English, data analysis can refer to two considerably different viewpoints: exploratory data analysis and data handling. For those who consider data analysis to consist of data handling, a viewpoint common among researchers in classical mathematical statistics, data are "handled" in order to apply existing theories. However, in our opinion, the result of this approach to data analysis has been an increasing divergence between theory and practice, a situation that is not particularly productive.

In the view of the editors, the term data analysis refers to thinking about approaches to analyzing events using data. In this view, the emphasis is upon how to collect the data and how to analyze it in order to extract the information it contains and thus clarify the events being studied. In this view, theory and practice become integrated and both theory and application develop from the data itself, with the result that applications of data analysis broaden and deepen. Viewed externally, theory and practice are fully interpenetrating. Under this definition of data analysis, methodological results include all stages up to the intertwined results, including, of course, the development and application of related computer software. Thus, we consider the term data analysis to include all of the concepts defined by the French term *analyse de données,* the English term data analysis, and the native Japanese term *deta kaiseki.*

This book brings together the results of clustering and multidimensional data analysis research conducted primarily in France and Japan. It is based upon papers given at the Japanese–French Scientific Seminar "Recent Developments in Clustering and Data Analysis" conducted on March 24–26, 1987, in Tokyo, Japan.

Most of the data analysis methods discussed in this book originated in France and Japan and have been developed through a close exchange of theory and practice between the two countries. In recent years, these methods have produced highly useful results in West Germany, the Netherlands, Spain, Italy, Africa, Eastern Europe, South America, and many other countries around the globe. Surprisingly, the theory and application of these methods is almost unknown in the English-speaking countries. At present, only two or three related books have been published in English.

As the papers will show, data analysis in Japan, with its emphasis on the importance of the data itself and on the "informatics" of the data, has much in common with the philosophy and methodology of *analyse de données* as it has developed in France. In addition, we may say that, in its approaches to data analysis, it shows a psychological kinship with the English term exploratory data analysis. In any case, this meaning of data analysis had its birth outside of the path of development of what is commonly called mathematical statistics. We can see that it developed with a reconsideration of the direction of developments in statistical mathematics as its driving force. It was not surprising then that the strong similarities in the approaches to data analysis in both France and Japan led to increasing calls for exchanges of information between the two countries, especially in the fields of clustering and multidimensional data analysis. Thus the Japanese–French Scientific Seminar was opened to create an opportunity for researchers from both countries to exchange the latest research results and to examine the application and effectiveness of data analysis as a supporting tool in the clarification of events in many scientific fields. The result was a stimulating exchange that demonstrated that the overall development of research results, both theoretical and practical, were as expected.

The contents of this Seminar are being published here for two reasons: first, to clarify recent developments in these data analysis methods, while also leaving a record of their foundation and development in English, where they are not well known; and second, to present the results of various classification techniques based on this philosophy, thus demonstrating their importance and practicality. Researchers and workers in fields from the behavioral sciences, biological sciences, medicine, and pattern recognition to the various industrial and agricultural sciences will find the clustering and multidimensional scaling methods described in this book of immense value.

Unfortunately, this book cannot include the various demonstrations of data analysis software that were presented at the Seminar. Software products such as these are essential to the practical application and extension of the clustering and multidimensional data analysis techniques described in this book and will certainly increase in importance as these analysis techniques are more widely accepted.

The editors wish to thank the Japan Association for the Promotion of Science (JAPS) and the Centre National de la Recherche Scientifique (CNRS) for funding this seminar. In addition, we would like to express our deep appreciation to the following organizations for their support and assistance: in Japan, the Institute of Statistical Mathematics (ISM), the Institute of Arteriosclerosis, the Health and Welfare Association, the Japan Marketing Association, and the Mitsubishi Foundation; in France, the Centre International de Statistique et d'Information Appliquées (CISIA), the Centre International d'Etudes des Télécommunications (CNET), and the Institut National de Recherche en Informatique et en Automatique (INRIA).

The Editors

Section 1

Data Analysis Techniques and Related Topics with Statistical Software

NEW DEVELOPMENTS IN MULTIDIMENSIONAL DATA ANALYSIS

Chikio Hayashi

University of the Air
11, Wakaba - 2, Chiba, Japan

I. INTRODUCTION

It is well-known that data analysis yields misleading results in complicated situations if multidimensional considerations are not taken into account. The importance of ideas for exploring phenomena in question is also emphasized. For this purpose, the following three phases in data analysis must be kept sight of: how to design the data, how to collect the data, and how to analyze the data. Of course, each of these entails the 'logic, methodology and philosophy', which I call the fundamental concept, of data analysis together with relevant concrete methods and theories (8).

Basic information is usually expressed as both quantitative and qualitative data. These multidimensional data are processed through various multidimensional methods under the guiding concepts of data analysis embodied in the three phases mentioned above to obtain useful information with validity. Needless to say, the properties inherent in the data must be taken into consideration in these three phases together with the background surrounding the data.

Here, apart from designing and collecting data, the quantification of qualitative data will be discussed as a method of analyzing statistically multidimensional data. In this case, the fundamental concept of data analysis plays a

ISBN 0-12-215485-1

particularly important role and provides the guiding concepts for both the development of method, or theory, and the design of computer software for data processing (8). These methods of quantifying qualitative data have been under development since 1948 in response to the demands of actual problem-solving. The theoretical system of quantifying qualitative data has evolved out of experience with 'data' itself.

Therefore, the wisest approach may be to explain heuristically in what situation the method was developed, when the past, present and future of quantification are discussed. The guiding concepts underlying quantification will be revealed through such an explanation, and an outline of the overall shape of the theoretical system of quantification will appear through the fog surrounding multidimensional data analyses.

II. BIRTH OF QUANTIFICATION OF QUALITATIVE DATA

The following examples (A, B, C) are found in the early stages of the theory of quantification of qualitative data.

A. Parole Prediction

There is a regime that allows a criminal who has finished more than one-third of his term to be paroled if he is fairly well educated and it is recognized as the result of investigation of his career and behaviour in prison that he will not commit a crime again in society.

We proceed in maximizing the precision of our prediction, that is, the success rate of prediction (in other words, the rate that the prediction "he will be good or bad in prognosis" turns out to be true) by quantifying the qualitative results of our investigations. It is at this point that the problem of quantifying the qualitative data arises. The contents of such data consist of items related to factors that are thought to be useful based on various analyses of past data in prediction of a criminal's prognosis in the society, for example, physical and hereditary evidence, character, family life, occupation, environment, financial situation, etc. However, human relations are much too complicated to be judged merely by these rather superficial factors. They must be contemplated from a higher point of view, and we have adopted some of the factors emerging from these considerations. For example, these include the growth pattern of a criminal from childhood to the present, his behaviour in prison, his psychological state at the time he committed the offence and the condition of the victims, his feelings in prison, desires, interest, attitude towards the society etc.

These factors are essentially qualitative. By studying these factors it is possible to ascertain the reactions of a criminal with respect to every item. These results are expressed in terms of a checklist of items in each category. In today's terminology, qualitative data were converted into categorical data. In the conventional treatment, categories were arbitrarily quantified in advance. For example, integers 1, 2, 3, 4 & 5 were assigned to each of the categories of an item, based on intuitive judgement. An ordinary linear discriminant analysis of these quantities was performed. This arbitrary quantification was extremely questionable for me, because the results differed depending upon how the numerical values were assigned to the categories. In particular, it was doubtful that a linear relation held in categories within an item and that equal spacing existed in them. Thus, the idea that quantification of categories is desirable only for our specific purpose (in this case, maximization of discriminant power) was born.

Quantifying these cagegories, we predicted the criminal's prognosis. The problem was to quantify these categories, and to synthesize the quantities obtained, to maximize the success rate of prediction. The success rate was calculated as a probability from the collective theory using the concept of statistical decision based on the Bayesian distribution concerning the occurrence of events. In order to maximize the probability of success in parole prediction, the categories in each item taken up must be quantified. This was mathematically possible (4).

Based on this idea, methods were developed for quantifying categories of items representing the above prediction factors in order to estimate literacy scores by several factors (5), and estimate the total attitude toward French or American culture using component questions (5) & (6).

B. Quantification in Sociometric Data

A matrix representation of human interaction patterns is used in a sociometric survey. e_{ij} means the attitude that the i-th person has toward the j-th person and expresses degree of nearness or intimacy between i and j. Let e_{ij} be expressed as a numerical value, even though it may be not represent a metrical value in the true sense.

Generally speaking, $e_{ij} \neq e_{ji}$, and e_{ii} can not be defined and thus need not be measured. The group structure of the persons involved can then be determined from the matrix representation of mutual human relations and interactions, where, e_{ij} being the larger, the degree of intimacy is higher. Of course, $e_{ij} \gtreqless 0$. For this purpose, the existence of subgroups must be ascertained.

In this case we may say that the sub-groups consist of persons who are intimate with each other. In one-dimensional treatment, it is natural for our purpose to rearrange the rows and columns of the matrix with non-specified diagonal elements, to bring the larger e's near the diagonal of the matrix. Assume that we give a numerical value x_i to the i-th person to analytically determine group structure. We quantify each person from the standpoint that persons with x values close to each other will be intimate with each other, and persons with distant x-values will be alienated from each other. Now, consider metric Q:

$$Q = -\sum_{i}\sum_{j} e_{ij} (x_i - x_j)^2 .$$

If e_{ij} is larger, it is desirable that x_i and x_j be close from the viewpoint of quantification, and if e_{ij} is smaller, that x_i and x_j be distant. This means that x values must be chosen so as to maximize the Q values under the condition that the variance of x equals 1. This metric Q has a very flexible property and is invariant under a linear transformation of e_{ij} with a positive multiplier. This point turns out to be very useful. This quantification of i is easily generalized to a multidimensional case (5), (10). It can be seen that this idea is quite similar to so-called multidimensional scaling (MDS), and gives a similar solution to that of MDS if e's can be assumed to be a rank order. But this was 'a future story'. This method was called e_{ij}-type quantification or the fourth-type of quantification.

C. Designing Labels for a Can of Tinned Food

This problem concerns a classification (based on the evaluation) of labels designed on a can of tinned food by individuals (testees). Various labels designed (objects) are shown to individuals (subjects). Individuals are instructed to choose all labels they like without restriction as to the number of labels chosen.

Here we classify labels depending on each individual's sense of taste. Refined labels are chosen by individuals of good taste and unrefined labels by individual of poor taste. Vice versa, individuals of good taste choose refined labels and individuals of poor taste choose unrefined labels. As formulated, this relationship is tautological and does not provide any solution to the problem. Naming of good-poor and refined-unrefined is only a matter of interpretation. It is essentially wiser to consider simultaneous classification of both objects and subjects. This idea leads to a rearrangement of subjects and objects to bring the choice responses closer to the diagonal of the rectangle. Suppose that we assign a numerical value x_j to the j-th object and y_i to the i-th subject to analytically determine a rearrangement of both objects and subjects. This suggests that x's and y's are quantified so as to maximize the correlation coefficient between x and y in the one-dimensional case. This is equivalent to maximizing the correlation ratio with respect to x's which is equal to the variance ratio, i.e. the between variance with respect to subjects divided by the total variance equals one minus the within variance based on the variance of x's within each subject divided by the total variance. This idea is easily written in an analytical form (7) and generalized to the multidimensional case. In the multidimensional case, a numerical vector is assigned to each object while another numerical vector is assigned to each subject so as to maximize a kind of correlation coefficient in the multidimensional case or a generalized multidimensional correlation ratio under some orthogonal conditions between mutual dimensions (9). This is a representation of spatial correspondence between subjects and chosen objects in each multidimensional space. The analytical realization of this idea is simply a mathematical problem.

Furthermore, this situation can be interpreted as a quantification of objects (subjects) in a matrix form, elements of which are given in terms of frequency of mutual choices or a correlation table between objects (subjects) (9). This is sometimes referred to as quantification of a matrix. This is synonymous with the quantification based on response pattern of subjects, because the matrix form is derived from response patterns of subjects in the process of calculation of x's. The numbers of mutual choices of objects by subjects become the elements of a correlation table for the matrix. The meaning of quantification in each case is quite the same. This suggests a wide range of applicability.

In addition to the problem of label design, this method was applied to various types of data, even in cases where a response pattern in items (reaction pattern of a subject to many objects (question items) in a survey or a test) is obtained as a large data set.

This method was similar to Guttman's scaling theory and equivalent to Benzécri's correspondence analysis. This was called quantification of response pattern or the third method of quantification afterwards.

These ideas in the early stage of quantification in practical problem-solving were based mainly on that of scaling of qualitative data.*

Example A was solved by quantifying the categories within items based on the outside variables. In this sense, it can

---

* At that time, the term "categorical data" was not yet used. Nevertheless, I place importance on the process of expressing qualitative data in terms of categorical data. This does not mean treating given categorical data but treating qualitative data, including how to express it in categorical form.

be said that the quantification of qualitative data grew out of the idea of scaling. Beginning with one-dimensional scaling, multidimensional scaling has naturally developed as a statistico-mathematical idea.

Example B is a problem of rearranging elements in a primitive sense of the one-demensional case. As this is too time-consuming and unwieldy when the number of elements is large, the idea of rearranging the elements using an analytical method was adopted and elements were scaled so as to maximize the appropriate measure. In this case there was no outside variable. It was natural that the rearrangement in a one-dimensional space was generalized to a rearrangement in a multidimensional space by using the concept of multidimensional scaling in a statistico-mathematical sense. This provided the solution to rearranging the elements through a scaling concept.

Example C has no outside variable. Simultaneous rearrangement of both subjects and objects for the purpose involved is the fundamental idea. This was analytically realized by maximizing the correlation ratio by quantifying (scaling) subjects and objects. Multidimensional treatment was done along the same line. The fundamental concept underlying all of these methods of quantification was the optimal scaling of qualitative information whether it is one-dimensional or multidimensional.

This fundamental concept has evolved gradually through the treatment of practical problems to be solved, following the advancement of statistico-mathematical logic, methodology, and philosophy (8).

III. OVERVIEW OF QUANTIFICATION METHODS

The methods developed mainly by the author together with other closely related methods is formally summarized in the following table in which the relationship among various methods is given.

Table. Method of quantification (or scaling) as one type of analysis of categorical data

---

I. Existence of outside variable
   (quantification or scaling of factors for estimating outside variable)
  A. numerical outside variable
    1. unidimensional
       1st method (one type of regression analysis)
    2. multidimensional
       1st method (by means of a vector correlation coefficient)
  B. categorical outside variable
    1. classification into 2 groups
      a. absolute inference
        (1) discrimination based on a measure of correlation ratio
            2nd method (one type of discriminant analysis)
        (2) discrimination based on a measure of success rate of estimation (or prediction)
      b. relative inference
         Guttman's quantification method of categorical factors in case of paired comparison
    2. classification into more than 3 groups
      a. absolute inference
        (1) unidimensional or ordered classification
            scaling by assignment of unidimensional value based on correlation ratio (2nd method mentioned above)
        (2) multidimensional or unordered classification
            (a) scaling by assignment of multidimensional numerical values based on generalized correlation ratio (2nd method mentioned above)
            (b) unidimensional scaling of multi-factors (multidimensional metrical space construction by multi-factors) based on generalized variance

b. relative inference
   (1) by paired comparison
       application of 2nd method or Guttman's method generalized
   (2) by simultaneous many objects comparison (for example ordering of N objects)
       application of 2nd method

II. Non-existence of outside variable
(quantification or scaling of factors for understanding their data structure)

A. data based on response pattern of elements
   1. representation of a degree of mutual dependence between 2 variables
      quantification of categorical variable by maximization of correlation coefficient
   2. quantification in multi-way classification (Iwatsubo)
   3. construction of spatial configuration of data based on relations among more than 3 variables
      3rd method (in the case of those variables being numerical, factor analysis or principal component analysis may be used under some strict conditions), Benzécri's analyse des données
      ⟷ Guttman's Scalogram Analysis, MSA, POSA, etc.

B. data based on relations between (among) elements
   1. numerical representation of similarity or dissimilarity
      a. between 2 elements
         (1) non-metrical treatment
             4th method (e_{ij} - type quantification by use of information of those relations with flexibility)
         (2) metrical treatment (if it is valid)
             K-L type quantification and Torgerson-Gower method
             ⟶ asymmetric case (Young, Chino)
      b. among more than 3 elements
         (1) non-metrical treatment (if it is valid)
             e_{ij} - type quantification (generalization of 4th method)
         (2) metrical treatment (if it is valid)
             (a) Harshman's PARAFAC-model
             (b) Tucker's three mode factor analysis
             (c) Torgerson's metrical multidimensional scaling (MDS)
   2. non-metrical representation
      a. representation of relations between 2 elements by an absolute judgement or criterion

(1) rank ordered representation of similarity or dissimilarity
non-metric MDS
(a) Shepard method
(b) Kruskal method
(c) smallest space analysis, SSA (Guttman, Lingoes, etc.)
(d) individual difference model (Carroll & Chang, Takane, Young, de Leeuw)
(e) asymmetric model (Young, Chino, etc.)
(2) belonging representation of similarity or dissimilarity to rank ordered group
non-metric MDS
Minimum Dimension Analysis (MDA), MDA-OR
(3) nominal classification
MDA-UO

b. representation of relations by a relative judgement
(1) by paired comparison
Hayashi's multidimensional unfolding method
(2) by simultaneous many objects comparison
Coomb's multidimensional unfolding method

After completing ordinary methods of quantifying qualitative data, MDS was advanced along my idea of data analysis. This entailed both MDA-OR (Minimum Dimension Analysis of Ordered Class Belonging) and MDA-UO (MDA of Un-Ordered Class Belonging) (10). This is a variation of e_{ij} - type (the fourth method of quantification as mentioned in II. B). This method can be considered a primitive prototype of MDS. A fool-proof solution to Harshman's PARAFAC-model was found using an idea similar to MDS-OR.

Based on the fourth method or KL-type quantification (10), the arrow and point method (APM) was developed (11). APM provides a representative method of rank order information by segmentation of the sample. For example it can be used for determining the relationship between the groups in question through the synthesis of marginal distributions on various questions shown by those groups. This method was utilized to summarize the great deal of simple information from simple tabulations of the many questions in those groups (12). Thus, APM appears to provide a means of summing up large amounts of simple information considered to be too complicated because of the excessive number of tabulations it would involve.

IV. FUTURE PROBLEMS

Numerous problems remain unsolved even within the scope of data analysis mentioned in this paper. Several of these problems are shown below.

i) Fuzzy quantification corresponding to fuzzy sets. How can a fuzzy data structure be represented in a multidimensional space which is not always Euclidean, but at the same time not so sophisticated that the data structure can not be understood intuitively. This problem has been treated in terms of probabilistic response. However this is only a temporary measure.

ii) How to uncover the data structure based on relational information involving three or more elements $R_{ijk\cdots}$.

iii) How to achieve independence from linear structure. Most methods of quantifying qualitative data entail some forms of statistical manipulation. The statistical methods stand, in principle, on a linear basis and the least squares method, which is efficient when there are linear relations in some sense. A new principle, which is not so sophisticated, must be explored. Generally, linear relations in the above meaning have been avoided by clustering elements. In other words, a method of quantification is applied in each group by clustering based on a number of concepts following an initial clustering of elements.

A more refined approach is needed.

iv) How to explore data structures by developing a procedure for finding both clusters and the data structure in each cluster simultaneously. Conventionally, clustering is done before the data structure is found. These clusters of elements are constructed intuitively or by trial and error in order that clear and distinct data structures can be uncovered, making different features among clusters extremely informative in relation with the information used in formation of clusters. Some kind of clear cut and refined method for executing a simultaneous procedure is desirable.

REFERENCES

1. Benzécri, J.P. et al. (1973). L'analyse des Données 1, 2, Dunod.
2. Guttman, L. (1946). An Approach for Quantifying Paired Comparisons and Rank Order, Annals of Mathematical Statistics, 17, 144-163.
3. Guttman, L. (1950). The Basis for Scalogram Analysis, Measurement and Prediction, (Stouffer, S.A. et al. (eds.)), 60-90.
4. Hayashi, C. (1950). On the Quantification of Qualitative Data from the Mathematico-statistical Point of View, Ann. Inst. Statist. Math., 2, 35-47.
5. Hayashi, C. (1952). On the Prediction of Phenomena from Qualitative Data and the Quantification of Qualitative Data from the Mathematico-statistical Point of View, Ann. Inst. Statist. Math., 3, 69-98.
6. Hayashi, C. (1954). Multidimensional Quantification I, II, Proc. Japan Acad., 30, 61-65, 165-169.

7. Hayashi, C. (1956). Theory and Example of Quantification (11), Proc. Inst. Statist. Math., 4 (2), 19-30 (in Japanese).
8. Hayashi, C. (1973). Methodological Problems in Mass Communications Research—from a Statistico-mathematical Standpoint, Studies of Broadcasting, 9, Nippon Hoso Kyokai, 121-151.
9. Hayashi, C. and Suzuki, T. (1974 & 1975). Quantitative Approach to a Cross-societal Research I and II: a comparative study of Japanese national character, Ann. Inst. Statist. Math., 26, 455-516 and 27, 1-32.
10. Hayashi, C. (1975). Minimum Dimension Analysis: MDA-OR and MDA-UO in Essays in Probability and Statistics, (Ikeda, S. et al. (eds.)), 395-412. Shinko Tsusho Co., Ltd. (1976).
11. Hayashi, C. (1979). Some Statistical Methods in Market Survey, Proceedings of the 42nd Session of the International Statistical Institute, (Manila, 1979), Bulletin of ISI, XLVIII (3), 51-70.
12. Hayashi, C., Hayashi, F., Suzuki, T. Lebart, L. and Kuroda, Y. (1985). Comparative Study of Quality of Life and Multidimensional Data Analysis: Japan, France and Hawaii, Data Analysis and Informatics (INRIA), 573-584.

INTERPRETATION OF SOME DATA ANALYSIS METHODS

Michel Jambu
Centre National d'Etudes des Télécommunications
PAA/TIM/ATL/AD
38 rue du Général Leclerc
92131 ISSY LES MOULINEAUX
FRANCE

I. INTRODUCTION

The scope of the paper is to propose rules of interpretation of correspondence cluster analysis by selecting explained classes and explaining variables involved in the algorithm of hierarchical classification. These rules of selection are based upon the contributions computations introduced a few years ago. The aim is to propose a computer aided correspondence cluster analysis. Graphics allow to highlight the interest of the rules.

II. CORRESPONDENCE CLUSTER ANALYSIS

A. Notations

Let I and J be two finite sets and let k_{IJ} be a data table of positive nombers associated with I and J

$$k_{IJ} = \{k(i,j) \; ; \; i \in I \; ; \; j \in J\}$$

The cloud of points used in correspondence cluster analysis is the following :

$$N_I(J) = \{(f_J^i, f_i) \; ; \; i \in I\} \subset R_J$$

f_J^i is called the profile of the element i on J

RECENT DEVELOPMENTS IN
CLUSTERING AND DATA ANALYSIS

ISBN 0-12-215485-1

The distance used in correspondence cluster analysis is the following :

$$d^2(i,i') = ||f_J^i - f_J^{i'}||^2_{f_J} = \sum_{i\in I}(f_j^i - f_j^{i'})^2/f_j$$

(χ^2 distance with centre f_j)

The correspondence cluster analysis determines a hierarchical classification associated with a correspondence data table and an algorithm derived from the distance used in correspondence analysis.
It is recalled the terminology used in hierarchical classification.

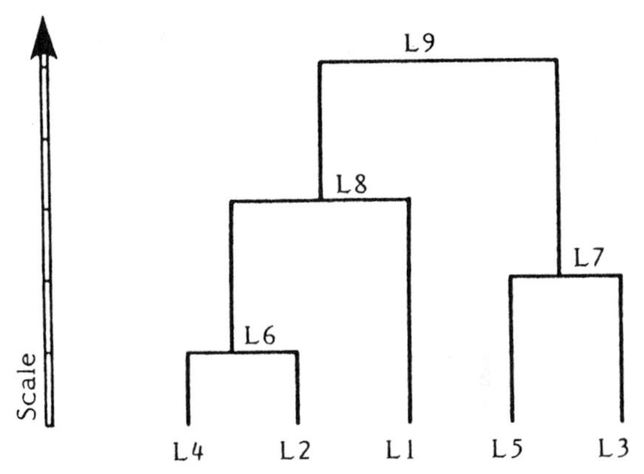

Fig. 1. A hierarchical classification

I = L1, L2, L3, L4, L5 set of basic elements on which the classification is built.

N = {L6, L7, L8, L9} set of nodes (or classes) which have been built

They are L6 = {L4, L2}, L8 = {L1, L2, L4}
L7 = {L5, L3}, L9 = {L1, L2, L3, L4, L5}.

Each node is formed by the union of two preexisting classes ; for each node, denoted by N, (except for the terminals) there are two sub-classes ; one is called the elder of N, denoted by A(N) ; the other is called the younger of N, denoted by B(N) : the couple {A(N), B(N)} is often called a subdivision of N or a dipole N (See Fig. 2).

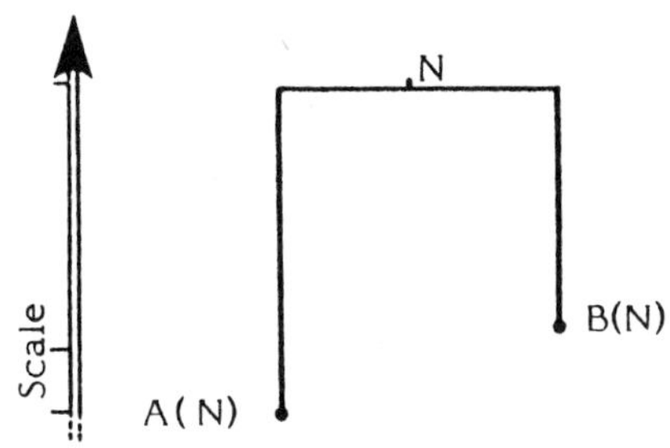

Fig. 2. A dipole N, A(N), B(N)

On the left there is a scale denoted by ν . This is the level measure of the classification. $\nu(N)$ is the value of the level measure of node N. The algorithm performed in correspondence cluster analysis is based upon the χ^2-metric and the decomposition of the variance of the cloud of elements associated with the correspondence data table (cf. Notations). The main result is the following :

$$\forall n \in N \quad \nu(n) = M^2(n) - M^2(A(N)) - M^2(B(N))$$
$$= f_{A(N)} \cdot f_{B(N)} / f_N \; || f_J^{A(n)} - f_J^{B(n)} ||^2_{f_J}$$

where $\quad M^2(n) = \sum_{i \in I} f_i \; || i - n ||^2$

$$M^2(I) = \sum_{i \in I} f_i \; || i - g ||^2$$

$$g(I) = \sum_{i \in I} f_i \; i$$

$||\cdot||^2$ is the χ^2-norm centred by f_J and the points i, n or g are identified by their profiles f_J^i, f_J^n, $f_J^g = f_J$. The algorithm performed is : at each step n, to find n such that $\nu(n)$ is minimum.

B. Decomposition of Variance in Hierarchical Classification

The variance, denoted by $M^2(I)$, can be decomposed according to a set of nodes of a h-classification whatever the algorithm performed to build it.

$$M^2(I) = \nu(I) + M^2(A(I)) + M^2(B(I))$$
$$M^2(A(I)) = \nu(A(I)) + M^2(A(A(I))) + M^2(B(A(I)))$$
$$\vdots \qquad \vdots \qquad \vdots$$

it means that $M^2(I) = \sum_{n\in N} \nu(n)$.

C. Level of the Classification

- $\nu(n)$ represents the part of the variance of the dipole n in the total variance $M^2(I)$
- $\nu(n)$ is the contribution of node n to the total variance $M^2(I)$
- $\tau(n) = \nu(n)/M^2(I)$ is the relative contribution of node n to the total variance $M^2(I)$.

The general shape of the classification depends on the decreasing of the value $\nu(n)$. We can understand a hierarchical classification as a serie of dipoles. The most important ones (in terms of variance) are on the upper levels ; the less ones are on the lower levels).

D. Contributions to the Deviation from the Center

Let consider a class n :

$$f_J^n = \{f_J^n \; ; \; j \in J\}$$

where $f_j^n = f_{nj}/f_n$ with $f_{nj} = \sum_{i \in n} f_{ij}$

and $f_n = \sum_{i \in n} f_i$.

Let consider $\rho^2(n) = ||f_j^n - f_j||^2_{f_J} = \sum_{j \in J} (f_j^n - f_j)^2 / f_j$

$$\rho^2(n) = \sum_{j \in J} \mu(n,j)$$

with $\mu(n,j) = (f_j^n - f_i^2)/f_j$

if $(f_j^n - f_j) \geq 0$ a sign + is assign to $\mu(n,j)$; it means that the contribution $\mu(n,j)$ is given because $f_j^n \geq f_j$ (greater than the average profile f_j)
if $(f_j^n - f_j) \geq 0$ a sign - is assign to $\mu(n,j)$ because $f_j^n \leq f_j$.

From geometric point of view the value $\mu(n,j)/\rho^2(n)$ represents the squared cosine of the angle formed by the axis j and the radius vector of the class n (identified as its center of gravity).

See Figure 3.

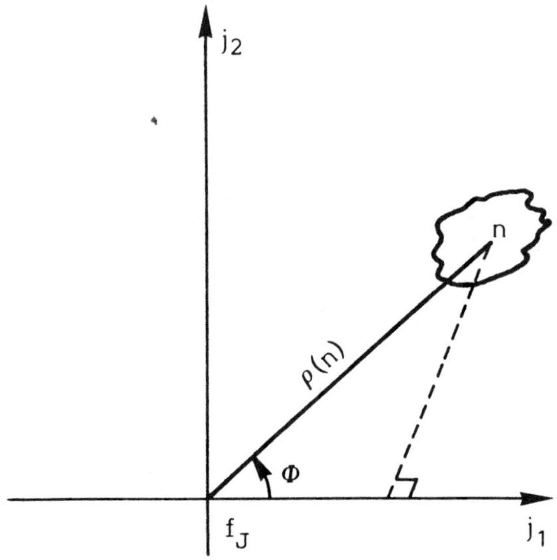

Fig. 3 Geometric representation on signed contributions

$$\mathrm{Cos}^2\phi = \nu(n,j)/\rho^2(n) = ((f_j^n - f_j)^2/f_j/\rho^2(n))$$

E. Contributions to the deviation between two classes

$\nu(n)$ represents the variance of the dipole n built with A(n) and B(n)

$\nu(n)$ can be decomposed as follows

$$\nu(n) = (f_{A(n)} \cdot f_{B(n)}/f_n) \| f_J^{A(n)} - f_J^{B(n)} \|^2 f_J$$

$$= (f_{A(n)} \cdot f_{B(n)}/f_n) \sum_{j \in J} (f_J^{A(n)} - f_J^{B(n)})^2/f_j$$

$$\nu(n) = \sum_{j \in J} \underbrace{(f_{A(n)} f_{B(n)}/f_n)(f_j^{A(n)} - f_j^{B(n)})^2/f_j}_{\nu(n,j)}$$

Since $\sum_{n \in N} \nu(n) = M^2(I)$

We have $M^2(I) = \sum_{n \in N} \sum_{j \in J} \nu(n,j)$.

The total variance is decomposed according to :

- the original axes j of J
- the set of nodes n

$\nu(n,j)$ is the mutual contribution of n and j to the total variance $M^2(I)$.

It is interesting to study the following values :

a) $\nu(n,j) / \sum_{n\in N} \nu(n,j)$

$$\sum_{n\in N} \nu(n,j) = In(j)$$

$\nu(n,j) / \sum_{n\in N} \nu(n,j)$ represents the part of node n in the variance of the axis j of J.

If $\nu(n,j)/In(j)$ close to 1 ; it means that the dispersion on the axis j of J is exclusively due to the points of the class n.

b) $\nu(n,j) / \sum_{n,j} \nu(n,j)$

$\nu(n,j)/M^2(I)$ represents the relative contribution of n and j to the total variance $M^2(I)$.

c) $\nu(n,j)/\nu(n)$ represents the relative contribution of the axis j of J to the separation of n into A(n) and B(n).

If $(\nu(n,j)/\nu(n)) \neq 1$; it means that the separation into two classes takes places exclusively in the direction of the axis j of J.

From geometric point of view $\nu(n,j)/\nu(n)$ represents the squared cosine of the angle formed by the axis j of J and the radius vector of the dipole $\{n, A(n), B(n)\}$

See Figure 4

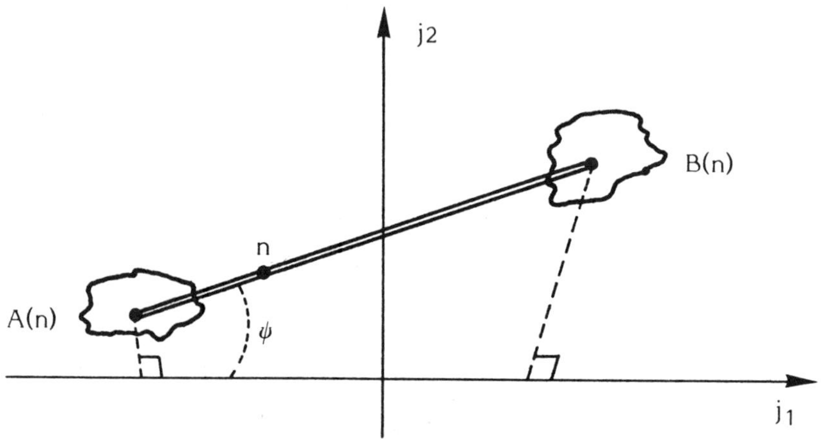

Fig. 4. : Geometric representation of contributions

Let consider the usual representation of the hierarchical clustering on the set I (cf. Fig. 5).
There are only the numbering of nodes, names of elements and level measures. With this graph, it is impossible to qualify the relationships between classes and to determine which elements j of J explain the formation of classes or the subdivisions into classes.

F. Graphics and Contributions Data Tables

With contributions data tables it is possible to improve the understanding of graphics in hierarchical clustering. Let consider the analysis of the semantic field associated with colours.
Values of contributions and explained j of J are put on the graphic.

II. RULES OF INTERPRETATION IN CORRESPONDENCE CLUSTER ANALYSIS

A. Problem

The problem is : how to select the information given in contribution data tables and level measures to have an explained classification ; it means a classification where the information is automatically select and put on the graph by the program. To do so, it is necessary to introduce <u>rules of selection</u> of nodes then rules of selection of explaining variables.

B. Selection of Nodes : First Class Relevant Subdivisions

<u>RULE 1</u> N is chosen such that :

$$\sum_{n=2*\text{CardI}-1,N} \tau(n) \geq 0.8 \text{ (for example)}$$

The given value 0.8 corresponds to a percentage of variance of nodes equal to 80 % of the total variance $M^2(I)$.

<u>RULE 2</u> N is chosen such that :

$\nu(N)/\nu(N+1)$ is maximum

N is chosen such that the difference of two successive levels is greater than the others.

<u>RULE 3</u> N is chosen such that

$\tau(N)$ is $\geq$ (1/CardI) representing the average value of $\tau(n)$.

<u>RULE 4</u> Let consider (n) and the sets

$A(N)$ and $B(N) \subset I$

with k_{IJ}, $k_{A(N)J}$, $k_{B(N)J}$ can be computed

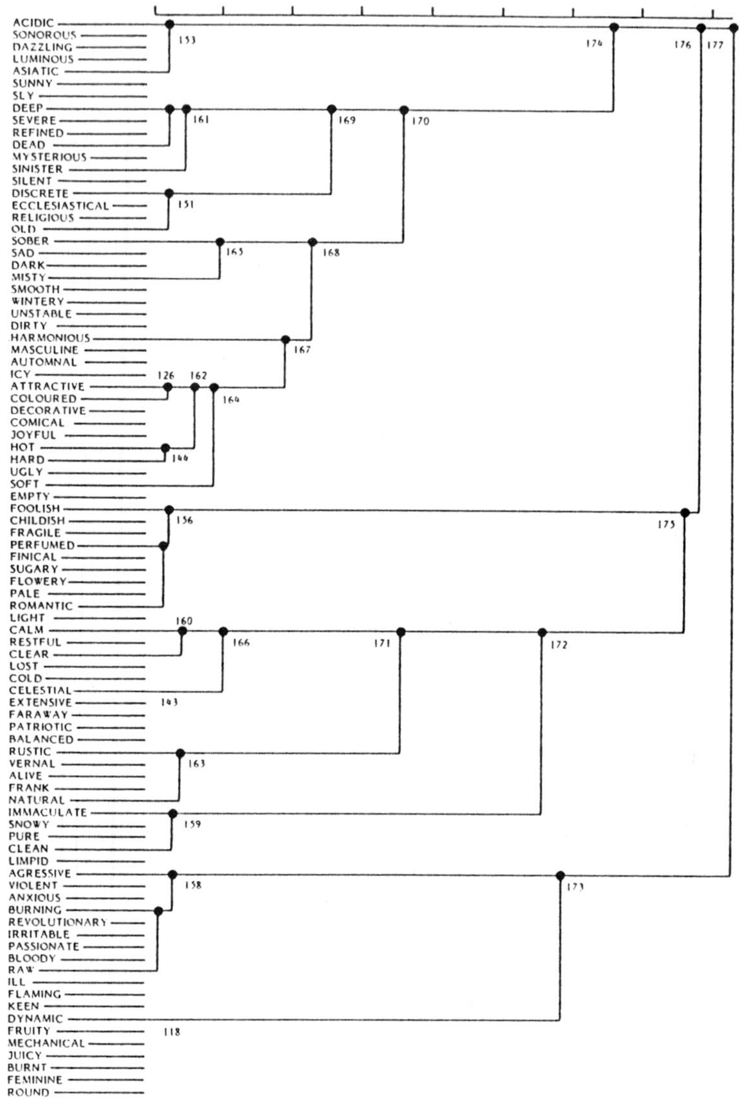

Fig. 5. : Usual representation of hierarchical classification. Analysis of the semantic field associated with colours (See. JAMBU, Cluster Analysis for Data Analysis, Chapter II, North Holland 1983).

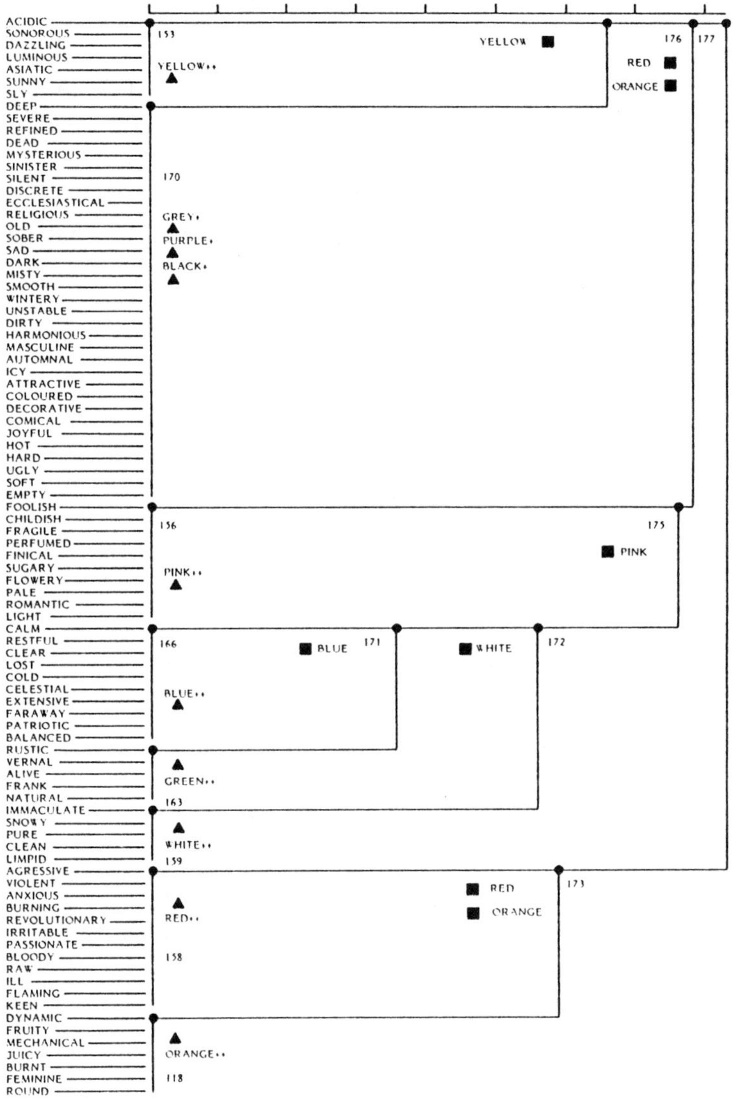

Fig. 6. : Representation of hierarchical classification with variables associated with contributions

■ J means the uppermost contributions of J to the subdivision into classes

▲ J means the uppermost contribution of J to the deviation from the center.

the value $\nu_{s(N)} = (f_{A(N)} \cdot f_{B(N)} / f_{(N)} || f_J^{A(N)} - f_J^{B(N)} ||^2_{f_J}$

where the elements j of J are randomly permuted, can be computed. If $\nu_{s(N)} \leq \nu_{(N)}$ N is chosen as a relevant subdivision.

Comments :

These rules can be applied to supress the lower branches of the hierarchical classification. But sometimes, it can be interesting to study lower or medium branches, because local relationships can be founded in hierarchical clustering. This introduces the following paragraph.

C. Selection of Nodes - Second Class Relevant Subdivisions

Local relationships can be obtained at the low level of the hierarchical classification. How to select (or to detect) the nodes involved with local relations.

RULE 5 $N' \neq N$ is chosen such that $\sum_{n \geq N'} \tau(n) \geq 0.9$ (for example

it means 90 % of the total variance $M^2(I)$.

RULE 6 The values $\nu(n,j)/M^2(I)$ are ordered by decreasing order

N' is chosen such that

$\nu(n,j)/M^2(I) \geq$ the average value of the total variance (1/n.j)

or $\sum_{n,j} \nu(n,j)/M^2(I) \geq 90$ % of the total variance.

Particular case :

$\rho^2(N') = 0$

and $\nu(N') = 0$ and $M^2(N') = 0$

with $N' \neq I$.

In this case N' represents a type with the same profil than the average profile.

D. Selection of Explaining Variables

To explain a classification means to give a meaning of subdivisions into classes in terms of percentage of explaining variables j of J. Variables are selected such that their contributions to the deviation from the center and/or the deviation into two classes are high. It is necessary to determine a threshold this is only made for relevant nodes or subdivisions N or N'.

<u>RULE 1</u> :

A variable j of J is an explaining variable

if $\mu(N,j)/\rho^2(N) \geq 0.5$

or if $\mu(N,j)/\rho^2(N) \geq$ the average value of the contributions (this explains the deviation from the center).

<u>RULE 2</u> :

A variable j of J is chosen :

if $\nu(N,j)/\ (N) \geq 0.5$

or if $\nu(N,j)/\ (N) \geq$ the average value of the contributions.

This allows to select variables which explain a subdivision N.

E. Explained hierarchical classifications

The principle of explained hierarchical classifications is to have graphical representations containing the information selected by the rules of selection of nodes and variables without consulting contributions data tables. New graphics are proposed (See Figs. 7, 8,9).

III. RESULTS

These rules have been applied to the example of colours. For the selection of nodes the following rules have been applied.

A. First Class Nodes

RULE 1 :

with 80 % of the total variance $M^2(I)$

N = 167

RULE 2 :

Maximum of $\nu(N)/\nu(N+1)$

N = 165

RULE 3 :

N = 164.

B. Second Class Nodes

RULE 6 :

N = 150.

C. Explaining Variables

The explaining variables are computed for relevant 1st class nodes only (from N = 177 to N = 164).

D. Explaining the Excentricity of Node N (using RULE 1 : average values)

Nodes	Explaining variables
N = 176	- RED (50%), - ORANGE (28%)
N = 175	RED (+20%), PINK (16%), GREEN (20%)
N = 174	- BLUE (11%), - RED (11%), GREY (12%), BLACK (12%), PURPLE (12%)
N = 173	RED (+50%), ORANGE (28%)
N = 172	BLUE (32%), GREEN (29%)
N = 171	BLUE (36%), GREEN (42%)
N = 170	GREY (16%), BROWN (148%), PURPLE (24%)
N = 169	PURPLE (46%), BLACK (31%)
N = 168	GREY (44%), BROWN (19%)
N = 167	- BLUE (12%), BROWN (45%)
N = 166	BLUE (71%)
N = 165	GREY (61%)
N = 164	ORANGE (44%), - BLUE (19%), - GREEN (13%).

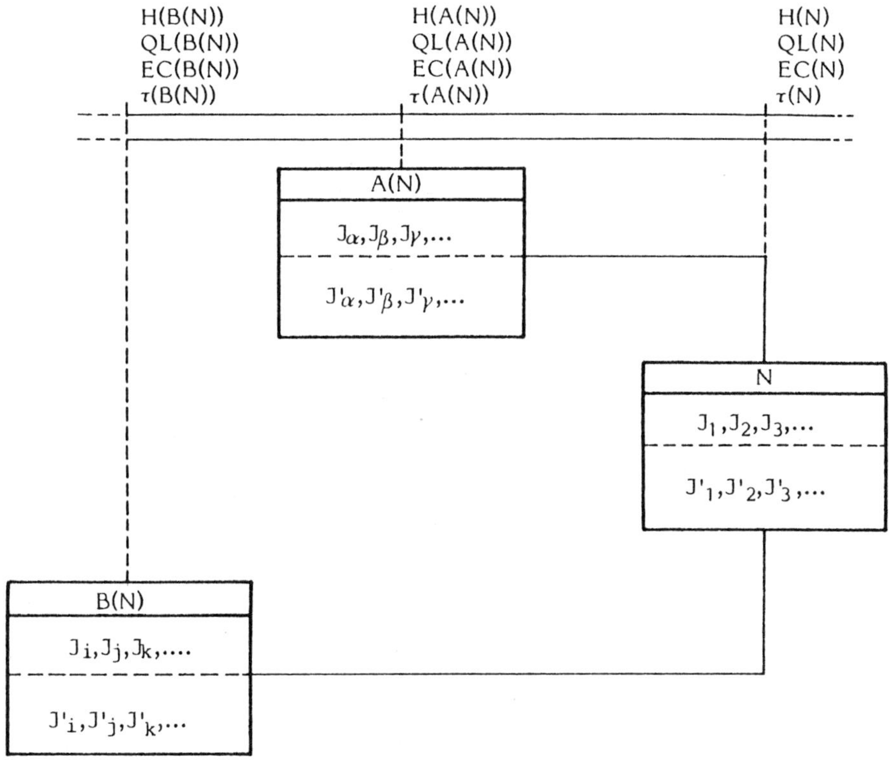

Fig.7 : Elements of representation of a dipole N, A(N), B(N)
N represents the number of node N
H(N) represents the homogeneity of node N
QL(N) represents the quality of the partition at level N
EC(N) represents the relative proportion of ρ^2 to ρ^2 maximum
$\tau(N)$ represents the proportion of variance of the subdivision N
J_1, J_2, J_3 represent the explaining variables of the variance of the subdivision of N into A(N) and B(N)
J_1', J_2', J_3' represent the explaining variables of the deviation from the center.

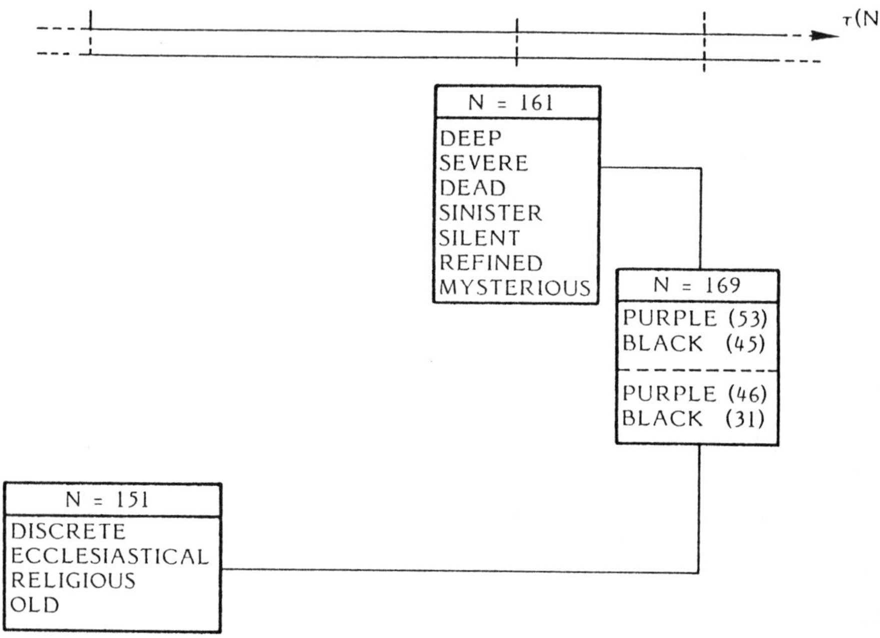

Fig. 8. : Interpretation of the dipole N = 169 by explaining variables.

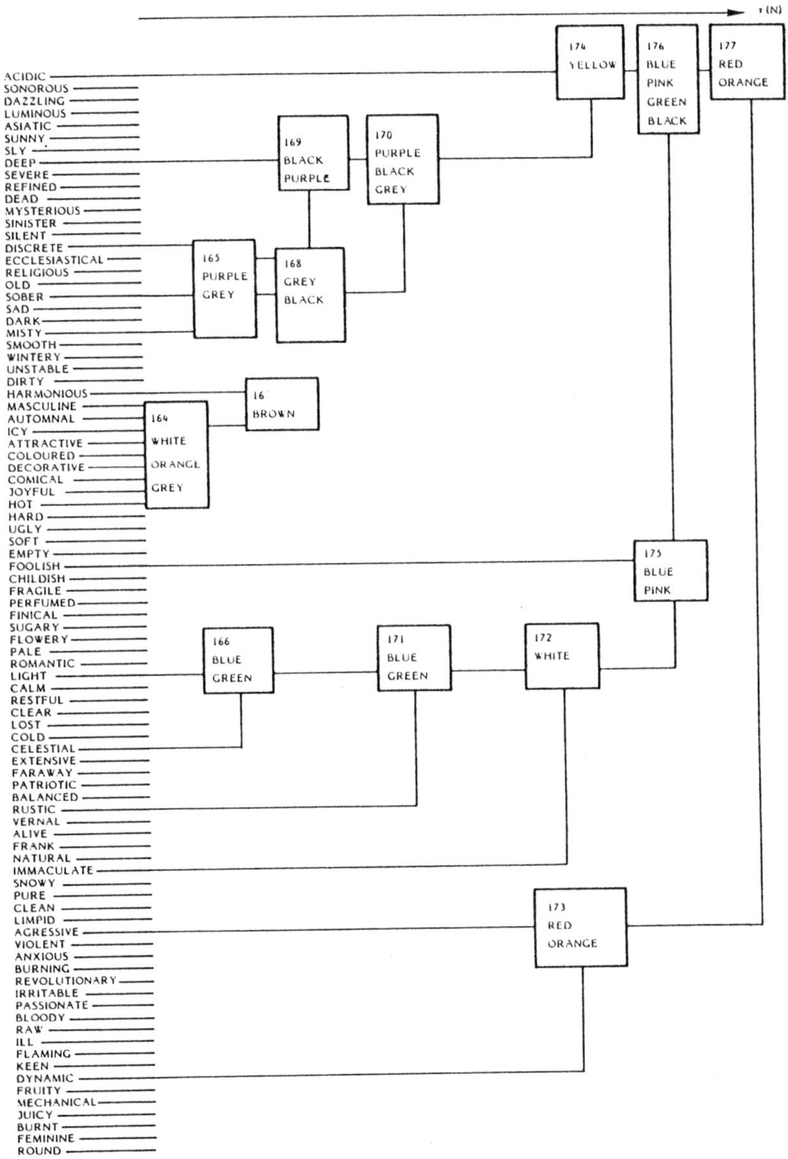

Fig. 9. : Summarized hierarchical classification with selection of relevant nodes and explaining variables.

E. Explaining the Subdivision N into Classes A(N) and B(N) ; (using RULE 1 : average value)

Nodes	Explaining variables
N = 177, A(N) = 176, B(N) = 173	RED (50%), ORANGE (28%)
N = 176, A(N) = 174, B(N) = 175	BLUE (20%), GREEN (19%) PINK (16%), BLACK (12%)
N = 175, A(N) = 156, B(N) = 172	BLUE (12%), PINK (73%)
N = 174, A(N) = 153, B(N) = 170	YELLOW (75%)
N = 173, A(N) = 158, B(N) = 118	RED (40%), ORANGE (50%)
N = 172, A(N) = 171, B(N) = 159	WHITE (72%)
N = 171, A(N) = 166, B(N) = 163	BLUE (44%), GREEN (51%)
N = 170, A(N) = 168, B(N) = 169	PURPLE (46%), BLACK (15%) GREY (17%)
N = 169, A(N) = 161, B(N) = 151	BLACK (45%), PURPLE (53%)
N = 168, A(N) = 167, B(N) = 165	GREY (62%), BLACK (15%) ORANGE (12%)
N = 167, A(N) = 164, B(N) = 120	BROWN (87%)
N = 166, A(N) = 143, B(N) = 160	BLUE (60%), GREEN (34%)
N = 165, A(N) = 141, B(N) = 128	PURPLE (62%), GREY (17%)
N = 164, A(N) = 144, B(N) = 162	WHITE (32%), ORANGE (33%) GREY (15%).

F. Graphics with Explaining Variables

There are two kinds of graphics. On one hand a succession of graphics representing a dipole N and its explaining variables for each subdivision N of the classification ; on the other hand a summarized classification with the whole information about nodes and explaining variables (see Fig. 8 and Fig. 9.).

V. CONCLUSION

The contributions in correspondence cluster analysis have been computed in LTSM software. The rules of selection of nodes and explaining variables will be introduced in the Data Analysis workstation named MULTISTAT[R] which will be presented out at the end of 1987. The principle of selection has been applied to correspondence factor analysis, to select factors and explaining elements of axes. These rules will be also introduced in MULTISTAT[R].

REFERENCES

M. Jambu : Cluster Analysis for Data Analysis - 1983 - North Holland.

MULTISTAT[R] is a registred trade mark of the Centre National d'Etudes des Télécommunications. FRANCE.

A GENERAL EUCLIDEAN APPROACH FOR MEASURING AND DESCRIBING ASSOCIATIONS BETWEEN SEVERAL SETS OF VARIABLES

Yves Schektman

Greco-CNRS 59, Université Paul Sabatier
118 route de Narbonne, 31062 Toulouse cedex, FRANCE

I. INTRODUCTION

Symmetric and dissymmetric association coefficients between several sets of variables are defined in the formal framework of euclidean spaces. They are equal to inertia of patterns of points in individual-space: more precisely, inertia of these patterns of points result from suitable choices of distances and orthogonal projections. So, concerning this last point, the results describe here are intimately connected to some Principal Component Analyses (P.C.A) under linear constraints (18). These aspects are described elsewhere (3,5,19,20) ; note that many methods, for example (1,16), are particular cases of these linear constrained P.C.A.

For many years, I thought that researches on distances are very important for future developments in multidimensional data analysis. An important point is to value conjoint effect of distances (21) and variances, or correlations, on inertia. So, I introduced distances "with relationship effects" and I have been developing, with my students (4,8,6,10), researches

RECENT DEVELOPMENTS IN
CLUSTERING AND DATA ANALYSIS

ISBN 0-12-215485-1

in this direction. Some of these results are described in section II.

General definitions of association coefficients, including classical ones, are given in section III. In a natural way, graphical analysis of these coefficients will consist in orthogonal projections of corresponding patterns of points on their principal planes. Of course, this approach generalizes partitions of association coefficients like (12) and classical techniques like correspondence analysis (J.P. BENZECRI).

The case of two sets of variables is described in detail. For more than two sets, the situation is more complex. However the general results presented in this paper should constitute guidelines for designing them: some reflexions on multiple correspondence analysis (14) or on measurement and description of association between one categorical variable and several others (15) are given in (21) and in (8).

II. SOME USEFUL DISTANCES

A. Distances with relationship effects

Consider a multivariate sample defined by n individuals and two sets of variables [xj] (j=1,p) and [yk] (k=1,q) . We suppose , without loss of generality , that all variables have zero means. Let xji be the value associated to the ith individual and to the jth variable . yki denotes the similar value but for yk. For any definition referring to x, generally a similar definition referring to y exists. Subsequently , we shall not give the definitions referring to y. All the values [xji] are regarded as forming a rectangular data matrix, denoted by X , with n rows and p columns and with xji in ith row and jth column .

Let $F = R^n$ be the variable-space and $E = R^{p+q} = Ex \oplus Ey$ be the individual-space , where Ex is the p-dimensional subspace associated to X . Let M and D be the euclidean distances in E and F , respectively . Generally, the matrix of D is a (n x n) diagonal one with the (i,i)th element equal to the weight wi of the ith individual, but D may be more complex. M may be supposed such that

$$\mathrm{Im}[\mathrm{Inx}(\mathrm{Inx}'\ M\ \mathrm{Inx})^{+}] = \mathrm{Sx} \subset \mathrm{Ex}$$

where Sx is the subspace spanned by the row vectors of the (n,p+q) matrix [X | 0] , Inx is the canonical injection from R^p into E, "+" denotes the Moore-Penrose Inverse (M.P.I) and "'" the transposition.

According to the variables partition in two sets and to the corresponding partition of E, let us denote:

$$V = \begin{bmatrix} Vx & Vxy \\ Vyx & Vy \end{bmatrix} \quad \text{and} \quad M = \begin{bmatrix} Mx & Mxy \\ Myx & My \end{bmatrix} \quad \text{where}$$

- V is the covariance-matrix of variables [x] and [y]
- $Vx = X'\,D\,X$, $Vy = Y'\,D\,Y$, $Vxy = X'\,D\,Y$.

<u>Definition 1</u> : M is with relationship effect for (Ex,Ey) if $Mxy = Mx\,[(Vx\ Mx)^{1/2}]^{+}\ Vxy\ My\,[(Vy\ My)^{1/2}]^{+}$ where $[(Vx\ Mx)^{1/2}]^{+}$ is the M.P.I of $(Vx\ Mx)^{1/2}$ with respect to Mx.

Let [uj(x) $\in$ Ex , Uj(x) $\in$ Fx] (j=1,p) be the principal vectors and components of the 3-uple [X,Mx,D], where Fx is the subspace spanned by the column vectors of X . Using the relation Uj(x) = X Mx Inx' uj(x) , it follows,

<u>Property 1</u> : M is with relationship effect for (Ex,Ey) , iff

$$\cos[uj(x)\ ,\ uk(y)] = 0 \quad \text{if } Uj(x)=0 \text{ or } Uk(y)=0$$
$$= r[Uj(x)\ ,\ Uk(y)] \quad \text{otherwise}$$

where r denotes the correlation coefficient.

Thus , the observed statistical relationships between variables [x] and [y] are carried, via principal elements, from F into E , in a geometrical way.

Denote by $V\bar{x}$ the matrices Vx^{-1} or $V\overset{+}{x}$ or $\chi^2 x$ (chi-square): note that they define useful distances when variables are linearly independent, dependent, or centered dummy-variables (D.V) of a categorical variable , respectively . It is easy to show that $Qx = X\ V\bar{x}\ X'\ D$ is the orthogonal projection onto Fx. If one supposes that $Im[Inx\ M\overset{+}{x}] = Sx$, for example $Mx = V\overset{+}{x}$, then $Px = Inx\ M\overset{+}{x}\ Inx'\ M$ is the orthogonal projection onto Sx. Moreover, if [x] are D.V of a categorical variable and if $Mx = \chi^2 x$, as $Inx\ (\chi^2 x)^{-1}\ Inx'\ M\ Iny = Inx\ Vx\ Inx'\ M\ Iny$, then the restriction to Ey of $Px = Inx\ (\chi^2 x)^{-}\ Inx'\ M$ is also the orthogonal projection onto Sx .

It is well known that Qx Qy and Qy Qx have the same non zero latent roots which are equal to the squared canonical correlation coefficients (C.C.C) of (Fx,Fy,D); it is easy to show the dual property in E :

<u>Property 2</u> : if M is with relationship effect for (Ex,Ey)
and if Mx is of full rank or equal to Vx or $V\overset{+}{x}$
and if My is of full rank or equal to Vy or $V\overset{+}{y}$
then Px Py , Py Px , Qx Qy and Qy Qx have the same non zero latent roots.

Consequently , (Ex,Ey,M) has the same non zero C.C.C as (Fx,Fy,D) . Note that if M is with relationship effect for (Ex,Ey) and if $Mx = V\bar{x}$ then $Mx[(Vx\ Mx)^{1/2}]^{+}\ X' = V\bar{x}\ X'$ and $Im[Mxy] \subset Im[V\overset{+}{x}]$. Let xi' be the ith row vector of $[\ X \mid 0\ ]$, one can show the following property which links the two systems of points $Nx = [xi\ (i=1,n)]$ and $Ny = [yi\ (i=1,n)]$

<u>Property 3</u> : if $Mx = V\bar{x}$, $My = V\bar{y}$
and if $Tr[Px\ Py] = Tr[Qx\ Qy]$
then M is with relationship effect for (Ex,Ey)
iff $\sum wi\ \|xi - yi\|^2$ is minimum .

Denote by (C) the three following conditions : M is with relationship effect for (Ex,Ey) , $Mx = \overline{Vx}$, $My = \overline{Vy}$. Thus , under conditions (C), if (Ex,Ey) and (Fx,Fy) have the same non zero C.C.C, the distance with relationship effect is the unique one so that Nx and Ny are the nearest. This property will be used to give, in section IV, an interesting geometrical illustration of the Regression in E. Note that if Mx = Vx and My = Vy then naturally M = V .

Finally, let us give the definition of distances which are opposed to distances with relationship effects .

<u>Definition 2</u> : M is with unrelationship effect for (Ex,Ey) if $Mxy = -\ Mx[(Vx\ Mx)^{1/2}]^{+}\ Vxy\ My[(Vy\ My)^{1/2}]^{+}$

The properties of these distances are the inverse of those of distances with relationship effects .

B. Distances - "Variability and association effects" .

If M is with relationship effect for (Ex,Ey) , then $Iy[Nx] = \sum wi\ \|Py(xi)\|^2$, i.e. the inertia of the orthogonal projection of Nx onto Sy , depends upon the statistical relationships between variables [x] and [y].

With regards to the approach presented in section I , it is natural to study the "variances and correlations effects" of variables [x] on the shape of Nx and consequently on the value of Iy[Nx] . More precisely , it seems reasonable to try to "eliminate" these effects : in this way, indeed, we may hope that the shape of "Py(Nx)", and therefore the value of Iy[Nx], will only depend upon the association between variables [x] and [y]. We propose measures of "variability and association effects" and some properties in (8,21,6) . In fact, for our subject, a simpler and more global definition will be enough.

<u>Definition 3</u> : In Sx , a distance eliminates simultaneously

"variability and association effects" (VAE) of variables [x] if $Ia[Nx] = 1$ for any line $\Delta a \subset Sx$.

One can show that $\chi^2 x$ is a full rank internal generalized inverse of Vx (4). Moreover, as $Vx\,\bar{Vx}$ is a projection matrix onto Sx, then Sx is the proper subspace of $Vx\,\bar{Vx}$ associated to the multiple latent root equal to 1. It follows,

<u>Property 4</u> : $Mx = \bar{Vx}$ eliminates the V.A.E of variables [x].

It is interesting to note that this result leads to another illustration of distances with relationship effects .

<u>Property 5</u> : the two following assertions are equivalent

(i) - M is with unrelationship effect for (Ex,Ey) ,

- $Mx = (Vx - Vxy\,Vy^{-1}\,Vyx)^{-1}$, $My = (Vy - Vyx\,Vx^{-1}\,Vxy)^{-1}$

(ii) $M = V^{-1}$.

Under the property 5 conditions , according to property 4, one may say that Mxy eliminates the part of V.A.E not eliminated by Mx and My , i.e : the "association effects" between variables [x] and [y] and the V.A.E part of variables [x] and [y] linearly depending on [y] and [x], respectively.

III. GENERAL ASSOCIATION INDICES .

A. Generalities

In this section , variables will be quantitative linearly independent , dependent , or D.V of a categorical variable, indifferently. From the following expression,

$$Ix[Ny] = Tr[Px\;Iny\;Vy\;Iny'\;M] = Tr[Mx^{+}\;Mxy\;Vy\;Myx]$$

it implies property 6 which shows that Ix[Ny] does not depend upon the choice of Mx.

Property 6 : if M is with relationship effect for (Ex,Ey)
and if Mx is full rank or equal to Vx or to Vx^{+}
then $Ix[Ny] = Tr[Vxy\ My\ Vyx\ Vx^{+}]$.

So, according to property 4, we shall see that $My = \bar{Vy}$ will lead to classical symmetrical association indices . On the contrary , we shall see that My = Iq will lead to classical dissymmetrical association indices .

B. Symmetrical association indices

Let [Rc] be the C.C.C of (Fx,Fy,D) , it is easy to show,

Property 7 : $\sum Rc^2 = Tr[Vxy\ \bar{Vy}\ Vyx\ \bar{Vx}]$.

As for D.V, $Qx = X\ Vx^{+}\ X'\ D = X\ \chi^2 x\ X'\ D$, thus properties 6 and 7 imply,

Property 8 : if M is with relationship effect for (Ex,Ey)
then (i) $My = \bar{Vy}$ ==> $Ix[Ny] = \sum Rc^2$,
(ii) $Mx = \bar{Vx}$ ==> $Iy[Nx] = \sum Rc^2$.

Thus,under property 8 conditions,Iy[Nx] and Ix[Ny] are equal to "Pearson coefficient of mean square contingency" when [x] and [y] are the D.V of two categorical variables, and to the "generalized correlation ratio" when [x] are quantitative and [y] the D.V of a categorical variable, respectively.

In conclusion , whatever the type of variables , property 8 points out that classical symmetrical association indices can be expressed by the same formula Ix[Ny] (or Iy[Nx]) .
This leads to the following general definition :

Definition 4: Symmetrical Relational Association Coefficients, between [x] and [y], are defined by

$$CARS[x,y/Mx,My\ ] = Ix[Ny]/(Ix[Nx]\ Iy[Ny])^{1/2}..$$

Note that this formula must be used with suitable distances; concerning Ix[Ny], for example, we must choose :

- $My = \chi^2 y$ when [y] are the D.V of a categorical variable,
- $My = V\overset{+}{y}$ when [y] are quantitative variables.

These choices are in agreement with property 4 and the remarks given in the beginning of section II- B . It is easy to verify that CARS[$x, y / \chi^2 x, \chi^2 y$] is equal to "Tschuprov coefficient" when x and y are categorical variables and more generally to "RV coefficient" (17) under conditions (C). Under (C), it may be useful to note (8) that

$$Ix[Ny] = (I[N/M] - I[N/Mo])/2$$

where N = [xi + yi (i=1,n)] , I[N/M] is the inertia of N with respect to M,and Mo = M when null association is assumed.

Finally , let us give a property which simplifies algorithms using Ix[Ny], when y is a categorical variable. Let [z] be the uncentered dummy-variables of y and REz be the restriction to Ez, as Vxz = Vxy when [x] are centered, property 9 shows that dummy-variables associated to y may be uncentered.

<u>Property 9</u> : if M is with relationship effect for (Ex,Ey)
and if $My = Mz = \chi^2 y \quad (= Vz^{-1})$
then Mxz = Mxy , REz[Px] = REy[Px]

C. Dissymmetrical association indices

Classical dissymmetrical association indices (7,22), from variables [x] to [y] can be defined by : $\sum \|Qx(yk)\|^2 / \sum \|yk\|^2$

From expression of Qx and property 6, it follows,

<u>Property 10</u> : $\sum \|Qx(yk)\|^2 = Tr[Vxy\ Vyx\ V\bar{x}]$.

<u>Property 11</u> : if M is with relationship effect for (Ex,Ey)
and if Mx is of full rank or Mx = Vx or $Mx = V\overset{+}{x}$
and if My = Iq
then $Ix[Ny] / Iy[Ny] = \sum \|Qx(yk)\|^2 / \sum \|yk\|^2$.

Definition 5: DisSymmetric Relational Association Coefficients from [x] to [y] are defined by

$$CARDS[x \to y \ / \ Mx, Iq] = Ix[Ny] \ / \ Iy[Ny].$$

Consequently ,

- "Goodman and Kruskal's tau" (for two categorical variables) and "Steward and Love's association" (for two sets of quantitative variables) are CARDS .
- It is natural to propose the CARDS[$y \to x$/My,Ip] for measuring dissymmetrical association from the qualitative factor y to the observed quantitative variables [x] (MANOVA for example), or the CARDS[$x \to y$/Mx,Iq] in the opposite case.
- Mx = Ip corresponds to classical dissymetrical association indices; so, other choices for Mx, more or less near to $\overline{Vx}$, will provide more or less dissymetrical indices.

D. Extensions

One may extend CARS and CARDS in two directions :

- definition for more than two sets of variables; properties of distances given above or in (18,8,4,21,6) will aid.
- "My" can be obtained by generators of distances (21): you may start from the symmetrical case (My = $\overline{Vy}$), then go to more or less dissymmetrical cases. Note that these generators, actually used to make relationship methods robust (6), are designed by using properties on elimination of V.E or V.A.E.

IV. SYNTHETIC RELATIONSHIP METHOD

The first kind of methods, briefly described here, deals with graphical descriptions in E of relationships between several sets of variables. For example, for two sets of variables, keeping in mind the results of section III, it is natural to want to see simultaneously the projections of

Px[Ny] and Nx onto the principal planes of Px[Ny] . This method, parametrized by suitable distances, is proposed as a general method for describing associations. Note that this practice provides also "principal partitions" of the CARS or CARDS concerned by the study. Obviously, one can show that classical methods like symmetrical (J.P.BENZECRI) or dissymmetrical (13) correspondence analyses, more generally canonical analysis, or multiple correspondence analysis (14) for more than two sets of variables,... are particular cases of this general method .

The following general property, connected to property 3, leads to a second kind of methods which deals with multidimensional linear fit.

<u>Property</u> <u>12</u> : (C) ==> [Py(x1)..Py(xn)]'=[0 | Qx(y1)..Qx(yq)].

Using this result, whatever the type of variables [x] and [y], property 3 provides, for q=1, a geometrical definition of the Regression in E (CARS[x,y] = r^2[y;Fx]) . What happens when one modifies Mx or My ?

V. SYNTHETIC CONCLUSIONS

- One program is enough for classical methods and the new ones proposed: for a given study, this program will determine the suitable patterns of points, distances and orthogonal projections .
- Users don't have to take into account the type of variables.
- Generators of distances can provide dynamical graphics according to the more or less dissymetrical feature of the the indices.
- Maximization or minimization of CARS or CARDS, relatively to an unknown categorical variable, leads to clustering methods (10,11), and we use them to propose a new valuation of

factors effects in MANOVA .

REFERENCES

1. BONIFAS,L. ESCOUFIER,Y. GONZALEZ,P.L. and SABATIER,R. (1984): Choix de variables en analyse en composantes principales. Rev.Stat.Ap, XXXII, 2, 5-15.
2. CAILLEZ,F. and PAGES, J.P. (1976): Introduction à l'analyse des données. SMASH-BURO-ASU, PARIS.
3. CARLIER,A. CROQUETTE,A and SCHEKTMAN,Y. (1983): Analyses en composantes principales sous contraintes linéaires. Rap.Tech. 8302, ENSAM-INRA-USTL, Montpellier, 25 p.
4. CROQUETTE,A. (1980): Quelques résultats synthétiques en analyse des données multidimensionnelles: optimalités et métriques à effets relationnels. Th. 3ème cycle, Toulouse.
5. CROQUETTE,A. and SCHEKTMAN,Y. (1985): Quelques propriétés extrémales des éléments propres des opérateurs non négatifs. Composantes principales sous contraintes. Rap.Tech., 8503 , ENSAM-INRA-USTL , Montpellier, 24 p.
6. FABRE,C. (1986): Contribution à la protection des méthodes relationnelles. Th. 3ème cycle, Univ.P.Sabatier, Toulouse.
7. GOODMAN,L.A. and KRUSKAL,W.H. (1954): Measures of association for cross-classifications. J.AMER. Statis. Assoc., 49, 732-764.
8. GRAU,D. (1983): Mesure des effets relationnels . Applications. Th. 3ème cycle, Univ.P.Sabatier, Toulouse.
9. GRAY,L.N. and WILLIAMS,J.S. (1975): Goodman and Kruskal's tau b: multiple and partial analogs. Proc. Soc. Statist. Assoc., Section of the Amer.Statis.Assoc.,444-448.
10. IBRAHIM,A. and SCHEKTMAN,Y. (1985): Analyses en partitions principales. Algorithmes et exemples. Actes des Journées de la Société Francophone de classification . La Grande Motte. Ed. C.PERRUCHET, Pub.CNET-PARIS , 61-89.

11. IBRAHIM,A. and SCHEKTMAN,Y. (1986): Principal Cluster analyses. In "Classification as a toal of research".W.GAUL and M.SCHADER Ed., El.Pub.B.V., North-Holland, 217-223.

12. LANCASTER,L. (1963): Canonical correlation and partitions of χ^2. Quant.Jouv.Math , 14 , 220-224.

13. LAURO,N. and D'AMBRA, L. (1983): L'analyse non symètrique des correspondances . In "Data analysis and Informatics", E.Diday and al.editors , North-Holland , 433-446 .

14. LEBART,L. MORINEAU,A and TABART,N. (1977): Technique de la description statistique . DUNOD .

15. LECLERC,A. (1976): Une étude de la relation entre une variable qualitative et un groupe de variables qualitatives. Inst.Stat.Rev., 44 , 2 , 241-248 .

16. RAO,C.R. (1964): The use and interpretation of principal component analysis in applied research. Sankhya , Ser.A , 26 ,329- 358.

17. ROBERT,P. and ESCOUFIER,Y. (1976): A unifying toal for linear multivariate statistical methods : the RV-coefficient. Appl.Statist., 25 , 3 , 257-265 .

18. SCHEKTMAN,Y. (1978): Contribution à la mesure en facteurs dans les sciences expérimentales et à la mise en oeuvre des calculs statistiques. Thèse d'état, Toulouse.

19. SCHEKTMAN,Y. (1982): Optimality and duality properties for defining constrained principal components. 4th southeast asian statistics seminar. Singapour , 12 p.

20. SCHEKTMAN,Y. and al.(1984): Analyses en composantes principales sous contraintes.Applications. In "Data analysis and informatics". E.Diday et al.ed., El.Sci. Pub.B.V. , North-Holland , 85-97 .

21. SCHEKTMAN,Y. (1984): Mètriques et analyse de données . Contributions et synthèse. 2ème colloque Franco-Soviétique Statistique et Analyse des Données. Issy-les-Moulineaux.

22. STEWART,D. and LOVE,W.(1968): A general canonical correlation index. Psychol.Bull., 70 .

DATA-ADAPTIVE METHODS IN MULTIVARIATE ANALYSIS

Takenobu Tasaki
Masashi Goto

SHIONOGI Kaiseki Center
1-22-41, Izumicho, Suita City, Osaka Japan

I. INTRODUCTION

In this paper, bootstrap and cross-validation methods are applied to the logistic and nonparametric regression analyses of ordered categorical responses. These methods are data-adaptive and intended to evaluate the accuracy of the statistic of interest only based on the given data. In order to illustrate our research, we adopt examples from the actual pharmacological studies on structure-activity relationships. Recently, in this field various multivariate analyses are most vigorously applied without taking care of such problems as outliers, influential observations and over-fitting. In order to examine such problems, it is necessary to diagnose the data and models and to evaluate the stability of obtained results.

The concept of the "bootstrap" method has been discovered by Efron (1979) and it is compared as a "Columbus's egg" in the field of statistics. Though traditionally statis-

ISBN 0-12-215485-1

ticians must develop sophisticated statistical procedure and its acompanying intensive computation based on the assumption of the underlying distribution of observations, in the bootstrap method, we have made computers to take over those hard tasks. We can say that both jackknife and cross-validation methods are also data-adaptive in the classical version.

II. DATA AND MODEL

Two sets of data obtained from the studies of structure-activity relationships of drugs are used here. The first set of data consists of the measurements of β- adrenergic blocking activities for the 37 derivatives of 1, 2-benzisoxazolyloxy-propanolamine, and of measurements of the four physicochemical indices, namely π_1, R_1, H_3 and LM_1, which had been considered to explain their bioactivities. The second set of data consists of the measurements of the photosynthesis inhibition activities of 40 benzene derivatives comprising mainly of phenol compounds, and of measurements of the two physicochemical indices, namely $logP$ and $-E\ (HOA)$. The β-adrenergic blocking and photosynthesis inhibition activities are classified into four ordered categories.

In practice, original measurements of these bioactivities were given as continuous variates. Actually we had many structural indicies other than stated above, which were not utilized in this paper. Yamakawa et al. (1986), Tasaki and Goto (1984) and Goto et al. (1986) have discussed on the selection of optimal subsets of the structural indices which most contribute to predict the bioactivities, and on the motives and effects of classifying the measurments of the bioactivities into ordered categories.

Logistic regression models to evaluate the dependence of categorical responses on explanatory variables have been discussed by Plackett (1974), McCullagh (1980) and Anderson & Philips (1981). Let us denote by X a vector of m explanatory variables (the structural indices), and by Y the response variable (the bioactivity) which is classified into one of the k ordered categories, respectively. We represent the ordered categories with integers from 1 to k. In the two examples mentioned above, $Y=1$ represents the weakest bioactivity, $Y=4$ the strongest and $Y=2$ and 3 the intermediate, respectively. The logistic regression model can be written as

$$prob\ (Y \leqq j \mid X=x) = \frac{exp(\theta_j - \beta^T x)}{1+exp(\theta_j - \beta^T x)}, \quad 1 \leqq j < k, \tag{1}$$

or equivalently as

$$prob\ (Y=j \mid X=x) = \begin{cases} \dfrac{exp(\theta_1 - \beta^T x)}{1+exp(\theta_1 - \beta^T x)}, & j=1, \\[2ex] \dfrac{exp(\theta_j - \beta^T x)}{1+exp(\theta_j - \beta^T x)} - \dfrac{exp(\theta_{j-1} - \beta^T x)}{1+exp(\theta_{j-1} - \beta^T x)}, & 1 < j < k, \\[2ex] 1 - \dfrac{exp(\theta_{k-1} - \beta^T x)}{1+exp(\theta_{k-1} - \beta^T x)}, & j=k. \end{cases} \tag{2}$$

The unknown parameters $\theta^T = (\theta_1, \ldots, \theta_{k-1})$ and $\beta^T = (\beta_1, \ldots, \beta_m)$ in Eq.'s (1) and (2) can be estimated by means of the maximum likelihood method.

Both π_1 in the first set of data and $logP$ in the second set denote the degree of hydrophobic property of a chemical compound. Chemicals of exceptionally high hydrophobic property (lipophilic property) will remain in the first hydrophobic layer (fat layer) and will not be distributed to the working area. Conversely, chemicals of very high hydrophilic property will remain in the first water layer. Therefore, it is considered that there will be the optimal values for either π_1 or $logP$. In consideration of this fact, the squared value of π_1 or $logP$, namely π_1^2 or $(logP)^2$ is normally added to the explanatory variables in the analysis on structure-activity relationships. In order to find out such nonlinear effect, a heuristic smoothing method proves to be effective. Copas (1983) has proposed the nonparametric regression method in the case of a binary response variable. We try to extend this method to the cases of ordered multi-categorical responses. Namely, the response variable Y can be identified with $k-1$ binary response variables Z_j as

$$Z_j = \begin{cases} 0 & (Y \leqq j) \\ 1 & (Y > J) \end{cases}, \quad j=1, \ldots, k-1. \tag{3}$$

In the nonparametric regression of Z_j on single explanatory variable X based on observations $\{(x_1, z_{1j}), \ldots, (x_n, Z_{nj})\}$ the response probability at the point $X=x$, say, $P_j(x)=prob\ (Z_j=1 \mid X=x)$, is estimated by

$$\hat{P}_j(x) = \sum_{i=1}^{n} Z_{ij}\Psi\{h^{-1}(x-x_i)\} / \sum_{i=1}^{n} \psi\{h^{-1}(x-x_i)\}, \tag{4}$$

where Ψ is a kernel function for probability estimation and h is a constant. In this paper we adopt $\Psi(u) = exp(-u^2/2)$

as the kernel function. The constant h plays an important role to adjust smoothness of the estimate $\hat{P}_j(x)$.

III. NONPARAMETRIC REGRESSION

In the nonparametric regression analysis, selection of an optimum h value in Eq. (4) is critical for our estimation procedure. If h is large, the estimate of the regression relationship has over-smoothing and disadvantage of high bias. On the other hand, if h is small, the estimate of the regression relationship has large dispersion and heavy up-and-down variations. To select the appropriate h value, we apply the cross-validation method. For this purpose, we calculate with fixed h the estimate of the response probability of the regression in the same way as Eq. (4) at the point x_i but based on the $n-1$ pairs of observations which are obtained by excluding (x_i, z_{ij}) from the whole n pairs of observations $\{(x_1, z_{1j}), \ldots, (x_n, z_{nj})\}$. The difference between the estimated value and z_{ij} is taken as a predictive residual. Hence, the predictive sum of squares (PSS) is obtained by summation of the squared values of n predictive residuals. We choose the value h which minimizes the PSS as the optimal one. For the first example, the PSSs with respect to the Z_3 component are plotted in Fig. 1 against h. From it, we see that the optimal value for h is 0.51.

Fig. 2 shows the estimated value of the nonparametric regression using the optimum h. In Fig. 2, the three dotted curves represent the estimated values of the nonparametric regression for $prob(Z_j=1|\cdot)$, $j=1, 2, 3$, while the broken curves represent the estimated values for $prob(Y \leqq j|\cdot)$, $j=1, 2, 3$, of the quadratic logistic regression model.

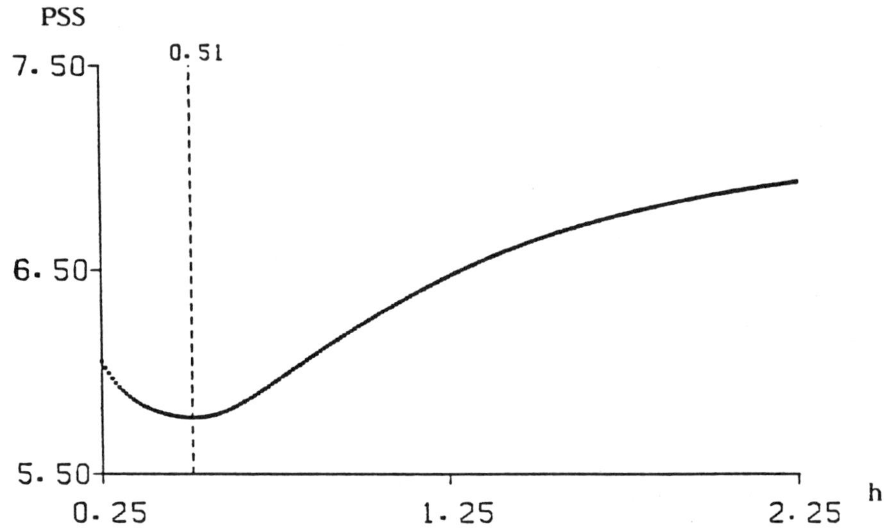

Figure 1 Plot of PSSs in cross-validation against h for the Z_3 component of β-blocking activity.

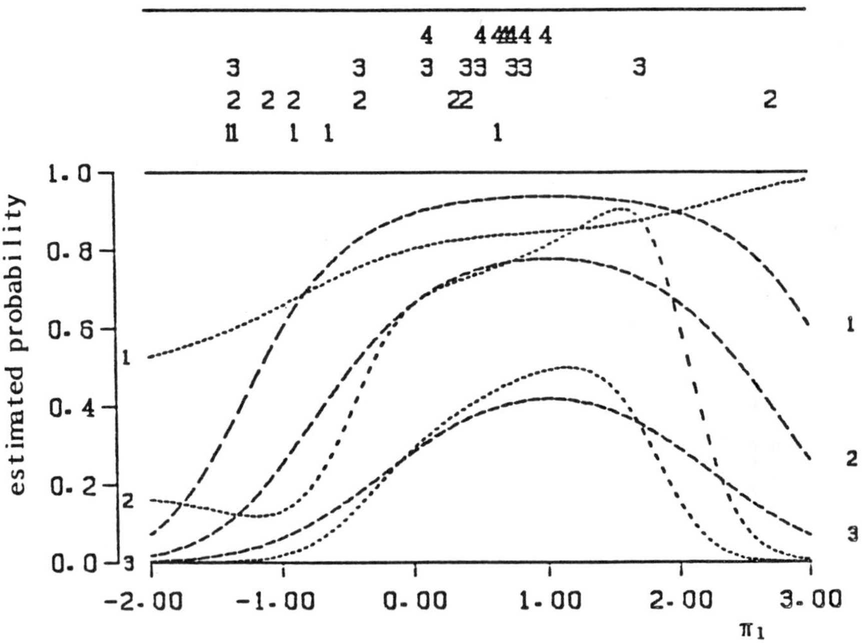

Figure 2 Nonparametric (dotted curves) and parametric (broken curves) estimates of regressions for β-blocking activities.

In Fig. 2, both the dotted curves and the broken curves behave similarly, as well as those to the Z_2 component. This is also true with respect to the Z_3 component. Therefore, it is seen that as far as the index π_1 is concerned, stronger β-blocking activities can be fully explained with the quardratic logistic regression model.

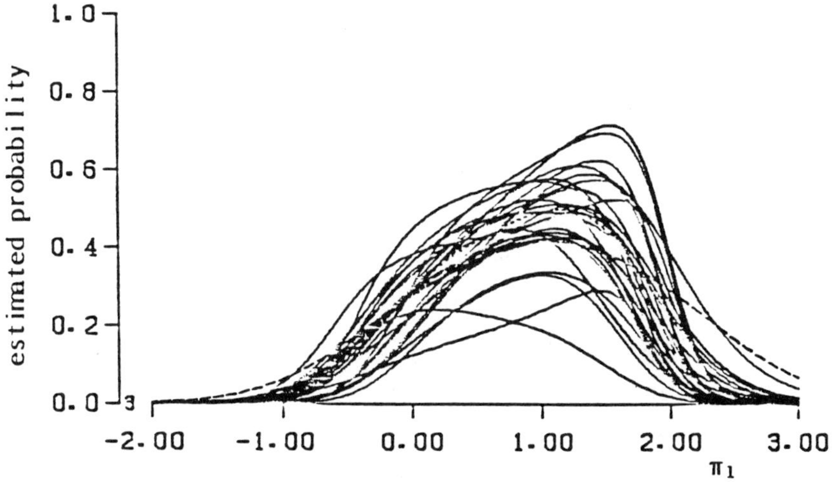

Figure 3 Bootstrapping of nonparametric estimates of the regression for the Z_3 component of β-blocking activity.

On interpreting the nonparametric estimates of regression, we have to take their reliability into consideration. One of the methods useful for assessing reliability is to evaluate replication stability by applying the bootstrap method. For this purpose, we draw independently n pairs of observations $\{(x_1^*, z_{1j}^*), \ldots, (x_n^*, z_{nj}^*)\}$ with replacement from our observations $\{(x_1, z_{1j}), \ldots, (x_n, z_{nj})\}$. The nonparametric regression model is then applied to the bootstrap samples. Fig. 3 shows the 25 regression curves obtained from the independent $B=25$ bootstrap samples, concerning the Z_3 component for the β-blocking activity.

Though the initially estimated values (the dotted curves) are completely enveloped within the bands of solid curves, the parametric estimates (the broken curves) deviate somewhat from the bands of the solid curves near its extreme right. This suggests that the parametric estimates for greater π_1 values are not so reliable.

IV. LOGISTIC REGRESSION ANALYSIS

The logistic regression model (1) was fitted by the maximum likelihood method, using $\{\pi^2{}_1, \pi_1, R_1, H_3, LM_1\}$ and $\{(logP)^2, logP, -E(HOA)\}$ as explanatory variables for the first and the second data set, respectively. The maximum likelihood estimates of θ and β, and the estimates of standard deviations of them based on the large-sample theory are shown in the 2nd and 3rd columns in Table 1.

Let $\hat{\theta}$ and $\hat{\beta}$ be the maximum likelihood estimates of θ and β, based on our data of size n, $(x_1, y_1),...(x_n, y_n)$. For convenience, let $prob\,(Y \leqq j\,|X=x)$ and $prob(Y=j \mid X=x)$ be expressed as $P_j\,(x)$ and $P_j\,(x)$ hereafter, respectively. Bioactivity y_* of a new compound with structure indices x_* can be predicted by the following two methods:

(1) to select the value j that maximizes the $\hat{p}_j\,(x_*)$ obtained by substituting $\hat{\theta}$ and $\hat{\beta}$ for θ and β in Eq. (2), and

(2) to select the value j so that $\hat{\theta}_{j-1} \leqq \hat{\beta}^T x_* < \hat{\theta}_j$, where for convenince, it is assumed that $\hat{\theta}_0 = -\infty$, $\hat{\theta}_k = +\infty$.

The apparent correct classification rate for the training data set and estimated one by the cross-validation method based on these two prediction rules, are shown in the 2nd and 3rd columns in Table 2, respectively.

Then, the bootstrap method was applied, regarding each

Table 1 Estimation of parameters of logistic regression models

1. β-blocking activities

parameter	MLE		Bootstrapping					
			(B=991: §4)			(B'=1000: §5)		
	Est.	S. D.	MEAN	S. D.	SKEW	MEAN	S. D.	SKEW
β_1	-1.785	0.458	-2.744	3.001	-6.964	-2.393	0.943	-1.838
β_2	2.578	0.639	4.030	3.738	9.532	3.343	1.189	1.443
β_3	7.305	2.411	9.453	11.65	11.12	9.635	4.106	1.584
β_4	-1.481	0.522	-1.910	1.638	-6.292	-1.945	0.861	-1.024
β_5	-3.284	0.844	-5.927	7.989	-17.97	-4.364	1.691	-2.420
θ_1	-21.52	4.956	-36.17	40.57	-15.70	-28.43	10.05	-2.206
θ_2	-18.77	4.591	-31.45	38.34	-17.00	-24.82	9.175	-2.144
θ_3	-15.89	4.219	-26.96	35.35	-17.50	-21.17	8.279	-2.219

2. photosynthesis inhibition activities

parameter	MLE		Bootstrapping					
			(B=1000: §4)			(B'=1000: §5)		
	Est.	S. D.	MEAN	S. D.	SKEW	MEAN	S. D.	SKEW
β_1	-0.421	0.203	-0.489	0.392	-0.768	-0.487	0.267	-0.599
β_2	3.895	1.453	4.579	3.022	1.193	4.517	1.934	0.824
β_3	-1.597	0.662	-2.202	1.490	-1.162	-1.801	0.819	-0.810
θ_1	-8.855	6.078	-13.39	11.25	-1.380	-9.709	7.667	-0.436
θ_2	-6.961	6.005	-11.11	10.69	-1.414	-7.572	7.566	-0.365
θ_3	-5.642	5.930	-9.637	10.54	-1.397	-6.095	7.491	-0.348

pair consisted of the response and the vector of explanatory variables as a sampling unit. Mean, standard deviation and skewness of the distribution of the bootstrap replications obtained from the $B = 1000$ bootstrap samples are shown in the 4th, 5th and 6th columns in Tables 1 and 2. With respect to 9 bootstrap samples in the first set of data, however, the maximum likelihood estimates were not obtained uniquely due to the "complete separation" (Anderson, 1974). For this reason, these 9 samples were excluded in the calculation of the typical values of the distribution.

Table 2 Estimation of correct classification rates for two prediction rules

1. β-blocking activities

rule	A	CV	Bootstrapping (B=991: §4)			Bootstrapping (B'=1000: §5)		
			MEAN	S. D.	SKEW	MEAN	S. D.	SKEW
(1)	62.16	48.65	62.67	4.705	-0.228	59.62	4.227	-0.134
(2)	62.16	54.05	62.71	4.593	-0.182	59.78	4.117	-0.079

2. photosynthesis inhibition activities

rule	A	CV	Bootstrapping (B=1000: §4)			Bootstrapping (B'=1000: §5)		
			MEAN	S. D.	SKEW	MEAN	S. D.	SKEW
(1)	57.50	50.00	53.53	5.607	-0.648	53.03	4.859	-0.120
(2)	50.00	45.00	48.86	7.112	-0.768	49.08	4.750	-0.187

A: apparent estimates
CV: cross-validation estimates
SKEW: skewness

V. RESIDUAL ANALYSIS

It would be natural to define the residuals of the logistic regression for the ordered categorical responses by $Z_j - \hat{P}_j(\bullet)$, where Z_j ($j=1, \ldots, k-1$) are binary component variables. In order to define a scalar residual, these $k-1$ residuals must be combined. Scalar residuals can also be defined by:

$$r = 1 - \hat{p}_j(\bullet), \text{ for } y = j. \tag{5}$$

In order to detect possible outliers using the scalar residuals, we try to extend the empirical probability plot proposed by Landwehr et al. (1984) as follows:

Step 1. Obtain residuals $\{r_i\}$ from the fitted model as in Eq. (5).

Step 2. Obtain $\{r_{(i)}\}$ by arranging $\{r_i\}$ in increasing order.

Step 3. Simulate data y_i^* from the k-nomial distribution whose cell probabilities are $\hat{p}_1(x_i), \ldots, \hat{p}_k(x_i)$ $(i=1, \ldots, n)$.

Step 4. Fit the same model to $(x_1, y_1^*), \ldots, (x_n, y_n^*)$ to obtain the bootstrap replications $\hat{\theta}^*$ and $\hat{\beta}^*$. Further, calculate $\hat{p}_j^*(x_i)$, $i=1, \ldots, n$, $j=1, \ldots, k$, using these values.

Step 5. Compute the residuals of y_i^* i.e. $\{r_i^*\}$ using Eq. (5), and order $\{r_i^*\}$ to obtain $\{r^*_{(i)}\}$.

Step 6. Repeat steps 3 to 5 B' times independently.

Step 7. Compute typical values of the ordered residuals over B' replications, say, $T_{(i)} = \text{median}\{r^*_{(i)}\}$, $i=1, \ldots, n$.

Step 8. Plot the ordered residuals $\{r_{(i)}\}$ from the original fit, against the typical values $T_{(i)}$ of the ordered residuals from the simulation, that is, plot $(T_{(i)}, r_{(i)})$, $i=1, \ldots, n$.

Step 9. Obtain and plot upper and lower confidence bounds at each of the $T_{(i)}$ by determining the upper and lower quantiles of B' values $\{ r^*_{(i)} \}$, which correspond to the prescribed confidence coefficents.

The empirical probability plot for the second set of data is shown in Fig. 4, where the number of the bootstrap samples is selected as $B'=100$. In Fig. 4, two residuals at the upper right part are located outside the confidence bounds. These largest residuals corresponded to the 6th and 22nd observations.

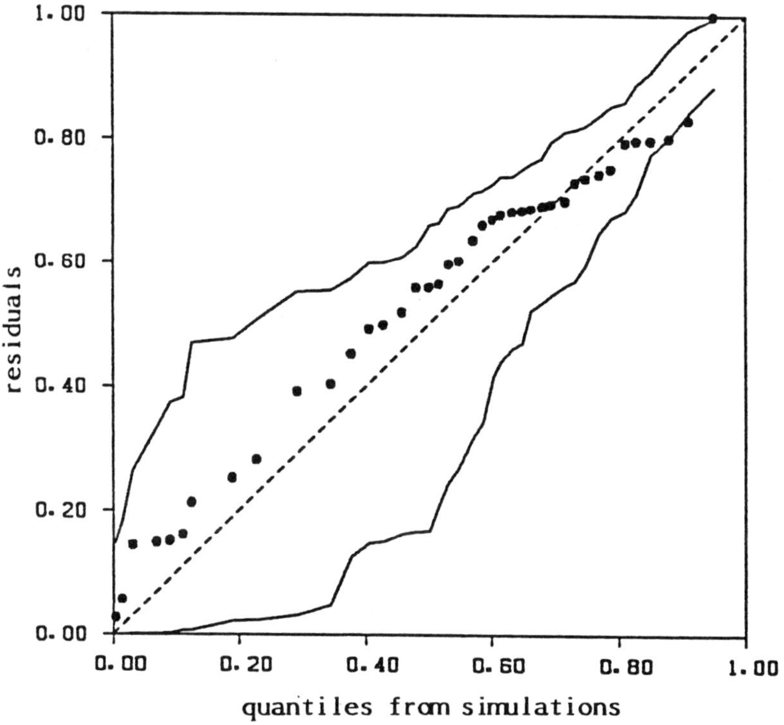

Figure 4 Empirical probability plot for the photosynthesis inhibition activities.

Only the 22nd observation was pointed out as an outlier by the sensitivity plot of maximum log likelihood based on the

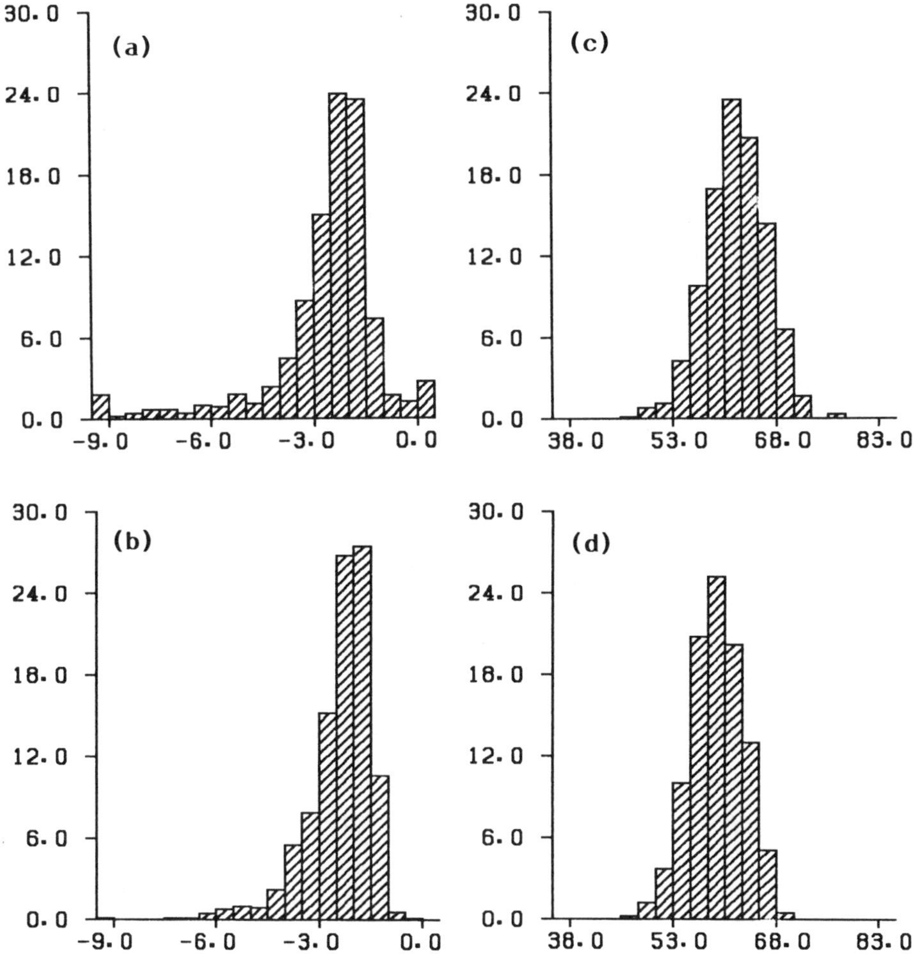

Figure 5 Distributions of bootstrap replications for β-blocking activities.

(a) Estimates of β_1 (Bootstrapping in §4)
(b) Estimates of β_1 (Bootstrapping in §5)
(c) Correct classification rate (Bootstrapping in §4)
(d) Correct classification rate (Bootstrapping in §5)

rule to exclude one observation at a time, though the plot is not shown here. Accordingly, the empirical probability plot was drawn, for *39* observations with the 22nd observation excluded, and it was found that all of the residuals obtained were located inside the confidence bounds.

The steps 3 to 6 suggest an alternative application of the bootstrap method in logistic regression analysis. Results obtained from the $B'=1000$ bootstrap samples generated as in the step 3, are shown in the 7th, 8th and 9th columns in Tables 1 and 2. The histograms for the $\hat{\beta}_1$ of the first set of data and for correct classification rates based on the prediction rule (1) are shown in Fig. 5. The values used in the histograms are replications obtained from two ways of bootstrapping.

From these results, it was suggested that the estimates of standard deviations for the maximum likelihood estimates based on the large-sample theory were too small, that the naive bootstrap procedure mentioned in section 4 provided greater deviations, and that distributions of $\hat{\beta}$ and $\hat{\theta}$ were skewed. It was also suggested that both bootstrap procedures could give more pessimistic estimates on the correct classification rates than the apparent rates.

VI. CONCLUDING REMARKS

If we want to obtain some productive findings in data analysis, instead of sticking to not carefully examined data, models and/or approaches set up at early stage of analysis, we should adjust them in the light of various phenomena in which they are involved in many cases. Information for such adjustments is usually obtained by means of stability analysis. Gifi (1981) has discussed on the subject of stabilities in multivariate analysis, by dividing them into three area like analytical, algebraic and replication stabilities. As the general approaches available in the evaluation of replication stability, he has introduced the delta method based on the asymptotic theory, and also the jackknife, the subsampling and the bootstrap methods. The last three methods reuse the given sample in the randomized way and therefore he called these the randomization methods.

Suppose that a sample of size n is given. Then by deleting successively one observation at a time, n sets of subsamples of size $n-1$ are generated in the normal jackknife method and correspondingly n pseudo-replications are obtained. In this case, by pseudo-relications we mean n statistics which are calculated based on those subsamples of size $n - 1$. Therefore, the number of different pseudo-replications is at most $\binom{n}{1}$. In the bootstrap method, on the other hand, overlapping of individual observations is allowed, and samples of size n are extracted with replacement. Let B be the repetition number in the bootstrap method. Then the corresponding B pseudo-replications are obtained. Accordingly, the number of different pseudo-values is at most $\binom{2n-1}{n}$. The two methods mentioned above give greatly different results in the case of estimating medians. In this case the jackknife method fails because most pseudo-values may be the same, however, the bootstrap method performs properly. If there were any outliers in the original sample, these two methods are influenced greatly by them. Impact on results from outliers depends on the two methods; for example, as far as inference of correlation is concerned the bootstrap method provides more resistant statistic against outliers, while on regression problem the jackknife method provides relatively higher resistance (see Tasaki & Goto, 1982, 1983).

We see that it is in the case of small samples for the bootstrap method to be superior among the other similar methods, say delta method and etc. Tasaki and Goto (1985) have discussed useful applications of the bootstrap method to the subjects on (1) test of dose-response relationships when observed response rates are close to eigher 0 or 1, (2) comparison of multivariate samples with smaller sample sizes compared to the number of variates, and (3) principal component analysis in small samples.

ACKNOWLEDGEMENTS

The authors would like to thank Prof. Takeaki Nagai of Oita University who had kindly read the first draft of this paper and gave them useful suggestions.

REFERENCES

1. Anderson, J.A. (1974). Diagnosis by logistic discriminant function: further practical problems and results. Appl. Statist., 23, 397-404.
2. Anderson, J.A. & Philips, P.R. (1981). Regression, discrimination and measurement models for ordered categorical variables. Appl. Statist., 30, 22-31.
3. Copas, J.B. (1983), Plotting p against x. Appl. Statist., 32, 25-31.
4. Efron, B. (1979). Bootstrap methods: another look at the jackknife. Ann. Statist., 7, 1-26.
5. Gifi, A. (1981). Nonlinear Multivariate Analysis.

University of Leiden, The Netherlands.

6. Goto, M., Inoue, T. & Tsuchiya, Y. (1986). Power-transformations for ordered categorical data. Behaviormetrika, 19, 17-39.

7. Landwehr, J.M., Pregibon, D. & Shoemaker, A. C. (1984). Graphical methods for assessing logistic regression models (with discussion). J. Amer. Statist. Assoc., 79, 61-83.

8. McCullagh, P. (1980). Regression models for ordinal data (with discussion). J. Roy. Statist. Soc., B42, 109-142.

9. Plackett, R. L. (1974). The Analysis of Categorical Data. London: Griffin.

10. Tasaki, T. & Goto, M. (1982). Bootstrap method and outliers. The Proceedings of 50th Conference of Japan Statistical Society, 63-64. (in Japanese.)

11. Tasaki, T. & Goto, M. (1983). Bootstrap method in regression analysis. The Proceedings of 51st Conference of Japan Statistical Society, 37-38. (in Japanese.)

12. Tasaki, T. & Goto, M. (1984). Logistic regression model for ordered categorical response and its applications to quantitative structure-activity relationships. Bull. of the Biometric Society of Japan, No. 5, 19-35.

13. Tasaki, T. & Goto, M. (1985). Bootstrap method in multivariate analysis for small samples. The Proceedings of 53rd Conference of Japan Statistical Society, 32-33. (in Japanese.)

14. Yamakawa, M., Ezumi, K. & Tasaki, T. (1986). Prediction of bioactivity with physicochemical indices. Japanese Journal of Applied Statistics, 15, 1-12. (in Japanese.)

SPECIFIC DESIGN OF A SOFTWARE FOR MULTIVARIATE DESCRIPTIVE STATISTICAL ANALYSIS THE CASE OF SPAD.N

Alain Morineau

CEPREMAP 142 rue du Chevaleret
75013 Paris FRANCE

I. GENERAL VIEW OF THE SOFTWARE

Multivariate descriptive statistical techniques, as developed for example by Benzécri (2), are divided into two main areas: principal axes methods and clustering techniques. Software SPAD.N is intended to implement these exploratory analyses for large arrays of data: several thousand rows or observations, and several hundred columns, mixing categorical and continuous variables.

Descriptive Principal Component Analysis performs the analysis of a set of continuous variables through correlation or covariance matrices. Two-way Correspondence Analysis is available for description of contingency tables, and Multiple Correspondence Analysis, for a set of categorical variables. These principal axes analyses describe the dispersion of a set of n points in p-dimensional space as points in low dimensional subspaces where we can appreciate dependences in terms of proximities.

ISBN 0-12-215485-1

Clustering techniques are exploratory methods usually used along with principal axes analyses. Some of these methods implemented in SPAD.N are described in (9).
In exploratory analysis, the statistician's behavior is radically different from that in mathematical statistics. An appropriate set of tools was to be designed. The problem actually was: - to work with groups of variables, which must be easily selected and handled; - to perform complex transformations on the variables; - to link heavy processings of data giving rise to voluminous intermediary results; - to describe the results themselves, which are often complex and require specific interpretation devices.

II. DESIGN OF THE SOFTWARE: MODULARITY

In developing the multivariate data analysis package, facility and efficiency must be balanced against one another. For convenience of use, an analysis is a sequence of "steps", called "procedures" (or PROC). The communication of the results from one step to another is automatically performed by means of external storage. Modularity contributes to the flexibility of the package: steps can be used in unforeseen sequences for particular applications. The user can select the most appropriate set of analytical steps for the data, and can save the results of any step for subsequent analysis. In practice each step is called by using a keyword and by defining a short sequence of parameters.
The following is an example of a sequence of steps (typical analysis of a set of categorical variables). Subsequent sections point out some of the pecularities in each part of the program.

PROC SELEC (see also section 3). This step allows selection of variables (columns of the table) and of observations (rows of the table) for the next step CORMU. We choose the "active" and "illustrative" variables among the categorical variables (continuous variables are also allowed). "Active" elements are those used to compute similarities and relationship, "illustrative" ones are a posteriori used to interpret the observed patterns.

PROC CORMU (see section 4). This is the main step for computation of multiple correspondence analysis. We get the eigenvalues and percentages of variance, the coordinates and contributions to variance for each element. "Test-values" (described lelow) are printed for each illustrative category in order to help the interpretation of the principal axes.

PROC GRAPH (see section 7). We may obtain many different types of graphical displays on any pair of axes. We can represent active categories, all or part of the illustrative categories, observations, plots mixing rows and columns, and graphical display of continuous variables whose coordinates are their correlation coefficients with the axes.

PROC RECIP (see section 5). Fast hierarchical algorithm on principal coordinates, one of the control parameters being the number of axes. The outputs include a graphical display of the dendrogram.

PROC PARTI. A final partition is chosen after visual inspection of the dendrogram. The cutting of the hierarchical tree is followed with a reallocation procedure in order to improve the quality of the final partition.

PROC DECLA (see section 6). This step provides a synthetic description of the classes using the basic continuous and categorical variables. It computes the contingency tables which cross-classify the partition with the categorical variables (and the chi-2 statistics). Then the classes are characterized by the most significant categories and continuous variables.

PROC GRAPH. Here the step is used to obtain graphical displays of cluster centers simultaneously with the categorical points, as well as density plots in which observations are identified by the digit of their cluster.

PROC ESCAL (see section 8). This step, which is also used to create new variables from old ones, is used here to save the main results of the analysis on a file, in such a way that they can be re-used for further interpretation.

III. EASY SELECTION OF RELEVANT DATA (PROC SELEC)

This part of the program selects "active" and "illustrative" variables and individuals, which is an essential operation in data analysis. The following example shows how the user specifies his command with an example.
The list of active categorical (or nominal) variables follows the key word NOMI ACT. Then the illustrative categorical variables are on two lines following NOMI ILL (the continuation mark is ">"). Continuous variables may only be illustrative (CONT ILL). The selection of individuals is announced by the command LSELI=FILT, which means that selection will be performed by "filtering" individuals through

logical conditions.
The conditions are expressed on the last line: select the males (value 1 for VII) who are workers (values 1, 2, 3 or 7 for (V24). Anything following ":" is a comment.

```
-------------------------------------------------------
  PROC SELEC
 ***  multiple correspondence on male workers  ***
:--- parameters (if not the default values)
LSELI = FILT
:--- retained variables
NOMI ACT 1, 2, 4--6, 8--10
NOMI ILL 12 17 19--22 25--27 30--32  34--36 38--40 >
             42  44  45  3  7
CONT ILL  37 43   46--59
END
:--- selection (VII=sex, V24=social category)
VII= 1 * V24 = 1,2,3,7
-------------------------------------------------------
```

The procedure SELEC allows the use of any variable as a weight for the individuals, or allows to choose a random weight if we want to check the stability of the results by simulations.

IV. SAVINGS IN THE COMPUTATIONAL STEP (PROC CORMU)

When categorical data array is in a complete disjunctive form, the amount of computation can be limited to the number of variables, whatever the number of categories. Considerable savings in computation costs occur since the transformation to binary coding is implicitly performed within the program.

Other technical features allow computation savings. For example the core diagonalization for multiple correspondence table is performed on a reduced matrix obtained after transformation of the table (the table is first expressed in a subspace corresponding to its rank dimensionality). When all responses are binary, the dimension of the table to be diagonalized is twice smaller.

For Correspondence Analysis (PROC CORBI) a program parameter allows the user to compute eigenvalues and eigenvectors without computing the matrix to be diagonalized. The algorithm allows the determination of the first eigenvectors by direct reading of the data (6), (5).

As we may have hundreds of illustrative categories likely to participate in the interpretation of the first principal axes, the step allows the user to select the relevant ones for each axis: the "test-values".

Let us suppose as a null hypothesis that the n(A) individuals in category A are drawn randomly (without replacement) from the n individuals. Then the expectation of the coordinate of point A (mean point of the individuals in category A) is 0 on any axis. Its variance is :

$$\mathrm{Var}[x(A)] = (n - n(A)) \; / \; (n - 1) \; n(A)$$

By application of the central limit theorem, the distribution of the point on an axis is approximated by the normal distribution, and :

$$t(A) = \sqrt{(n - 1) \; n(A) \; / \; (n - n(A))} \; x \; (A)$$

has an approximate standardized normal distribution under the null hypothesis. This value is called the test-value of point A on the principal axis. The greater the test-value is, the more significant the category is for the interpretation of the axis (11).

V. SPECIFIC TOOLS FOR CLUSTERING (PROC RECIP)

Classifying a large data set is a complex task. Generaly neither the distance matrix nor the data array can be stored in fast memory. Distances must be computed when needed and algorithms must proceed by reading the data file row by row. Computation time however has to remain within reasonable limits. One possible strategy provided by the software is to proceed in the following way (12).

First we restrict the data to the first principal coordinates. Usually the ten or twenty first principal coordinates can reconstitute most of the non-random part of data. Distances are computed with substantial saving, and the array can be stored in fast memory.

We build a complete hierarchical tree using the Reciprocal Neighbors Chain Search Algorithm (3). It is a fast algorithm for which maximal cost is proportional to n^2 , where n is the number of objects. We use Ward's criterion for aggregation, which is a minimum variance criterion that ensures the consistency with principal axes methods.

Inspection of the dendrogram gives hints about suitable partitions. However any partition obtained by cutting a hierarchical tree is expected to be sub-optimal due to the constraint that the k-partition wust contain the (k+1)-partition and be included in the (k-1)-partition.

A Moving-Center Aggregation (k-means type algorithm) improves the initial partition. Iterations stop when the percentage gain is below a threshold (e.g. 0.05%). It usually needs less than 10 iterations (PROC PARTI).

VI. AUTOMATIC CHARACTERIZATION OF CLASSES (PROC DECLA)

For exploratory or descriptive purposes, the main interest is to obtain interpretable classes. Each class is described with the help of all the original variables. The continuous variables can be ranked by comparing their means within the class, to their overall means. To rank the importance of variable x, one can compare the actual mean with the expected mean assuming that individuals were allocated at random (without replacement) into the class. For class k, we use the "test-value":

$$t = (\bar{x}(k)-\bar{x})/s(x,k) \quad \text{with} \quad s^2(x,k)=(n-n(k))s^2(x)/(n-1)n(k)$$

The categories of categorical variables can be ordered in a similar way. If there are n(j) individuals in category j, and n(j,k) individuals who are also in class k, we compute :

$$p(j, k) = \text{Prob}(N \geq n(j, k))$$

N is a hypergeometric variable with parameters n, n(j) and n(j,k). The hypergeometric null hypothesis expresses random allocation in class k. Test-values are the normal values giving the same probabilities, and are used to sort categories according to their interest for describing class k (11).

VII. THE GRAPHICAL TOOLS (PROC GRAPH)

Graphical pictorialization is certainely an important step in multivariate data analysis. The package must provide powerful and flexible facilities to make the graphical operations on principal axes easy to perform. The characteristics of each graph are determined by the set of 21 parameters, most of them can take a default value. Any command statement for a graph can specify:

- the type of points to be included: individual points, variable points, categorical points, cluster points;
- the coordinate axes: any principal axis or quantitative variable; the direction of an axis may be reversed for easier comparison with others graphs;
- the identifiers: type and length of the identifiers of the points (it is also possible to choose arbitrarily a character to differenciate a category of individuals);
- the selection of points: the user may request automatic selection of the points according to their variance on the axes, or to their squared cosines, or both;
- the processing of superimposed points: either ask the program to shift slightly any point which would otherwise be hidden, or to list the coordinates of hidden points;
- the size of the graph printout: it is determined in height and width, either by a number of characters and lines, or by a number of pages;
- the graph scope: a parameter eliminates (or returns to the margin) all points which are too far away from the centre; otherwise the user may impose the scales (for example, to superimpose graphs); etc.

VIII. CREATING AND COPYING VARIABLES (PROC ESCAL)

This procedure enables the user to copy results from previous analyses (principal coordinates, and clusters), and to create new variables using existing variables (4),(5). Operations to be performed on the data are specified in a "program" written by the user, with a language easy to use and versatile. In addition to standard operations (arithmetical and logical operators, usual statistical transformations, random values, partition of continuous variables, cross-tabulation of two variables,...), the language includes possibility of loops, jumps, "if-then-else", etc.

Some functions in the language are specifically dedicated to preprocessing the data. For example: NITEM(V) gives the number of categories of variable V; SUMIT(3,V) gives the number of individuals in the third category of V; COUNT(2,list...) gives the frequency of value 2 in the given list of variables; BINCOD creates the binary coding of a categorical variable.

IX. EXTENSIONS OF THE SOFTWARE

Such a modular software may easily develop, since the addition of new procedures is straight-forward. Here are two of the next extensions.

Exploratory analysis of textual data: the texts of responses to open-ended questions or interview transcripts are submitted to analysis without any preliminary coding (8). Synthetic graphical displays, connected mainly with Correspondence Analysis, show relationships between the words most frequently encountered.

Statistical criteria can select the most typical words of any category of respondents, and the most characteristic sentences in order to insert words in their context (10). The emphasis is put on the connection between textual and numerical information in the data: words can be located on planes describing relationship between categories; similarly categories can be located on planes describing relationship between words.

Local and partial analyses: contiguity constraints can be introduced in the analysis in order to describe the local (or partial) relationship between individuals (1),(7). One may be interested in keeping constant the "graph effect" in order to visualize the discrepancies between local level and global level, and show which level is responsible of the observed patterns.

REFERENCES

1. Aluja, T., Lebart, L. (1984): Local and Partial Principal Component Analysis and Correspondence Analysis, In "COMPSTAT 1984", Physica Verlag, 113-118.
2. Benzécri, J.P. (1973): L'Analyse des Données, Tome 1 (La Taxinomie), Tome 2 (L'Analyse des Correspondances). Dunod, Paris.
3. Benzécri, J.P. (1982): Construction d'une Classification Ascendante Hiérarchique par la Recherche en Chaine des Voisins Réciproques. Cah. Anal. Donn., 7, 209-218.
4. Lambert, T. (1984): Présentation d'un Système Portable de Prétraitement, L'Etape Escal. Bull. Techn. CESIA, 2, 28-43.
5. Lambert, T. (1986): Réalisation d'un Logiciel d'Analyse de Données. Thesis, Université Paris-Sud.

6. Lebart, L. (1974): On the Benzecri's Method for Finding Eigenvectors by Stochastic Approximation, In "COMPSTAT 1974", Physica Verlag, 202-211.

7. Lebart, L. (1978): Programme d'Agrégation avec Contraintes. Cah. Anal. Donn., 3, 3, 275-287.

8. Lebart, L. (1982): Exploratory Analysis of Large Sparse Matrices with Application to Textual Data, In "COMPSTAT 1982", Physica Verlag, 67-76.

9. Lebart, L., Morineau, A., Warwick, K. (1984): Multivariate Descriptive Statistical Analysis, Correspondence Analysis and Related Techniques for Large Matrices. Wiley, New York.

10. Morineau, A. (1984): Computational and Statistical Methods for Exploratory Analysis of Textual Data, In "COMPSTAT 1984", Physica Verlag, 372-377.

11. Morineau, A. (1986): Inferential Techniques Following a Multivariate Descriptive Statistical Analysis, In "Proceeding 2d Catalan International Symposium on Statistics Barcelona", 321-329.

12. Morineau, A., Lebart, L. (1986): Specific Clustering Algorithms for Large Data Sets and Implementation in SPAD Software. In "Classification as a Tool of Research", (W. Gaul and M. Schader, eds), Elsevier North-Holland, 321-329.

A TEST OF GOODNESS OF FIT BASED ON GRAPHICAL METHOD

Kazumasa Wakimoto

Okayama University
2-1-1, Tsushima-Naka, Okayama 700, JAPAN

I. INTRODUCTION

In the previous papers([1] and [2]), we proposed a new test statistic for testing the goodness of fit in the case of the continuous and the multinomial distributions. In here, we propose more another test statistic in the case of the multinomial distribution and give the similar discussion as the previous paper[2] for it.

Let us consider an experiment in which there are k mutually exclusive possible outcomes $A_1, A_2, \ldots\ldots, A_k$. Let p_i be the probability that event A_i will occur at a trial of the experiment and let n trials be made. Let n_i be the observed value of the event A_i which is a random variable such that $E(n_i)=np_i$, i=1, 2,...., k (see Table 1-1).

Table 1-1

	A_1 A_2 ...... A_k	Total
Observed value	n_1 n_2 ...... n_k	n
Expected value	np_1 np_2...... np_k	n

ISBN 0-12-215485-1

Then we consider the test of the null hypothesis

$$H_0 : p_i = p_i^* , \ i=1, 2, \ldots, k.$$

where the p_i^* is postulated value of the probability.

For this test, the traditional test statistics are mainly T_1, T_2, T_3 and W (see[2]).

We discuss a new test statistic using the linked line chart, and calculate the exact percent point for some values of n, k in the case of $p_1=p_2 \cdots\cdots = p_k$. We also illustrate the comparison of the power for some specified value of p_i between the proposed test statistic and the traditional test statistics.

II. GRAPHICAL REPRESENTATION AND A TEST STATISTIC

Let

$$d_i = \frac{(n_i - np_i)^2}{np_i}, \quad i=1,2,\ldots,k.$$

Then we draw a linked line chart as following steps:

Step 1. Rearrange the values $d_1, d_2, \ldots, d_k$ of size k in the order of magnitudes as follows:

$$d_{(1)} \leqq d_{(2)} \leqq \cdots\cdots \leqq d_{(k)} .$$

Step 2. Draw the x-axis and the y-axis intersecting at right angles at the origin O. Draw a line $\overline{OP_1}$ of length $d_{(k)}$ in the direction of the x-axis starting from the origin O. Next, draw another line $\overline{P_1P_2}$ of the length $d_{(k-1)}$ in the direction of $\pi/2(2k-1)$ radians from the line $\overline{OP_1}$ starting from the point P_1. Continue in the same manner and draw other lines $\overline{P_2P_3}, \ldots, \overline{P_{k-1}P_k}, \ldots, \overline{P_{2k-1}P_{2k}}$ of the lengths $d_{(k-1)}, \ldots, d_{(1)}, d_{(1)}, \ldots, d_{(k)}$ in the direction of $\pi/2(2k-1)$

radians from the former lines(see Fig. 1).

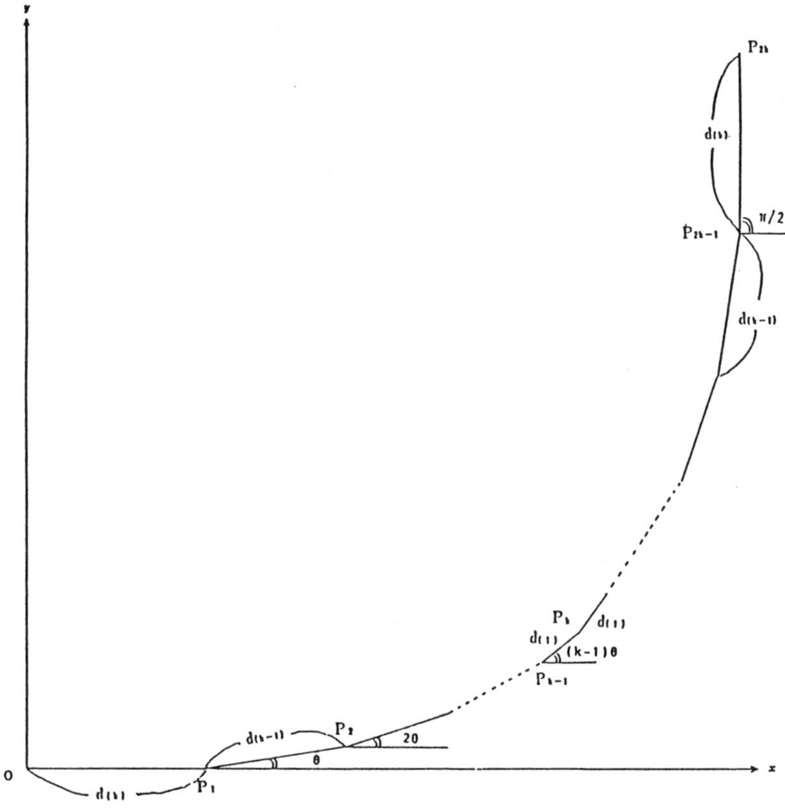

Fig. 1

In Fig. 1, let us denote the length of $\overline{OP}_{2k}$ by L_n^2 .

Then we have

$$L_n^2 = x^2 + y^2,$$

where

$$x = \sum_{i=1}^{k} D_{(k-i+1)} \cos\frac{i\pi}{2(2k-1)} + \sum_{i=k+1}^{2k} D_{(i-k)} \cos\frac{i\pi}{2(2k-1)}$$

$$y = \sum_{i=1}^{k} D_{(k-i+1)} \sin\frac{i\pi}{2(2k-1)} + \sum_{i=k+1}^{2k} D_{(i-k)} \sin\frac{i\pi}{2(2k-1)}.$$

$$L_w = \frac{k}{n} L_n^2 .$$

III. CALCULATION OF THE EXACT PERCENT POINTS

We use usually the approximate percent points based on the asymptotic χ^2 distribution for testing the hypothesis H_0. Therefore we have some errors compare with the test using the exact percent points, for example, as shown in Table 3-1, where

$$T_1 = \sum_{i=1}^{k}(n_i - np_i)^2/np_i ,$$

$$T_2 = 2\sum_{i=1}^{k} n_i \log(n_i/np_i) ,$$

$$T_3 = 4\sum_{i=1}^{k}(\sqrt{n_i} - \sqrt{np_i})^2 .$$

Table 3-1 The exact probabilities for 95 percent approximate point 9.49 in the case of k=5

n	10	30	50	70	90
$P(T_1 \leqq 9.49)$	96.0	95.4	95.4	95.0	95.2
$P(T_2 \leqq 9.49)$	92.5	94.5	94.6	94.6	94.7
$P(T_3 \leqq 9.49)$	63.9	91.6	92.9	94.4	94.3

So, we obtained the exact percent points of L_W by calculating all possible combinations with the probability

$$\frac{n!}{n_1!\, n_2!\, \ldots\, n_k!}\, p_1^{n_1}\, p_2^{n_2} \,\ldots\, p_k^{n_k}$$

for $(n_1, n_2, \ldots, n_k)$ satisfying the constraint

$$\sum_{i=1}^{k} n_i = n$$

under the hypothesis H_0: $p_1 = p_2 = \ldots = p_k$.

Example. Exact percent points of L_w are obtained as shown in Table 3-2 and Table 3-3.

Table 3-2 Exact 95 percent point of L_w

n \ k	3	4	5	6	7	8
10	28.04 (0.9410 ∼0.9776)	53.37 (0.9485 ∼0.9629)	101.2 (0.9447 ∼0.9602)	172.4 (0.9461 ∼0.9555)	239.4 (0.9411 ∼0.9616)	402.3 (0.9458 ∼0.9655)
20	14.03 (0.9445 ∼0.9549)	27.05 (0.9378 ∼0.9501)	51.86 (0.9497 ∼0.9519)	88.91 (0.9498 ∼0.9504)	127.0 (0.9491 ∼0.9513)	189.0 (0.9467 ∼0.9501)
30	7.626 (0.9318 ∼0.9522)	19.68 (0.9477 ∼0.9510)	36.60 (0.9499 ∼0.9527)	58.21 (0.9499 ∼0.9512)	90.53 (0.9499 ∼0.9503)	127.0 (0.9493 ∼0.9503)

Table 3-3 Exact 99 percent point of L_w

n \ k	3	4	5	6	7	8
10	70.93 (0.9898 ∼0.9944)	123.7 (0.9863 ∼0.9946)	203.6 (0.9898 ∼0.9950)	308.7 (0.9853 ∼0.9928)	483.8 (0.9897 ∼0.9916)	646.0 (0.9896 ∼0.9903)
20	30.29 (0.9864 ∼0.9911)	58.56 (0.9893 ∼0.9911)	102.2 (0.9896 ∼0.9901)	155.6 (0.9897 ∼0.9904)	240.8 (0.9899 ∼0.9906)	327.0 (0.9896 ∼0.9900)
30	18.34 (0.9860 ∼0.9901)	38.38 (0.9893 ∼0.9904)	70.58 (0.9897 ∼0.9901)	109.2 (0.9898 ∼0.9901)	157.7 (0.9899 ∼0.9901)	

IV. COMPARISONS OF THE POWER AMONG T_1, T_2, T_3, W AND L_w

Let us consider the alternative hypothesis H_1: $p_1=p_2=\ldots=p_4=\theta$, $p_5=1-4\theta$ in the case of k=5. Then we can calculate the exact power of the test statistics ; T_1, T_2, T_3, W and L_w for $\theta=0.01, 0.02, \ldots.. 0.50$ using the exact percent points. We show the power and its range for some values of θ ; $\theta=0.14, 0.18, 0.22, 0.24$ in Table 4-1.

Table 4-1 Exact power and its range

θ		Power	Range
0.14	T_1	0.6785 ∿ 0.6964	0.0179
	T_2	0.6311 ∿ 0.6314	0.0003
	T_3	0.5589 ∿ 0.5613	0.0024
	W	0.6876 ∿ 0.6896	0.0020
	L_w	0.6718 ∿ 0.6821	0.0103
0.18	T_1	0.1131 ∿ 0.1263	0.0132
	T_2	0.1104 ∿ 0.1117	0.0013
	T_3	0.0998 ∿ 0.1026	0.0028
	W	0.1196 ∿ 0.1218	0.0022
	L_w	0.1140 ∿ 0.1190	0.0050
0.22	T_1	0.1007 ∿ 0.1151	0.0144
	T_2	0.1186 ∿ 0.1210	0.0024
	T_3	0.1231 ∿ 0.1261	0.0030
	W	0.1071 ∿ 0.1104	0.0033
	L_w	0.1038 ∿ 0.1077	0.0039
0.24	T_1	0.3485 ∿ 0.3902	0.0417
	T_2	0.5127 ∿ 0.5212	0.0085
	T_3	0.5689 ∿ 0.5714	0.0025
	W	0.3644 ∿ 0.3758	0.0114
	L_w	0.3533 ∿ 0.3566	0.0033

From the values of exact power for θ=0.01, 0.02,...., 0.05, we find that the powers of T_1, W and L_w are larger than those of T_2 and T_3 for $\theta<0.2$. and the powers of T_2 and T_3 are larger than those of T_1, W and L_w for $\theta>0.2$. We also find that though the power of L_w is very similar to those of T_1 and W, the range of power L_w is different from those of T_1 and W(see Table 4-1). If we change the values of k and n under the conditions; $p_1=p_2=....=p_{k-1}=\theta$, $p_k=1-(k-1)\theta$, we obtain the similar results.

REFERENCES

1. Shirahata, S., Wakimoto, K. and Tarumi, T. (1985). A Goodness of Fit Test based on the Linked Line Chart, Austral. J. Statist. 27, 163-171.

2. Wakimoto, K., Odaka, Y. and Kang, L. (1987). Testing the Goodness of Fit of the Multinomial Distribution based on Graphical Representation, Computational Statistics and Data Analysis, 5, 137-147.

GRAPHICAL ANALYSIS OF RANKS

Yasumasa Baba

Institute of Statistical Mathematics
4-6-7, Minami-Azabu, Minato-ku, Tokyo, JAPAN

I. INTRODUCTION

In various practical fields, ranking plays an important role as a measure to evaluate something which cannot be measured on any objective scale. For instance, in the field of sensory evaluation ranking is indispensable.

In this paper a graphical method for analysis of ranks is presented, which is based on the rank graph. In the following section rank graph and linked rank graph are illustrated briefly according to the present author's paper (2). A graphical method for rank test is shown and examined by Monte Carlo method. Comparing the graphical method with Kramer's rank test by using an example, it is shown that the graphical method has advantages.

II. GRAPHICAL REPRESENTATION OF RANKS

In this section let us illustrate the rank graph and linked rank graph. Let us call the individual doing the ranking judge and the material to be ordered item. Let us assume n sets of rankings of k items are given by n judges as follows:

$$\begin{array}{l} R_{11}, R_{12}, \ldots, R_{1k} \\ R_{21}, R_{22}, \ldots, R_{2k} \\ \vdots \quad \vdots \quad\quad \vdots \\ R_{n1}, R_{n2}, \ldots, R_{nk} \end{array}$$

ISBN 0-12-215485-1

where $(R_{\alpha 1}, R_{\alpha 2}, \ldots, R_{\alpha k})$ is a set of rankings given by the α-th judge. Let the frequency of rank ℓ given to the j-th item by n judges be $f_{j\ell}$. That is,

$$f_{j\ell} = \sum_{\alpha=1}^{n} \#(R_{\alpha j}=\ell). \qquad (1)$$

Then

$$n = f_{j1} + f_{j2} + \ldots + f_{jk} \quad (j=1,2,\ldots,k)$$

holds. Let

$$p_{j\ell} = f_{j\ell} / n \qquad (j=1,2,\ldots,k;\ell=1,2,\ldots,k) \qquad (2)$$

Then $p_{j\ell}$ will be the relative frequency or ratio of the ℓ-th rank assigned to the j-th item. Define the angles corresponding to rank ℓ by

$$\theta_\ell = \frac{\ell - 1}{k - 1}\pi \qquad (\ell=1,2\ldots,k) \qquad (3)$$

The direction corresponding to the first rank is 0, the direction corresponding to the second rank is $\pi/(k-1)$, and so on. The final is therefore π.

We will show the procedure for constructing a linked rank graph and a rank graph as follows:

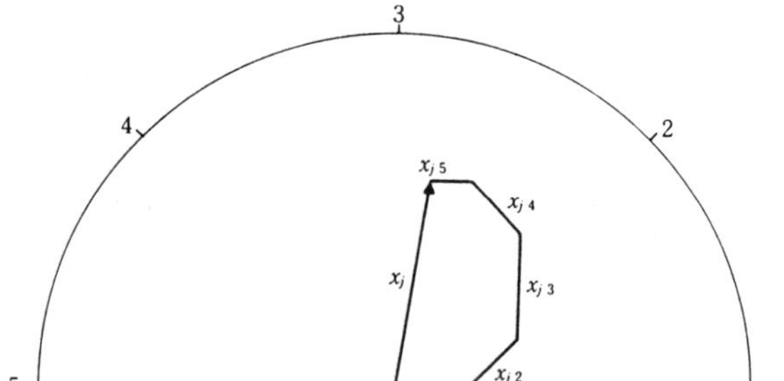

Fig.1 Illustration of the way to present rank data. The example shows a case with 5 rankings.

Linked rank graph

(1) Draw a semicircle with the radius of unity and scale marks corresponding to θ_ℓ. Then θ_ℓ represents the direction assigned to rank order ℓ (see Figure 1).

(2) Put

$$x_{j\ell} = (p_{j\ell}\cos\theta_\ell \; , \; p_{j\ell}\sin\theta_\ell \;), \tag{4}$$

then draw a linked line formed by connecting the vectors $x_{j1}, x_{j2}, \ldots, x_{jk}$ in this order.

The graph which consists of such linked lines for all items is called "linked rank graph". This graph represents the distribution of the ranks given to the item j by the judges. An example of linked rank graph is shown in Figure 2. It should be noted that the total length of a linked line is unity and it is always inside the semicircle, because

$$p_{j1} + p_{j2} + \ldots + p_{jk} = 1 \qquad (j=1,2,\ldots,k)$$

holds.

Rank graph

(1) Draw a semicircle with the radius of unity and scale marks corresponding to θ_ℓ as in the linked rank graph.

(2) Define the vector x_j by

$$x_j = \sum_{\ell=1}^{k} x_{j\ell}$$

$$= (\sum_{\ell} p_{j\ell}\cos\theta_\ell \; , \; \sum_{\ell} p_{j\ell}\sin\theta_\ell) \; (j=1,\ldots,k) \tag{5}$$

which is the resultant vector of $x_{j1}, x_{j2}, \ldots, x_{jk}$.

For the sake of convenience, we will call x_j the item vector hereafter. Draw the item vectors corresponding to each item. We call this "rank graph". An example of rank graph is shown in Figure 3.

III. GRAPHICAL TEST OF RANKS

Let the frequency of the rank ℓ given to an item by n judges be f_ℓ, which satisfies the following relation.

$$n = f_1 + \ldots + f_k$$

The item vector is defined as follows:

$$x = (c, s), \tag{6}$$

where

$$c = \frac{1}{n} \sum_{\ell=1}^{k} f_\ell \cos\theta_\ell$$

and

$$s = \frac{1}{n} \sum_{\ell=1}^{k} f_\ell \sin\theta_\ell .$$

respectively. Here θ_ℓ's have been defined already by (3).

If there is no difference between k items, then those will be ranked at random by n judges. Then the ranks given to each item may be considered as a random sample of size n from a multinomial distribution with the probabilities $p_\ell = 1/k$ $(\ell=1,2,\ldots,k)$. Hence the null hypothesis for one item is

$$H_0: p_\ell = 1/k \quad (\ell=1,2,\ldots,k)$$

The distribution of an item vector under the hypothesis can be obtained by counting the combination of a set of f_1, $f_2,\ldots,f_k$. Since the counting procedure is not realistic when n is large, a large sample theory is required. It was shown by the present author that the asymptotic distribution of (c,s) is the 2-dimensional normal distribution. As is well known, an equi-probability curve of a 2-dimensional normal distribution is an ellipse (3). Therefore we can obtain a confidence ellipse or an elliptic critical region.

Let (c,s) denote the coordinate of the item vector. Under the null hypothesis, we can obtain the following critical ellipse in large sample case (2).

$$\frac{c^2}{a^2} + \frac{(s - \mu_0)^2}{b^2} = 1 \tag{7}$$

where

$$a^2 = -\frac{k+1}{nk}\log\alpha ,$$

$$b^2 = -\frac{k - 1 - 2k\mu_0^{\,2}}{nk}\log\alpha$$

and

$$\mu_0 = k^{-1}\cot\{\pi/2(k-1)\},$$

respectively. Here α is a level of significance or the probability that an item vector locates out of the ellipse under the null hypothesis. By describing the ellipse on a rank graph, we can visually test the null hypothesis.

We examined the asymptotic distribution with various n and k by using the Monte Carlo method. Table 1 shows the results of α=0.05. It seems that the approximate distribution is useful for practical use.

IV. EXAMPLE

Suppose that it was required to evaluate two kinds of new products of something like orange juice. Let us call those A and B. Suppose that twenty judges ordered five kinds of orange juices including A and B according to their preference by means of a blindfold test and Table 2 was obtained. Let us call other three items C, D and E, which played a role of control.

Table 1. The results of Monte Carlo simulation. Numerals in the table show the percentage that the item vector located out of the critical ellipse of $\alpha=0.05$ within 25×10^4 repetitions.

n \ k	5	8	10
10	4.60	4.51	4.43
20	4.70	4.79	4.79
30	4.81	4.86	4.85
40	4.81	4.89	4.88
50	4.83	4.93	4.92
60	4.96	4.90	4.89
70	5.01	4.91	4.98
80	4.99	4.97	4.89
90	4.94	4.91	4.89
100	4.96	5.02	4.97

Table 2. Frequencies of ranks given to five items.

rank \ item	A	B	C	D	E
1	8	10	1	0	1
2	2	7	2	3	6
3	0	3	6	7	4
4	2	0	7	6	5
5	8	0	4	4	4
S_i	60	33	71	71	65

S_i : rank sum

The linked rank graph and rank graph drawn from the table are shown in Figure 2 and Figure 3. The ellipse in Figure 3 shows the critical boundary with $\alpha=0.05$. From the graphs we can obtain various information: The item B is more preferred; the average of the ranks given to the item A is equal to the expected value of random ranking but the distribution of the ranks is dispersed.

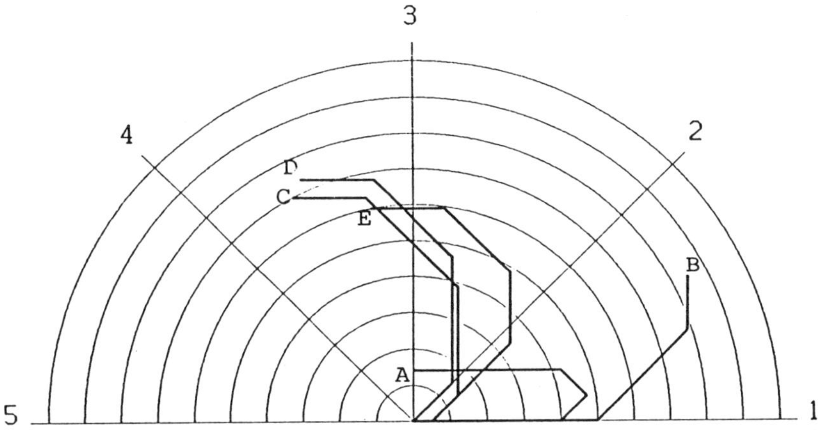

Fig.2 An example of a linked rank graph. The marks on the circumference of the outside semi-circle show the direction corresponding to each rank. The semi-circles represent the degree of concordance of ranks given to each item.

V. DISCUSSION

Kramer's method is familiar as a test for random ranking (4), (5). In his method the rank sum is used as a test statistic. For the above mentioned problem the upper and lower boundary of 0.05 significance level are 45 and 75, respectively. From Table 2 the null hypothesis is accepted for item A, C, D and E and rejected for B. Comparing with this, the items A and B are rejected by the present graphical test.

Here let us consider the difference between the present graphical test and Kramer's test. In the Kramer's test the null hypothesis that k items are ranked at random is tested by a rank sum. Therefore it is equivalent to a test by the average rank of each item. The graphical method, on the other hand, can be regarded as a test about an average rank and a degree of concordance because the coordinate of rank graph consists of average rank and degree of concordance (see Appendix). We can find from Figure 3 that for the item A the null hypothesis was rejected because of low degree of concordance.

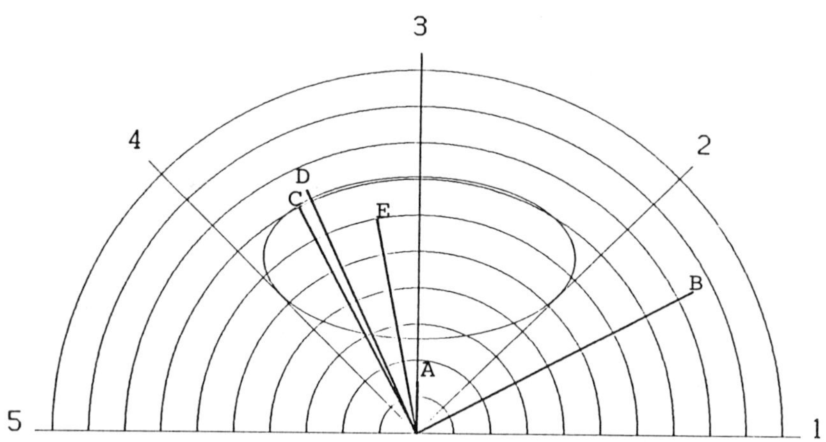

Fig.3 An example of a rank graph.
The marks of the circumference of the outside semi-circle show the direction corresponding to each rank. The semi-circles represent the degree of concordance of ranks given to each item. The ellipse represents the critical boundary of α=0.05.

SUMMARY

The graphical method presented here has advantages as follows: (1) The average rank and degree of concordance of each item can be visually grasped. (2) Therefore, the difference in rankings for each item by classified sample can be intuitively grasped. (3) In the case of a large sample, the null hypothesis for each item can be tested visually, therefore without the bother of using a table.

APPENDIX

Let us consider the mathematical properties of the item vectors. Let us denote the length and direction of item vector x_j as w_j and ϕ_j respectively. That is,

$$w_j = |x_j| \quad \text{and} \quad \phi_j = \arg(x_j)$$

From the viewpoint of the statistics of directional data (1), (6), we know that ϕ_j and w_j have the meanings of mean direction and circular variance of angles which are induced from rankings. It should be noted that when the ranks given to the j-th item consist of only the first and/or last ranks, ϕ_j has no meaning as the mean direction because the final point of the resultant vector x_j is located on the horizontal line. Under the condition $w_j \neq 0$, by replacing the ℓ and θ_ℓ of (3) with $\bar{R}_j$ and ϕ_j, respectively, and then solving it for $\bar{R}_j$, we have the average rank of the j-th item as follows

$$\bar{R}_j = \frac{k - 1}{\pi} \phi_j + 1.$$

When all ranks assigned to the j-th item have the same value, w_j is equal to unity, but when they have different ranks, w_j is less than unity. Therefore w_j is considered as a quantity of the degree of concordance of rankings. So w_j is called a coefficient of the concordance of the item.

Accordingly, the direction and the length of the vector x_j on a rank graph represent the average rank and degree of concordance, respectively, for the j-th item. Hence the coordinate of an item vector consists of average and degree of concordance of ranks.

Let

$$v_j = 1 - w_j \ .$$

Then v_j has the meaning of a circular variance. Accordingly, in wrapped normal distribution model (6), a conventional standard deviation of the rank of the j-th item may be defined as follows:

$$s_j = \frac{k-1}{\pi}\sqrt{-2\log_e(1-v_j)} = \frac{k-1}{\pi}\sqrt{-2\log_e w_j} \ .$$

REFERENCES

1. Baba, Y. (1981). Statistics of angular data : wrapped normal distribution model (in Japanese). Proceedings of the Institute of Statistical Mathematics, 28, 1,41-54.

2. Baba, Y. (1986). Graphical analysis of rank data, Behaviormetrika, No.19, 1-15.

3. Cramér, H. (1946). The Mathematical Methods of Statistics, Princeton University Press

4. Kramer, A. (1960). A rapid method for determing significance of differences from rank sums, Food Technol., 14, 576-581.

5. Kramer, A. (1963). Errata, Food Technol., 17, 1596-1597.

6. Mardia, K. V. (1972). Statistics of Directional Data, Academic Press, New York.

A UNIFIED STUDY OF MULTIVARIATE DATA ANALYSIS METHODS BY NONLINEAR FORMULATIONS AND UNDERLYING PROBABILISTIC STRUCTURES

Nobuyuki Otsu Takio Kurita Hideki Asoh

Electrotechnical Laboratory
1–1–4 Umezono, Sakura-mura, Niihari-gun, Ibaraki, 305 JAPAN

I. INTRODUCTION

In multivariate data analyses most of the methods are formulated in linear forms, and less attention is paid to such problems: what are the nonlinear extensions in general?, and how related are those methods?

Linear formulations utilize only the statistics up to the second order; viz., means and covariances, and which makes the methods practical. However, the essential meanings of such statistics, or in more strict saying, what underlying probabilistic structures they reflect, remain obscure. If we extend linear methods to nonlinear forms, we need assume the higher order statistics (moments) of data, and in the limit, whole knowledge of underlying probability distributions. Such general nonlinear extensions will, however, provide us with deeper understanding of each method and clear relationship of the methods.

From this theoretical standpoint, it has been shown that the general nonlinear extension of the linear discriminant analysis is closely related to a Bayesian frame (inference) [1] and so is the nonlinear multiple regression analysis in the context of pattern recognition [2]. In both cases, the "intersection coefficients" play important roles.

Along this line, we are conducting a unified study, including other methods of multivariate data analysis. In what follows, we shall briefly summarize the recent developments and results of the research: see [3] and [4] for detail.

ISBN 0-12-215485-1

II. DEFINITIONS AND BASIC CONCEPTS

Let $A = \{a_i\}_{i=1}^{M}$ and $B = \{b_j\}_{j=1}^{N}$ be two sets of events, and $\mathbf{u}_i \in R^m$ and $\mathbf{v}_j \in R^n$ be representations of a_i and b_j in topological vector spaces R^m and R^n, respectively. In matrix forms; $U = [\mathbf{u}_1, \ldots, \mathbf{u}_M]\,(m\times M)$ and $V = [\mathbf{v}_1, \ldots, \mathbf{v}_N]\,(n\times N)$. We assume all the probabilities. Those are represented in matrix forms;

$$P_A = [p(a_i)\delta_{ij}]\,(M\times M\text{diagonal}),\; P_B = [p(b_j)\delta_{ij}]\,(N\times N\text{diagonal}),$$
$$P = [p(a_i\wedge b_j)]\,(M\times N), P_{A|B} = [p(a_i|b_j)]\,(M\times N), P_{B|A} = [p(b_i|a_j)]\,(N\times M),$$

and $\mathbf{p}_A = P_A\mathbf{1}_M$ and $\mathbf{p}_B = P_B\mathbf{1}_N$ in vector forms, where $\mathbf{1}_M$ is an M-dimensional vector consisting of unit elements. Then, $\mathbf{u}_i$ and $\mathbf{v}_j$ are observed as random variables (data), and the statistics (up to the second order) can be defined in the usual ways; for example, $\bar{\mathbf{u}} = U\,\mathbf{p}_A$ and $\Sigma_U = UP_AU' - \bar{\mathbf{u}}\bar{\mathbf{u}}'$, where $'$ represents the transpose.

It is noted that in case A and/or B are continuous, we use the density distributions of $\mathbf{u}$ and/or $\mathbf{v}$; for example, $p(\mathbf{u})$ and $p(\mathbf{u}|b_j)$ in case A is continuous.

Consider mappings $\Phi : \mathbf{u}_i \in R^m \mapsto \mathbf{x}_i \in R^K$ and $\Psi : \mathbf{v}_j \in R^n \mapsto \mathbf{y}_j \in R^L$; $X = [\mathbf{x}_1, \ldots, \mathbf{x}_M]$ and $Y = [\mathbf{y}_1, \ldots, \mathbf{y}_N]$ in matrix forms. Then, a method of multivariate data analysis can be viewed as seeking for optimal spatial configurations of A and/or B in R^K and/or R^L, namely X and/or Y, by optimizing the mapping Φ and/or Ψ under some criterion which is defined on the basis of the statistics in the mapped spaces. The usual linear formulations assume (confine) the mappings Φ and/or Ψ to be linear of $\mathbf{u}$ and/or $\mathbf{v}$, respectively. In general nonlinear formulations, we directly apply the variational calculus to seek for optimal Φ and/or Ψ.

For the later discussions, we also define the following measures which represent probabilistic relations induced in A (or B) by averaging the conditional probabilities over B (or A). The latter S_A and S_B were called the intersection coefficients in [1].

$$\Gamma_A = [\gamma(a_i \wedge a_j)] = [\sum_{k=1}^{N} p(b_k)p(a_i|b_k)p(a_j|b_k)] = P_{A|B}P_BP'_{A|B} \quad (M\times M),$$
$$\text{similarly } \Gamma_B = [\gamma(b_i \wedge b_j)] = P_{B|A}P_AP'_{B|A} \quad (N\times N). \qquad (1)$$

$$S_A = [s(a_i|a_j)] = [\sum_{k=1}^{N} p(a_i|b_k)p(b_k|a_j)] = P_{A|B}P_{B|A} \quad (M\times M),$$

$$\text{similarly} \quad S_B = [s(b_i|b_j)] = P_{B|A}P_{A|B} \quad (N\times N). \tag{2}$$

where we observe $\Gamma_A = S_A P_A$ and $\Gamma_B = S_B P_B$.

In continuous case, we also define,

$$\gamma(\mathbf{u} \wedge \tilde{\mathbf{u}}) = \int p(\mathbf{u}|\mathbf{v})p(\tilde{\mathbf{u}}|\mathbf{v})p(\mathbf{v})\, d\mathbf{v}, \tag{3}$$

$$s(\mathbf{u} \mid \tilde{\mathbf{u}}) = \int p(\mathbf{u}|\mathbf{v})p(\mathbf{v}|\tilde{\mathbf{u}})\, d\mathbf{v}. \tag{4}$$

III. NONLINEAR EXTENSIONS OF LINEAR METHODS

A. Nonlinear multiple regression analysis (N-MRA)

Let $\mathbf{u} \in R^m$ and $\mathbf{v} \in R^n$ be continuous random variables, and $p(\mathbf{u}\wedge\mathbf{v})$ denotes the probability density function. Then, the problem of N-MRA from A to B is stated as :

FIND such a mapping $\Phi : R^m \to R^n$; for $\mathbf{x} = \Phi(\mathbf{u})$ that minimizes

$$\varepsilon^2 = \int\int \|\mathbf{v} - \mathbf{x}(\mathbf{u})\|^2 \, p(\mathbf{v}|\mathbf{u})d\mathbf{v}\, p(\mathbf{u})d\mathbf{u}. \tag{5}$$

The solution of this problem is,

$$\mathbf{x} = \Phi(\mathbf{u}) = \int \mathbf{v}\,(p(\mathbf{v}|\mathbf{u}) - p(\mathbf{v}))d\mathbf{v} + \bar{\mathbf{v}}. \tag{6}$$

B. Nonlinear canonical correlation analysis (N-CCA)

Let $\mathbf{u} \in R^m, \mathbf{v} \in R^n$ be continuous random variables, and $p(\mathbf{u} \wedge \mathbf{v})$ denotes the probability density function. Then, the problem of N-CCA is formalized as :

FIND such mappings $\Phi : R^m \to R^L$; $\mathbf{x} = \Phi(\mathbf{u})$ and
$\Psi : R^n \to R^L$; $\mathbf{y} = \Psi(\mathbf{v})$ that satisfy the following.

1) $\bar{\mathbf{x}} = \bar{\mathbf{y}} = \mathbf{0}$ (zero means)
2) $\Sigma_X = \Sigma_Y = I_L$. (unit covariance matrices)
3) $tr\,(\Sigma_{XY}) \to$ max.

Solving this problem turns out to be reduced to solving next simultaneous integral equation system for $\mathbf{x}$ and $\mathbf{y}$.

$$\Lambda\mathbf{x}(\mathbf{u}) = \int \mathbf{y}(\mathbf{v})\,(p(\mathbf{v}|\mathbf{u}) - p(\mathbf{v}))\,d\mathbf{v} \tag{7}$$

$$\Lambda\mathbf{y}(\mathbf{v}) = \int \mathbf{x}(\mathbf{u})\,(p(\mathbf{u}|\mathbf{v}) - p(\mathbf{u}))\,d\mathbf{u}. \tag{8}$$

Where Λ is a diagonal matrix (Lagrange multipliers). Note the similarity between (6) and (7). Eliminating $\mathbf{x}$ from (7) and (8), we obtain the following integral eigen equation for $\mathbf{y}$.

$$\Lambda^2\mathbf{y}(\mathbf{v}) = \int \mathbf{y}(\tilde{\mathbf{v}})(s(\tilde{\mathbf{v}}\mid\mathbf{v}) - p(\tilde{\mathbf{v}}))\,d\tilde{\mathbf{v}}. \tag{9}$$

We can obtain the solution of N-CCA from (7) and (9).

C. Nonlinear discriminant analysis (N-DA)

In linear formulation, Discriminant Analysis is a special case of Canonical Correlation Analysis. This is also true for nonlinear case.

Let $\mathbf{u} \in R^m$ be continuous random variable (data), and b_j be random event (category, class). Probability density function $p(\mathbf{u}|b_j)$ is assumed. Then the problem of N-DA is to find $\Phi : R^m \to R^L$ such that $\mathbf{x} = \Phi(\mathbf{u})$ maximizes the discriminant criterion $tr(\Sigma_t^{-1}\Sigma_b)$, where Σ_t and Σ_b are the total covariance matrix and the between-class covariance matrix of $\mathbf{x}$, respectively.

To see the relation between N-DA and N-CCA ,we regard the class average of $\mathbf{x}$ as the representation of class b_j, that is,

$$\mathbf{y}_j = \bar{\mathbf{x}}(b_j) = \int \mathbf{x}(\mathbf{u})p(\mathbf{u}|b_j)\,d\mathbf{u}. \tag{10}$$

Then, we have the relation between covariance matrices : $\Sigma_b = \Sigma_{XY} = \Sigma_Y$. Hence, under (10), maximizing the discriminant criterion is essentially equivalent to maximizing N-CCA's criterion $tr(\Sigma_{XY})$ subject to the constraint $\Sigma_X = I_L$. In fact, we have

$$\mathbf{x} = \Lambda^{-2}\sum_{j=1}^{N} p(b_j|\mathbf{u})\mathbf{y}_j, \quad [S'_B - \mathbf{1}_N\mathbf{p}'_B]Y' = Y'\Lambda^2. \tag{11}$$

This is essentially same as N-DA solution which has already obtained in [1]. Note the simirality between (11) and the set of (7) and (9).

IV. INTERPRETATION OF QUANTIFICATION METHODS

A. Quantification method 2 (QM2) and 3(QM3)

Hayashi's QM2 [5] is the discriminant analysis for qualitative data. We shall consider nonlinear extension of QM2 (N-QM2).

Let $U = \{\mathbf{u}_i \in \{0,1\}^m\}_{i=1}^M$ be a finite set of qualitative data, and B a set of classes. An optimal nonlinear mapping $\mathbf{x}_i = \Phi(\mathbf{u}_i)$ that maximizes the discriminant criterion $tr(\Sigma_t^{-1}\Sigma_b)$ is given by

$$X = \Lambda^{-1} Y P_{B|A}, \qquad [\Gamma_B - \mathbf{p}_B \mathbf{p}'_B] Y' = P_B Y' \Lambda^2, \tag{12}$$

where matrix Y can be interpreted as a set of representative vectors of each class. Notice the correspondence between (12) and (11) in N-DA.

On the other hand, regarding a contingency table as joint probabilities of A and B, the normal equation system of Hayashi's QM3 [5] is given by

$$[\Gamma_A - \mathbf{p}_A \mathbf{p}'_A] X' = P_A X' \Lambda^2, \quad Y = \Lambda^{-1} X P_{A|B} \tag{13}$$

$$[\Gamma_B - \mathbf{p}_B \mathbf{p}'_B] Y' = P_B Y' \Lambda^2, \quad X = \Lambda^{-1} Y P_{B|A}. \tag{14}$$

Notice that (14) corresponds to (7) and (9) in N-CCA, which shows QM3 is essencially N-CCA. Further, the equivalence between (12) and (14) reveals the close relationship between N-QM2 and QM3.

The singular value decomposition of probability matrix P is also reduced to (13) or (14). This means that N-CCA and/or QM3 has a close relation to SVD approximation of matrix P. In detail, see [4].

B. Relationship to quantification method 4 (QM4)

Hayashi's QM4 [5] is multidimensional scaling based on similarity. Assume the degree of similarity between a_i and a_j is given by $\gamma(a_i \wedge a_j)$, then the normal equation is obtained as

$$[\Gamma_A - \mathbf{p}_A \mathbf{p}'_A] X' = P_A X'(I - H/2) \tag{15}$$

where H is a diagonal matrix (Lagrange multipliers).

This equation is the same as the eigen equation (13). The result explicitly shows that QM4 is closely related to N-QM2 or QM3, reflecting the underlying probabilistic structure.

C. Properties of distances defined on the scores

Instead of the usual Euclidean distance, we shall consider a distance between the scores $\mathbf{x}(a_i)$ and $\mathbf{x}(a_j)$ which is defined by an weighted sum with eigenvalues in the eigen equation (13). The distance is given by

$$[\mathbf{x}(a_i) - \mathbf{x}(a_j)]'\Lambda[\mathbf{x}(a_i) - \mathbf{x}(a_j)] = \sum_{k=1}^{N}\{q(a_i|b_k) - q(a_j|b_k)\}^2 p(b_k) \quad (16)$$

where $q(a_i|b_k) = \{p(a_i|b_k) - p(a_i)\}/p(a_i)$. This means that the weighted distances essentially reflect the differences between q's defined on the underlying probabilities. Moreover, we can rewrite (16) as follows by using Bayes formulus.

$$\sum_{k=1}^{N} \frac{1}{p(b_k)} \left\{ \frac{p(a_i \wedge b_k)}{p(a_i)} - \frac{p(a_j \wedge b_k)}{p(a_j)} \right\}^2 \quad (17)$$

which is essentially equivalent to the distance used in Analyse des Correspondances [6].

REFERENCES

1. Otsu,N. (1975). Nonlinear Discriminant Analysis as a Natural Extension of the Linear Case, *Behaviormetrika*, **2**, 45-59.
2. Otsu,N. (1982). Optimal Linear and Nonlinear Solutions for Least-square Discriminant Feature Extraction, *Proc. of 6th Int. Conf. on Pattern Recognition*, 557-560.
3. Kurita,T., Asoh,H., and Otsu,N. (1986). Probability-theoretical Interpretation of the Quantification Methods, (in Japanese), *IECE Tech. Report*, **PUR86-17**, 57-66.
4. Asoh,H., Kurita,T., and Otsu,N. (1987). An Interpretation of Canonical Correlation Analysis and Discriminant Analysis by Nonlinear Extensions, (in Japanese), submitted to *The Japanese Journal of Behaviormetrics.*
5. Hayashi,C. ,Higuchi,I, and Komazawa, T. (1970). *Information Processing and Mathematical Statistics*, (in Japanese), Sangyo-Tosho.
6. Benzécri,J.-P. (1973). *L'Analyse des Données II*, BORDAS.

OPTIMUM CLASSIFICATION BOUNDARIES BASED ON A CONCOMITANT VARIABLE IN SAMPLE SURVEY --- SOME APPLICATION TO THE CURRENT STATISTICS OF COMMERCE ---

Shidou Sai*
Masaaki Taguri**

*) Okayama College of Commerce, 2-10-1, Tsushimakyou-machi, Okayama 700, JAPAN
**) Chiba University, Faculty of Science, 1-33, Yayoi-cho, Chiba 260, JAPAN

I. INTRODUCTION

In practical situation, it is impossible to stratify the population based on the information of the objective variable Y itself whose mean should be estimated. It is then usual to carry out the stratification based on the information of a concomitant variable X which has considerably high correlation with Y; for example, the information of past sample surveys or a pilot survey is sometimes available.

In this paper we propose a stratified random sampling procedure using a concomitant variable for two sample allocation methods; Equal Allocation (EA) and Neyman Allocation (NA). Some kinds of robustness in the proposed procedure are also investigated. The analysis of these facets of robustness must give effect to the procedure in practical situations. Finally the procedure is applied to some practical example; the Current Statistics of Commerce in Japan, which shows great improvement of the precision in estimation.

RECENT DEVELOPMENTS IN
CLUSTERING AND DATA ANALYSIS

ISBN 0-12-215485-1

II. OPTIMUM STRATIFICATION BASED ON A CONCOMITANT VARIABLE

Throughout this paper a stratification method is limited to interval division for a concomitant variable X. The population is divided into L strata, and the i-th stratification point is denoted by x_i. A random sample with size n_i is taken from each stratum, and the total sample size is $n = \sum_{i=1}^{L} n_i$. Let f(x,y) be the joint probability density function (p.d.f.) of X and Y. Suppose the regression function of Y on X, $Z=\eta(X)$, is strictly monotone increasing (or decreasing), and h(z) is the p.d.f. of Z. The weight, mean and variance of the i-th stratum are denoted by w_i, μ_{Zi} and σ_{Zi}^2 respectively, where the first suffix Z means the variable (the similar notations are used hereinafter). As an estimator for the population mean μ_Y, the unbiased estimator $\overline{Y} = \sum_{i=1}^{L} w_i \overline{Y}_i$ is taken up, where $\overline{Y}_i$ is the sample mean of the objective variable Y in the i-th stratum; that is, $\overline{Y}_i = \sum_{j=1}^{n_i} Y_{ij}/n_i$.

Referring to the results for PA given by Taga (1), let us investigate the cases of EA ($n_i = n / L$) and NA ($n_i = n w_i \sigma_{Yi} / \sum_{j=1}^{L} w_j \sigma_{Yj}$). Only the results for NA will be shown to save spaces, though the results for EA are also obtainable by quite similar discussion. The variance of the estimator $\overline{Y}$ can be represented using $\theta(x)$, the conditional variance of Y for given x. Let us now investigate the following three cases in which the conditions for $\theta(x)$ may approximately satisfy in some practical situations:

[Case 1] The conditional coefficient of variation is constant for any fixed x; $\theta(x)/\eta^2(x) = C$ ($C \geqq 0$).

[Case 2] The joint distribution of X and Y is bivariate normal; $\eta(x) = \mu_Y + \rho\sigma_Y/\sigma_X(x - \mu_X)$; $\theta(x) = (1 - \rho^2)\sigma_Y^2$.

[Case 3] More general case; $\theta(x) = C\eta^{\alpha}(x)$, where C and α are non-negative constants.

As [Case 3] is a natural extension of the previous two cases, it is enough to show the result for this case. A necessary condition for the optimum stratification points z_i^* is

$$w_i[Cz_i^{*\alpha} + C\nu_{Zi}^{(\alpha)} + (z_i^* - \mu_{Zi})^2 + \sigma_{Zi}^2]/n_i$$
$$= w_{i+1}[Cz_i^{*\alpha} + C\nu_{Zi+1}^{(\alpha)} + (z_i^* - \mu_{Zi+1})^2 + \sigma_{Zi+1}^2]/n_{i+1}, \quad (1)$$

where $\nu_{Zi}^{(\alpha)}$ is the α-th moment about the origin in the i-th stratum. The optimum stratification points for X can be obtained by the inverse transformation $x_i^* = \eta^{-1}(z_i^*)$. Note that this result of course includes that of the optimum stratification by the objective variable itself as a special case.

III. ROBUSTNESS ON A REGRESSION FUNCTION AND THE CONSTANT C

A regression function and the constant value C should be estimated in practical situation. It must be, therefore, important to investigate the influence on the variance $V(\overline{Y})$ by small variation of such quantities. Suppose that the stratification points z_i' are slightly different from the corresponding optimum points z_i^*; $z_i' = z_i^* + \Delta z_i$ $(i=1,2,\ldots,L-1)$. The weight, mean and variance of Z in the i-th stratum for the stratification points z_i' are denoted by w_i, m_{Zi} and s_{Zi}^2 respectively, and Δw_i is defined by $h(z_i^*)\Delta z_i$. $V'(\overline{Z})$ and $V^*(\overline{Z})$ are the variances of the estimator $\overline{Z} = \sum_{i=1}^{L} w_i \sum_{j=1}^{n_i} \eta(X_{ij})/n_i$ corresponding to the stratification points z_i' and z_i^*, respectively. In the following calculations the higher order of $\Delta^3 w_i$ and $\Delta^3 z_i$ are neglected, and n_i's are assumed not to vary for NA even if the stratification points change.

(A) Robustness on the Constant C

If the estimated value of C is different from the actual (true) value C^* by ΔC; $\hat{C} = C^* + \Delta C$, then the increase of the variance of $\overline{Y}$ can be given by

$$V'(\overline{Y}) - V^*(\overline{Y}) = \Delta C/\hat{C}\,[V'(\overline{Z}) - V^*(\overline{Z})]$$
$$= \sum_{i=1}^{L-1} \Delta w_i/2 \cdot [w_i'\{(z_i' - m_{Zi})^2 + s_{Zi}^2\}/n_i - w_{i+1}'\{(z_i' - m_{Zi+1})^2 + s_{Zi+1}^2\}/n_{i+1}]. \quad (2)$$

This difference might not be so large for small change of C referring to the result by Taguri (2); especially for NA.

(B) Robustness on a Regression Function

If the estimated regression function $\hat{\eta}(x)$ is different from the actual (true) one $\eta^*(x)$ by $\zeta(x)$; $\hat{\eta}(x) = \eta^*(x) + \zeta(x)$, then the increase of the variance of $\overline{Y}$ can be given by

$$V'(\overline{Y}) - V^*(\overline{Y}) = \sum_{i=1}^{L-1} \Delta w_i/2 \cdot [w_i'\{(z_i' - m_{Zi})^2 + s_{Zi}^2\}/n_i - w_{i+1}'\{(z_i' - m_{Zi+1})^2 + s_{Zi+1}^2\}/n_{i+1}] \cdot \delta_2(x_i')/[1 + \delta_1(x_i')], \quad (3)$$

where δ_1 and δ_2 are the quantities of order $O(\zeta(x)/\eta^*(x))$. If $\zeta(x_i')/\eta^*(x_i')$ is small enough, then $\delta_2(x_i')/[1+\delta_1(x_i')]$ is not so large; whereas the right-hand side of the above formula excluding that term is nothing but $V'(\overline{Z}) - V^*(\overline{Z})$. Thus the increase of the variance may not be so large for small change of a regression function.

IV. SOME NUMERICAL EXAMPLE

Let us apply the above results to "The Current Statistics of Commerce" in Japan. This statistics provides the frequency distribution of the number of employees x in an establishment, which is adopted as a concomitant variable in this example. In addition, the total and the sum of squares of annual sales for each x are given by the result of a sample survey. Our objective is to estimate the mean of the annual sales, which is an objective variable in this case. From the graph of the sample mean and standard deviation, it was reasonable to limit the regression function to a linear function through the origin. The value of α is put to 2, because any conditional coefficients of variation of Y for fixed x were almost equal to each other. The estimated regression function and the value of C were $z = \hat{\eta}(x) = 9907.166\ x$ and $\hat{C} = 1.62881$. The sample sizes in respective strata for NA were estimated using the results of the sample survey.

Table 1 shows the stratification points $\{x_i\}$ and the variances $V(\overline{Y})$ of the estimator $\overline{Y}$ which were obtained by Taga's method for PA and by the proposed method for EA and NA, in case of L = 2, 3 and 4. In addition to these quantities, we computed the relative efficiencies e(L) of stratified random sampling with L strata to simple random sampling for the three sample allocation methods. For reference, we also give the optimum stratification points and the corresponding optimum variances which could be obtained by computing for all possible combinations of stratification points. From this table, the following results could be obtained:

1. Stratified random sampling with L=4 decreases the variance of $\overline{Y}$ by about 25% for PA and by about 70% for EA and NA compared with simple random sampling.
2. The variances of $\overline{Y}$ for EA and NA are about half or less than half in comparison with those for PA, independently of the number of strata.
3. The stratification points for the three sample allocation

Table 1. Stratum Π_i, Variance $V(\overline{Y})$, Efficiency e(L) for PA, EA, NA by the Proposed and the Optimal method

L	Allocation	method	Π_1	Π_2	Π_3	Π_4	$V(\overline{Y})$	e(L)
1							26.634	1.0000
2	PA	Taga	1-15	16-50			21.821	.8193
		Optimal	1-20	21-50			21.605	.8112
	EA	Proposed	1-10	11-50			11.400	.4280
		Optimal	1-10	11-50			11.400	.4280
	NA	Proposed	1-9	10-50			11.529	.4329
		Optimal	1-10	11-50			11.335	.4256
3	PA	Taga	1-8	9-23	24-50		20.540	.7712
		Optimal	1-9	10-26	27-50		20.355	.7643
	EA	Proposed	1-6	7-18	19-50		10.009	.3758
		Optimal	1-6	7-20	21-50		9.414	.3535
	NA	Proposed	1-5	6-16	17-50		9.823	.3692
		Optimal	1-4	5-20	21-50		9.130	.3428
4	PA	Taga	1-3	4-11	12-26	27-50	20.085	.7541
		Optimal	1-4	5-11	12-26	27-50	20.017	.7516
	EA	Proposed	1-4	5- 9	10-21	22-50	8.522	.3200
		Optimal	1-4	5-10	11-26	27-50	8.159	.3063
	NA	Proposed	1-3	4- 8	9-20	21-50	8.431	.3166
		Optimal	1-4	5-10	11-26	27-50	7.998	.3003

methods are not so different from the corresponding optimum points. Thus the variances of $\overline{Y}$ are quite similar to those for the optimum stratification. These results indicate that the assumption on the value of α and the estimation of the regression function and the value of C have had small influence on the variance of $\overline{Y}$.

REFERENCES

1. Taga, Y. (1967). On optimum stratification for the objective variable based on concomitant variables using prior information, Ann.Inst. Statist. Math. 19, 101-129.
2. Taguri, M. (1982). Optimum stratification and its robustness (IV) --- Robustness on stratification points, Rep. Statist. Appl. Res., JUSE, 29, 32-41.

ON THE IDENTIFICATION PROBLEM OF AGE-PERIOD-COHORT ANALYSIS

Toshiro Tango

Division of Clinical Epidemiology
The Tokyo Metropolitan Institute of Medical Science
3-18-22, Honkomagome, Bunkyo-ku, Tokyo 113, JAPAN

I. INTRODUCTION

In epidemiological studies, temporal trends of incidence and mortality from a particular disease often need to be analyzed in terms of three types of effects: 1) Age effects due to the process of biological aging, 2) Period effects due to the epoch of death (or diagnosis), and 3) Cohort effects associated with the period of birth.

An available data set is often restricted to those tabulated in the form of a two-way (age x period) contingency table where the width of age group is equal to the length of period. Let d_{ij} and N_{ij} denote the number of events and the population at risk in the ith age group and jth period, called (i,j) cell here, i =1,...,I; j =1,...,J, respectively, and let $r_{ij} = d_{ij}/ N_{ij}$. To this data, the following Poisson regression model, generally referred to as "age-period-cohort model (APC model)", has been applied:

$$\text{Log } E(r_{ij}) = \mu + \alpha_i + \beta_j + \gamma_k, \ \Sigma\, \alpha_i = \Sigma\, \beta_j = \Sigma\, \gamma_k = 0 \qquad (1)$$

ISBN 0-12-215485-1

where α_i , β_j , and γ_k represent parameters of effects due to age, period and cohort, respectively.

It is, however, well known that simultaneous estimation of all these three parameters is not possible because of the linear interdependency among i,j and k:

$$k = I - i + j \quad (= 1,\ldots,I+J-1 = K). \tag{2}$$

The following system summarizes the identification problem of the APC model (Tango,1985). For any constant t and any set of parameters (α_i^* , β_j^* , γ_k^*), another set of parameter computed from the equations:

$$\begin{aligned} \alpha_i &= \alpha_i^* + t\,L(i,I) \\ \beta_j &= \beta_j^* - t\,L(j,J) \\ \gamma_k &= \gamma_k^* + t\,L(k,K) \end{aligned} \tag{3}$$

also satisfies the APC model, where $L(x,y)=x - (y+1)/2$.

In this paper, we critically review previous suggestions made for elimination of the identification problem and suggest methods for presenting the results of analyses based upon this type of model which minimize the serious risk of misleading interpretations.

II. ELIMINATION OF NON-UNIQUENESS

A. Subjectively Reasonable Constraint

Many researchers have tried to eliminate this source of non-uniqueness by imposing an arbitrary, though subjectively reasonable, constraint on the parameter of the model. In practice, however, there seems to be few occasions when just one specific constraint appears reasonable. Further, while the choice of constraint does not affect the goodness-of-fit of the model to the data, it does affect the estimates of the parameters and their trend.

For example, Mason and Smith(1985), in the analysis of lung tuberculosis mortality data in the U.S., imposed $\alpha_1 = \alpha_2$ based upon "close observation" of two figures which arrayed age-specific mortality rates by period and by cohort. They observed that "the rates for the early ages are fairly close, whether looked at from a cohort or a period perspective". This speculation is somewhat misleading because their observation leads directly to the following two assumptions, not to the above equality:

$$E(r_{1j}) = E(r_{2j}) \quad \text{for all } j,$$

$$E(r_{1j}) = E(r_{2\ j+1}) \text{ for all } k.$$

However, these constraints are also contradictory since it is derived that

$$\alpha_1 - \alpha_2 = \gamma_{k-1} - \gamma_k , \qquad \text{for all } k,$$

$$\alpha_1 - \alpha_2 = \beta_{j+1} - \beta_j , \qquad \text{for all } j,$$

indicating that all the first difference of period and cohort effects are equal, which is obviously strange.

B. Apparently Objective Constraint

Some other approaches to eliminate the non-uniquness of the model have also been suggested. Osmond and Gardner(1982) and Nakamura(1982) proposed a mathmatical constraint in the model such that, if the age-cohort model is better than the age-period model, then their solution tends to put more weight on the cohort effects. Essentially, they choose one of the infinitely many possible solutions on the ground that it has certain good properties. Such a strategy can only be justified if the property which identifies the unique solution has any biological basis but no such justification has been offered. Their solution is, therefore, totally arbitrary.

On the other hand, Tango(1984) and Boyle and Robertoson, independently, suggested the finer grouping of data by noting that the diagonal lines deliminate birth cohorts with grouping interval equal to that used for age and period. When individual records are available, we can devide the cell (i,j) into two right-angled triangular cells indexed as (i,j;s), s =0,1, so that the cell (i,j;0) and (i,j;1) constitute two successive non-overlapping cohorts. Then by introducing non-overlapping cohort effects $\{\gamma'_k\}$ instead of overlapping cohort effects $\{\gamma_k\}$, an alternative APC model can be suggested:

$$\text{Log } E(r_{ij}^{(s)}) = \mu + \alpha_i + \beta_j + \gamma'_k \quad , \quad k = I - i + j + s \qquad (4)$$

where $r_{ij}^{(s)}$ denote the observed rate in the (i,j;s) triangular cell. This model is easily shown to be fully identifiable. Boyle and Robertosn(1987) insisted that the analysis using the model (4) can proceed without arbitrarily setting two cohort effects to be equal, or any other such scheme.

However, if you examine closely data structure comprising 2IJ triangular cells, it is revealed that the more appropriate but still unidentifiable APC modelling is, for each s,

$$\text{Log } E(r_{ij}^{(s)}) = \mu + \alpha_i^{(s)} + \beta_j^{(s)} + \gamma_k^{(s)} \quad , \quad k = I - i + j \qquad (5)$$

where $\alpha_i^{(s)}$, $\beta_j^{(s)}$ and $\gamma_k^{(s)}$ denote the parameters of effects relating to the triangular cell (i,j;s) where the corresponding parameters, for example, $\alpha_i^{(0)}$ and $\alpha_i^{(1)}$, are generally not equal. But if we can reasonably assume that for all i,j and k,

$$\alpha_i^{(0)} = \alpha_i^{(1)} \quad , \quad \beta_j^{(0)} = \beta_j^{(1)} \quad , \quad \gamma_{k-1}^{(0)} = \gamma_k^{(1)} \qquad (6)$$

then the model (4) is derived. Therefore the truth of conclusion to which one is led by using the model (4) depends upon the validity of the constraint (6). Namely, model (4) can be considered as one of procedures which impose some constraints on the model parameters whether they are reasonable or not.

C. Comparison of Two-Effects Models

Different alternatives which have been done many times in literatures is to consider models which include only two of the effects, for example, age-cohort model and age-period model, and to compare their goodness-of-fit. This is also shown to be seriously misleading since the two-effects model is equivalent to the three-effects model where the effects excluded from the model has no curvature components but may have "inestimable" linear component. For example, consider age-cohort model with parameters (α_i^* , γ_k^*). This model is clearly equivalent to the APC model with parameter values (α_i^* , 0 , γ_k^*) and also to the APC model with parameter values (α_i , β_j , γ_k) which can be calculated from the system of equations (3):

$$\begin{aligned} \alpha_i &= \alpha_i^* + t\,L(i,I) \\ \beta_j &= 0 - t\,L(j,J) \\ \gamma_k &= \gamma_k^* + t\,L(k,K) \end{aligned} \tag{7}$$

where the case of t = 0 corresponds to the age-cohort model. Therefore, within the additive three-effects model, hypothesis that one effect is absent can only be tested for the curvature component, such as

$$H_0 : \tilde{\beta}_1 = \tilde{\beta}_2 = \cdots = \tilde{\beta}_J = 0$$

where $\{\tilde{\beta}_j\}$ are curvature components for period effects given by

$$\tilde{\beta}_j = \beta_j - s_\beta L(j,J) \tag{8}$$

$$s_\beta = \sum_{j=1}^{J} \beta_j L(j,J) \Big/ \sum_{j=1}^{J} L(j,J)^2 \tag{9}$$

where s_β represents the inestimable slope for linear trend of period effects. Analogous expressions for the curvature components $\tilde{\alpha}_i$ and $\tilde{\gamma}_k$ can be similarily given.

III. USE OF ESTIMABLE FUNCTIONS OF PARAMETERS

It is clear from the previous section that there has been no satisfactory resolution of the problem of identifiability of APC models. This led Kupper et al.(1985) to conclude that at present, such models offer little or no advantage over simple graphical methods. I do not think so. A better statistical analysis will not only summarize the data in a succinct manner, but will also make clear its limitations. No sophistication of method can create information where that is lacking.

We would like to emphasize, however, that there are other important aspects of APC models which can be identified, and that models can still provide a more parsimonious representation of the data than simply graphing the full data. Holford(1983) and Tango(1985) suggested the decomposition of these unidentifiable effects into two orthogonal components: inestimable linear component and estimable curvature component which has useful information, given by (8). An even more simple alternative is based upon the second differences: for example,

$$\Delta^2\gamma_k = 2\,\tilde{\gamma}_k - \tilde{\gamma}_{k-1} - \tilde{\gamma}_{k+1} = 2\,\gamma_k - \gamma_{k-1} - \gamma_{k+1} \tag{10}$$

and

$$E_k = \Delta^2\gamma_k + \Delta^2\gamma_{k+1} \tag{11}$$

These contrasts can be considered as local curvature components in comparison with the overall curvature components $\tilde{\gamma}_k$. The usefulness of these curvatures are investigated and illustrated with several examples recently by Tango(1987), in which the contrast E_k has been introduced by considering the nature of overlapping cohort effects of APC model. Contrasts based on second differences might have the practival advantage that the value taken is affected by only neighbouring data.

Depending on the problem, another comprehensible parameterization of the model which is identifiable may be considered but the reliability of these estimates should be discussed.

REFERENCES

1. Boyle,P. and Robertson,C.(1987). Statistical modelling of lung cancer and laryngeal cancer incidence in Scotland, 1960-1979, American Journal of Epidemiology, 125, 731-744.
2. Holford, T.R.(1983). The estimation of age, period and cohort effects for vital rates, Biometrics, 39, 311-324.
3. Kupper, L.L., Janis,J.M., Karmous,A. and Greenberg,B.G. (1984). Variance and dissent age-period-cohort analysis: A review and critique, Journal of Chronic Diseases, 38, 811-830.
4. Mason, W.M. and Smith,H.L.(1985). Age-period-cohort analysis and the study of deaths from pulmonary tuberculosis. In "Cohort Analysis in Social Research: Beyond the Identification Problem", (Mason and Fienberg, eds), 125-227 Springer-Verlag.
5. Nakamura,T.(1982). A Bayesian cohort model for standard cohort table analysis (in Japanese), Proceedings of the Institute of Statistical Mathematics, 29, 77-97.
6. Osmond,C. and Gardner,M.J.(1982). Age, period, and cohort

models applied to cancer mortality rates, Statistics in Medicine, 1, 245-259.

7. Tango,T.(1984). The estimation of age, period and birth cohort effects in the cohort analysis (in Japanese), Japanese Journal of Applied Statistics, 13, 11-26.
8. Tango,T.(1985). Estimation of age, period and cohort effect : decomposition into linear trend and curvature components (in Japanese), Japanese Journal of Applied Statistics, 14, 45-49.
9. Tango,T. and Kurashina,S.(1987). Age, period, and cohort analysis of trends in mortality from major diseases in Japan, 1955-1979: Peculiarity of the cohort born in the early Showa Era, Statistics in Medicine (In Press).

Section 2

Automatic Classification and Related Techniques

SOME RECENT ADVANCES IN CLUSTERING

Edwin Diday

INRIA
Domaine de Voluceau - Rocquencourt
B.P. 105 - 78153 LE CHESNAY CEDEX (FRANCE)

Université Paris IX-Dauphine
Place de Lattre de Tassigny
75775 PARIS CEDEX 16

I. INTRODUCTION

The main aim of clustering is to give an inter-class structure as a partition or a hierarchy which are the most common (see for instance, Sneath and Sokhal (1973), Anderberg (1973), Benzecri (1973), Diday, Lemaire, Pouget, Testu (1982), Jambu (1978), Lerman (1980)). Other kinds of inter-class structure may be used such as overlapping clusters and more recently pyramids (Diday (1984), Diday, Bertrand (1985)). We give first a general definition of a "Classification Space" $\mathbb{P}$ which covers all these inter-class structures. Usually a cluster is said to be "good" if its objects are not far from its center of gravity, a seed point, a kernel, etc. ...

RECENT DEVELOPMENTS IN
CLUSTERING AND DATA ANALYSIS

ISBN 0-12-215485-1

We have generalized this notion of representation of a cluster by introducing the general definition of what we call a "representation space" denoted by $\mathbb{L}$.

When the size of the population to classify gets large, a given statistical model may be wrong for the whole population, but very accurate, for certain of its parts. To approach this problem, we have used the so-called method of "nuées dynamiques", which consists of defining first a criterion W(L,P), which measures the fit between a representation $L \in \mathbb{L}$ called "Kernel", and a classification $P \in \mathbb{P}$, (generally, a partition). Then, an algorithm is defined, based on the choice of two functions g and f, characterizing the representation and allocation processes. If g and f are chosen properly, it can be proved that the criterion W improves at each iteration and converges to a stable position. The main interest of this method is to unify a large variety of problems, as well as to give algorithms leading to a (locally optimal) solution. We give six examples which illustrate this approach which has been widely developed in Diday et al.(1980,85). The second part of the paper concerns "Inferential Clustering". More practically, if the user has some ideas on the kind of clusters that he wishes to obtain and if he is able to express them on a sample by a hierarchy, we show that this information makes it possible to "infer" the parameters $a_1,\ldots,a_4$ of the Lance and Williams formula (1967) under some constraints in such a way as to obtain a hierarchy (on the complete population) which will have a good fit with the partial hierarchy given by the user, on a sample.

The last part of the paper is devoted to the graphical representation of clusters. Pyramids allow a graphical representation of overlapping clusters, that induce a special dissimilarity index closer to the data than ultrametrics. We introduce the notion of "coherent" data for a dissimilarity index which is "compatible" with an order on the population and is well represented by an indexed hierarchy; on the other hand, we show that for some kind of data associated to a dissimilarity index and called "chaotic", an indexed hierarchy and any ultrametric will have a very bad fit with the data. That is why we introduce the notion of "ultramine" and its anti-indexed hierarchical representation.

II. CLASSIFICATION SPACE AND REPRESENTATION SPACE

A. Definitions

The most classical classification spaces which are used in pratice are "partitions" and "hierarchies". Let Ω be the set of objects, it is well known that a partition P is a set of subsets of Ω $P = (P_1,\ldots,P_k)$ such that $\Omega = \bigcup_i P_i$ and $P_i \cap P_j = \emptyset$ $\forall$ $i,j = 1,\ldots,k$. If i and j exist such that $P_i \cap P_j \neq \emptyset$ we say that P is a covering of Ω. A hierarchy H is a set of subsets of Ω which contains the single elements and Ω and which satisfies the following property : $h,h' \in H \Rightarrow h \cap h' = \emptyset$ if $h \subset h'$ or $h' \subset h$. The hierarchies have been extended to pyramids which allows the graphical representation of overlapping classes (see figure 3); a pyramid P is a set of subsets of Ω which contains the single elements and Ω and which satisfies the following property : $h,h' \in P \Rightarrow h \cap h' = \emptyset$ or $h \cap h' \in P$ and for which there is an order $\mathcal{O}$ on Ω such that the elements of any $h \in P$ constitute an interval of $\mathcal{O}$ (for more details on pyramids see §3). It is possible to formalize more generaly the notion of "classification space" by the following definitions.

Definition of a "Classification Space" :

The set $\mathbb{P}$ is a classification space on Ω iff it satisfies the four following properties :

a) $\mathbb{P} \subset P(P(\Omega))$ where $P(\Omega)$ is the set of subsets of Ω

b) $\forall\ s \in \mathbb{P}$, $s = \{P_1,\ldots,P_k\} \Rightarrow \bigcup_i P_i \equiv \Omega$

c) $\exists\ s \in \mathbb{P}$, such that $\Omega \in s$.

d) $\exists\ s \in \mathbb{P}$, such that $\forall\ \omega \in \Omega$ $\{\omega\} \in s$.

It may be shown that partitions, coverings, hierarchies and pyramids are special kinds of classification space $\mathbb{P}$.

To summarize each class of a given classification numerous kinds of representations may be used, for instance, the center of gravity is one of the most popular and we will see in 2.2 other kinds of class-representation.

It is possible to give a general definition of this "representation" notion.

Definition of a representation structure associated to Ω for a classification space $\mathbb{P}$.

Ω has a "representation structure" iff there exists

a) a set L and a mapping $D : P(\Omega) \times L \rightarrow \mathbb{R}^+$

b) a set $\mathbb{L}_k$ and a mapping $W : \mathbb{P} \times \mathbb{L}_k \rightarrow \mathbb{R}^+$.

The set L is the "representation space" of each class and the set $\mathbb{L}$ is the "representation space" of each element of the classification space $\mathbb{P}$.

Usually : $D(P_i,L_i)$ where $(P_i,L_i) \in P(\Omega) \times L$ measures the fit between the class P_i and the representation L_i often called "kernel" or "seed" of P_i.

$$W(P,L) = \sum_{i=1}^{k} D(P_i, L_i)$$ and measures the fit between the classification $P = (P_1, \ldots, P_k) \in \mathbb{P}$ and $L = (L_1, \ldots, L_k) \in \mathbb{L}$.

B. The MND method (MND means in French "Méthode des nuées Dynamiques").

The main characteristic of the MND method is that it defines a new and large family of problems. This concerns optimizing a criterion that expresses the fit between a classification of objects and a way of representing the classes of this classification. The problem of optimization arises then as the *simultaneous seeking of the classification and the representation of this classification from among a set of possible classifications* $\mathbb{P}$ *and representations* $\mathbb{L}$ *which optimize the criterion*. More precisely, this problem may be expressed in the following way :

find a couple $(L^*, P^*) \in \mathbb{L} \times \mathbb{P}$

where $\mathbb{L}$ is a representation space that minimizes a criterion

$$W : \mathbb{L} \times \mathbb{P} \to \mathbb{R}^+ ;$$

in other words such that :

$$W(L^*, P^*) = \operatorname{Min}\{W(L,P) / L \in \mathbb{L}, P \in \mathbb{P}\} .$$

In order to minimize this criterion, we mainly use alternative steps of representation and allocation mapping until we reach the convergence which gives a locally optimal solution to the problem in question.

The MND method, as it is generally used, consists of :

1) Choosing a classification space $\mathbb{P}$ (this space is generally the set of partitions).

2) Choosing a representation space $\mathbb{L}$.

3) Defining a criterion W from $\mathbb{L} \times \mathbb{P} \to \mathbb{R}^+$.

   This criterion must make it possible to measure the fit between every classification $P \in \mathbb{P}$ and every representation $L \in \mathbb{L}$ of this classification.

4) Defining an optimization problem whose aim is to minimize the criterion W. This may be stated in the following way: *Seek simultaneously the classification* $P \in \mathbb{P}' \subset \mathbb{P}$ *and a representation* L *of this classification in such a way that* P *and* L *have the best possible fit in terms of the criterion* W.

   ($\mathbb{P}$ is, for example, the set of partitions and $\mathbb{P}'$ the set of partitions in k classes).

5) Constructing an algorithm (known as "nuées dynamiques") to solve this problem; this algorithm consists of using alternatively a representation mapping $g : \mathbb{P} \to \mathbb{L}$ and an allocation mapping $f : \mathbb{L} \to \mathbb{P}$ (see figure 1).

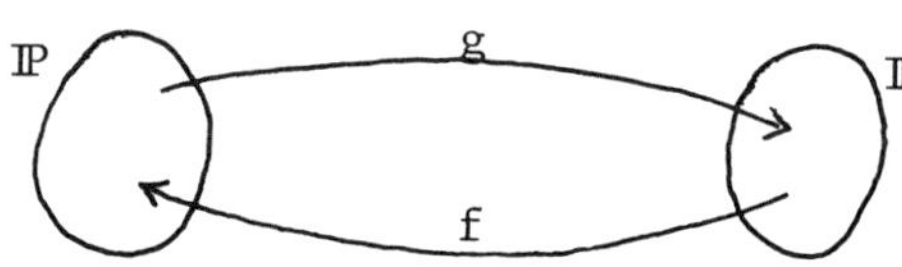

Figure 1

   This is initialized using a classification $P^0 \in \mathbb{P}$ or a representation $L^{(0)} \in \mathbb{L}$ which may be estimated or drawn at random.

6) Study the convergence properties of this algorithm : it is possible to define a sequence $u_n = W(v_n)$ with

$v_n = (L^{(n)}, P^{(n)})$, where $P^{(n)} \in \mathbb{P}$ and $L^{(n)} \in \mathbb{L}$ is the representation of $P^{(n)}$ obtained by applying g.
If the allocation mapping (making it possible to go from $L^{(n)}$ to $P^{(n+1)}$) and the representation mapping (making it possible to go from $P^{(n+1)}$ to $L^{(n+1)}$) have been well chosen, it can generally be shown that the sequences v_n and u_n are convergent and that the sequence u_n is decreasing. In other words, it can be shown that the algorithm makes the criterion decrease at each application of the functions g and f until a stable position is reached.

C. Some applications of the MND method

The MND method makes it possible to express and to give a local optimal solution for a large family of problems. If $\Omega \subset \mathbb{R}^n$, $\mathbb{P}' = \mathbb{P}_k$ the set of the partitions in k classes of Ω and $\mathbb{L} = \mathbb{R}^n$ it comes about that the MND is reduced to a classical approach which comes within the framework of K-means algorithms (see Mac Queen (1967)) the representation of each class is its center of gravity and the criterion $W(L,P)$ expresses the sum of the inertia of each class.

The MND formalism has made it possible to solve many other less classical problems with a lot of practical applications. For instance :

. Typological selection of variables : find simultaneously a partition and a given number of selected variables which characterize as well as possible each class of this partition (see Diday (1976) and Meunier et al. (1986)).

. Adaptative distances : find simultaneously a partition and a distance (local or global) among a family of admissible distances such that the objects be the most similar in each class of the partition according to the associated distance (see Diday, Govaert, Lemoine (1978)).

- Crossing clustering : find simultaneously a partition on the objects and a partition on the variables such that the value taken in each crossing :((class of objects) × (class of variables)) be as homogeneous as possible (see Diday et al. (1980 & 1985), Govaert (1983)).
- Canonical analysis : when data arrays have a large size, it is legitimate to look for linear combinations depending on the local tendencies which may appear in the population. Both these tendencies and the canonical components which are associated to them in the "best" possible way may be simultaneously detected. We present in Diday (1987) several algorithms in order to minimize the criterion, in both the case where the data are centered and the case where they are not. In the case where all the variables are qualitative, the problem is posed in terms of the factor analysis of correspondance and it amounts to looking for the clusters of objects, that induce the largest χ^2 on the contingency between the variables. If one of the two variable blocks is composed of incidence variables, we find interesting methods of local discriminant analysis.
- Clustering and regressions : find simultaneously a partition $P = (P_1,\ldots,P_k)$ and its representation $L = (L_1,\ldots,L_k)$ where the L_i are the coefficients of the local regression associated to P_i. The criterion measures for instance the sum of the mean square error of k local regressions (see Diday et al (1980,1985)). The local regressions obtained by the associated MND algorithm are generally better than the global one.
- Typological factor analysis : in Diday, Simon (1976) a generalization of the classical factor analysis to the simultaneous determination of k local factorial planes, instead

of one, is given. We look for a partition $P = (P_1,...,P_k)$ and $L = (L_1,...,L_k)$, where L_i is a plane which defines the representation of P_i. The associated MND algorithm converges towards a solution which tends to minimize the local inertia with respect to the local plane L_i defined by the subset P_i. The local factorial planes (or axes) obtained, explain a greater part of the inertia than the global factorial planes.

. Clustering and discriminant factor analysis : the problem consists of looking for the best discriminant factors of the clusters that we have to determine simultaneously.

The MND has been applied to many other problems for instance for clustering objects defined only by their distances, for finding clusters having a good fit with a curve taken in a family of admissible curves (the space of representation), in mixture decomposition etc... (see Diday et al. (1980,1985)).

LEARNING HIERARCHICAL CLUSTERING FROM EXAMPLES

A classical problem of Pattern recognition consists, of looking for an operator of classification (a "classifier") induced from a learning set on which classes are known. A problem frequently encountered in practice is the one of looking for an operator of clustering (a "clusterfier", as opposed to "classifier") induced from a learning set of which classes are also known. In the first case, we have to find an operator which allocates each new object to one of the classes defined by the learning set. In the second case, we have to find an operator which detects classes in the complete population, taking into account as much as possible the information given by the classes of the learning set. To illustrate this general problem a simple example is given in figure 2.

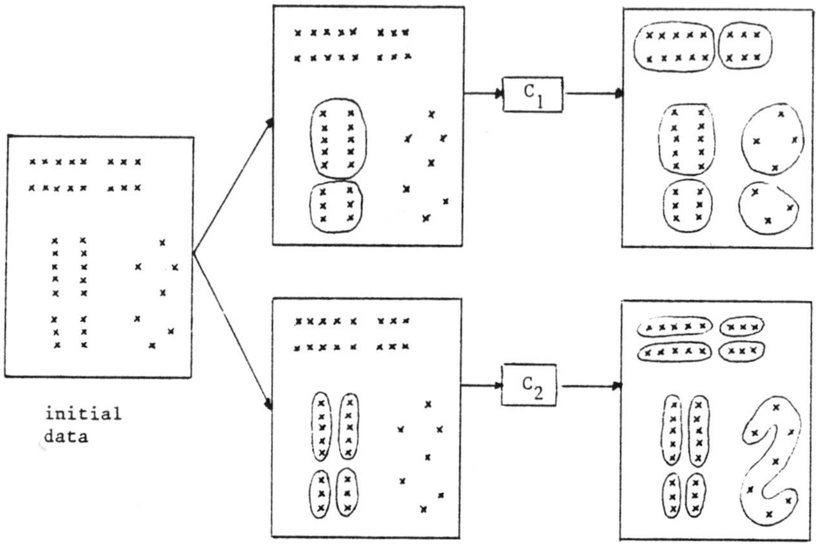

Figure 2 : The classes defined on a subset induces the clusterfier C_1 or C_2 which gives clusters on the whole initial set.

If we wish to find a clusterfier which gives an algorithm of partitioning (i.e. a partitioning clusterfier) we have to induce for instance the allocation and representation mapping of the MND method from a partition given on a sample. As a partition may be considered as a particular case of a hierarchy (we obtain exactly a hierarchy by adding to a partition the singletons and the whole set) the problem of finding a "partitioning clusterfier" is embedded in the problem of finding a "hierarchical clusterfier". More precisely a "hierarchical clusterfier" is an operator which, starting from a known hierarchy on a sample, produces an algorithm which gives a hierarchy on the whole set as close as possible to the given hierarchy on the sample. The most popular algorithm of hierarchical clustering depends mainly on the way of aggregation of two clusters h_i, h_j (denoted $\delta(h_i, h_j)$). Therefore the hierarchical clusterfier will have to include a measure of aggregation from a hierarchy given on a sample. Lance and Williams (1967) have proved that the most classical measure of aggregation δ (as the single, complete, average link, etc...) may be obtained from a good choice of the parameters a_i of their well known recurence formula :

$$\delta(h, h_1 \cup h_2) = a_1\ \delta(h, h_1) + a_2\ \delta(h, h_2) + a_3\ \delta(h_1, h_2) + a_4\ |\delta(h, h_2) - \delta(h, h_1)|\ .$$

Therefore to induce a measure of aggregation δ from a hierarchy H given on a sample it suffices to compute the value of the parameters a_i. It is possible to do so by setting $f(h) = \delta(h_i, h_j)$ = the height of the lowest class h of H containing $h_i, h_j \in H$.

With the help of these quantities and the general recurence formula, one obtains a system of equations with 4 unknown and $(m-1)(m-2)/2$ if m is the number of objects in the sample. Knowing the a_i by solving these equations and the dissimilarity index between the objects, it is possible to build the hierarchy on the whole set by using the recurence formula. Unfortunately, such a hierarchy may have "inversions" which means that $f(h) < f(h')$ may occur although $h' \subset h$. In Diday and Moreau (1984) we provide simple conditions on the a_i which insure the non-existence of inversions. Therefore we have finally to solve a system of m equations under constraints on the unknowns. What often happens in practice is that a user wishing to make a hierarchical classification does not know which of the usual panoply of dissimilarity indices will be the best one for his data; it can also happen that none of these indices satisfies the data that he must deal with; such a problem arises with regard to the choice of one of the known indices and possibly, the creation of new indices.

This method has made it possible to find aggregation indices adapted to particular learning sets (elongated class, spherical class with central kernel, half spherical class with central kernel, noising elongated class...). Some of these new indices make it possible to recognize more specific classes than the usual indices.

IV. NEW KINDS OF GRAPHICAL REPRESENTATION IN CLUSTERING

A. From hierarchies to pyramids

We have already given the definition of a hierarchy and a pyramid. If two clusters overlap one must be included in the other for a hierarchy but this is not necessary for a pyramid. It is possible to show that pyramids may be considered as a natural extension of hierarchies. To do so we need some recall :
An ultrametric δ is a dissimilarity index which satisfies the following property : for any triple $x,y,z \in \Omega$ we have

$$\delta(x,z) \leq \text{Max}(\delta(x,y),\delta(y,z)) .$$

(It may be shown that δ is an ultrametric iff all the triangles are isoceles with the base smaller than the sides).
A dissimilarity s and an order $\mathcal{O}$ are compatible iff for any triple x,y,z ordered according to $\mathcal{O}$ we have

$$s(x,z) \geq \text{Min}(s(x,y),s(y,z)) .$$

A pyramidal index s is a dissimilarity index for which there is an order $\mathcal{O}$ such that s and $\mathcal{O}$ are compatible.

We have then the following properties : the pyramids contain the hierarchies and allows the representation of overlapping clusters by a "pyramidal dendrogram" (see figure 4). Moreover the ultrametrics are contained in the pyramidal indices. An indexed classification is a couple (C,f) where C is a set of subsets of Ω and f is a map $C \to \mathbb{R}^+$ such that

.$f(h) = 0$ iff h is a single element of Ω

.$h,h' \in C$ and $h \subset h'$ strictly, implies $f(h) < f(h')$ (i.e. no inversions).

Indexed hierarchies (H,f) and indexed pyramids (P,f) are special cases of indexed classification. It may be shown that the set of indexed hierarchies is in bijection with the set of ultrametrics and that there also exists a bijection between the set of pyramids and the set of pyramidal indices. Let $M(s,\mathcal{O}) = [s(\omega_i,\omega_j)]$ be the n×n matrix whose terms are the dissimilarity index $s(\omega_i,\omega_j)$ values, and where rows and columns are ordered according to $\mathcal{O}$. It may be shown that for any ultrametric δ there exists an order $\mathcal{O}$ such that δ and $\mathcal{O}$ be compatible. We say that a matrix is robinson iff the terms of the rows and columns never decrease when moving away, in either direction, from the main diagonal. It may be shown that $M(\delta,\mathcal{O})$ is robinson if $\mathcal{O}$ is compatible with δ and that there exists a bijection between the robinson matrices and the pyramidal indices. All these results are summarized in figure 3. In figure 4 we give an example of pyramid. For more details on pyramids, see Diday (1984) and Diday, Bertrand (1986).

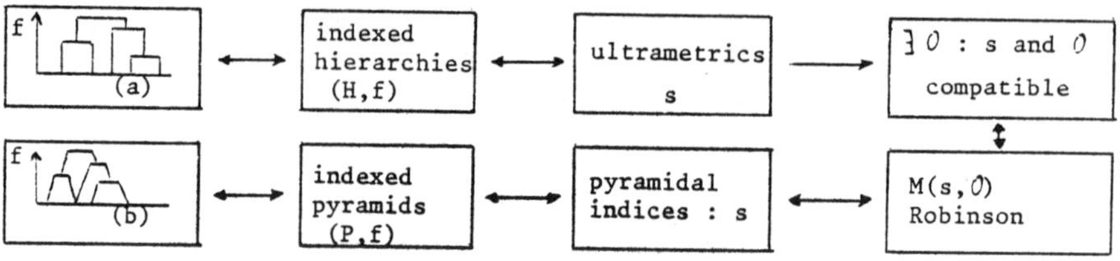

Figure 3 : (a) is an hierarchical dendrogram (H,f);
(b) is a pyramidal dendrogram (P,f);
A↔B means "bijection" between A and B
A→B means : A implies B .

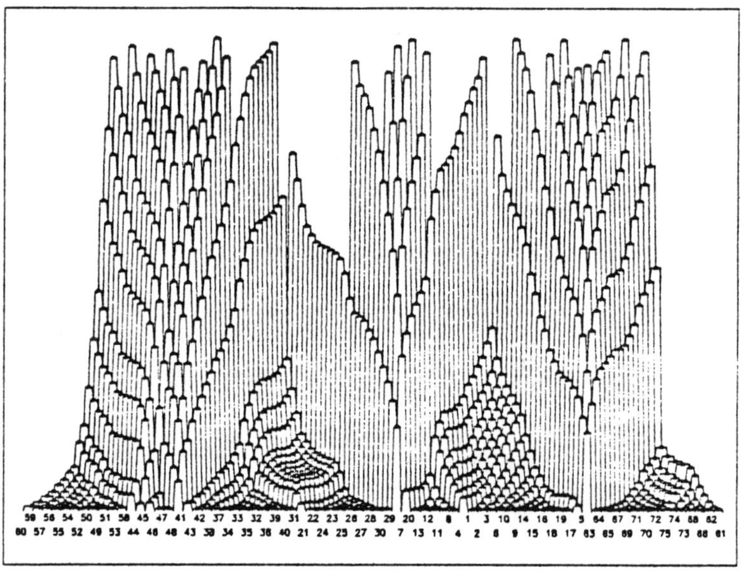

Figure 4 : An example of pyramid

B. Representation of "chaotic" data by "ultramines"

We have seen that there exists a bijection between hierarchies and ultrametrics. Therefore when we represent data by a hierarchy (i.e. dendrogram) it means that we represent the given initial dissimilarity index by an ultrametric. The result is that the fit between the dissimilarity index and a hierarchy will be that much better the more it satisfies the following properties of an ultrametric :

1) The dissimilarity index satisfies the ultrametric inequality.
2) There exists an order which is compatible (see § 4.1) with the dissimilarity index.

If these two conditions are not satisfied at all the fit between the data and any dendrogram (i.e. between a dissimilarity s and any ultrametric δ) will be very bad. That is why we have introduced the notion of "ultramine" which will have a good fit with that kind of data. We say that a dissimilarity index is an ultramine iff all the triangles are isoceles with the base larger than the sides. Then, it may be shown that there exists an order $\mathcal{O}$ (termed "anti-compatible") with any ultramine u such that for any triple $x,y,z \in \Omega$ ordered according to $\mathcal{O}$ we have $u(x,z) \leq \text{Max}\{u(x,y),u(y,z)\}$. It may be shown (see Diday (1986)) that if for more than four couples the dissimilarity s takes distinct values then there may not exist simultaneously an order compatible and an order anti-compatible with s. In the first case we will say that the data are "coherent" and in the second case that they are "chaotic" for the given dissimilarity index s. If they are "chaotic" they may be represented by an anti-indexed hierarchy which will have a good fit with the largest distances and for which the classes represent "oppositions" (see Diday (1986)) Coherent data may be represented by indexed hierarchies which will have a good fit with similar objects and for which the classes represent "chains" or "clusters" of close objects.

In practice the data are neither completely chaotic nor completely coherent. There are many ways of defining a criterion which measures if the data are more or less chaotic or coherent for a given dissimilarity index s by comparing the number of triangles which are "almost" isosceles with the base larger than the sides to the number of triangles which are "almost" isosceles with the base smaller than the sides. According to a given dissimilarity s, let T be the set of all the triangles of Ω. Let E(t) be the difference between the smallest sides and the greater sides of $t \in T$. Let N be the number of triangles for which $E(t) \neq 0$. If $N > 0$ we have, for instance, the following criterion :

$$W(\Omega,s) = \frac{1}{N} \sum_{t \in T} \operatorname{sign}(E(t)) .$$

If s is ultramine $W(\Omega,s) = -1$, if it is ultrametric $W(\Omega,s) = +1$ and $W(\Omega,s) = 0$ if the number of isosceles triangles for which the base is larger then the sides is equal to the number of isoceles triangles for which the base is smaller than the sides when all the triangles are isosceles.

C. Crossing, orders and ultrametrics : application to vizualisation of consensus for comparing classifications

The visual representation of a hierarchy induces an order on the objects. If we want to represent the same hierarchy taking into account some external constraints (for example an order on the objects induced by another hierarchy, a partition, etc...) crossings may appear. There is a crossing in the hierarchy when the vertical branch associated with a class of the hierarchy is cut by a horizontal branch.

We use the notion of compatibility between an order and an index of dissimilarity and we introduce the notions of semi-compatibility and weak-compatibility. We study the matrix characterizations which arise from a generalization of Robinson matrices. We relate these notions to the notions of minimum spanning path in the sense of the dissimilarity index chosen. When this index is an ultrametric we obtain interesting results which relate the order of the objects corresponding to a visualization of a hierarchy and the ultrametric induced by this hierarchy. Based upon these results several algorithms are developed which make it possible to build hierarchies which are visually similar, hierarchies which realize consensus etc... For more details see Diday (1986).

REFERENCES

1. Andenberg (1973). Cluster analysis for applications, Academic Press.
2. Benzecri, J.P. (1973). L'analyse des données, Tomes 1 et 2, Dunod (Paris).
3. Diday, E. (1976). Typological selection of parameters, Pattern Recognition and artificial Intelligence, Academic Press. Also, 4th IJCPR, Kyoto.
4. Diday, E., Simon, J.C. (1976). Clustering analysis, Digital Pattern Recognition, Edit. by K.S. Fu, Springer Verlag.
5. Diday, E. et al. (1980). Optimisation en classification automatique, a book of 900 pages published by INRIA.
6. Diday,E. et al. (1985). Optimisation en classification automatique, a book of 360 pages in Russian published by Economica.
7. Diday, E., Govaert, G., Lemoine, Y. (1978). New Kind of representation in clustering. Proc. 4th IJCPR Kyoto Japan.
8. Diday, E. (1985). Croisements ordres et ultramétriques : applications à la recherche de consensus en classification automatique, Metron, Vol. XLIII, n° 3-4.
9. Diday, E. (1982). Crossing orders and ultrametrics : application to visualization of consensus for comparing classification, Proceed. of Compstat Physica Verlag.

10. Diday, E. (1984). Une représentation des classes empiétantes : les pyramides. Rapport de recherche n° 291 available at INRIA in English.

11. Diday, E., Bertrand, P. (1986). An extension of hierarchical clustering : the pyramids. Pattern recognition in practice. 2. North Holland.

12. Diday, E. (1981). New kinds of graphical representation in clustering. Compstat, Physica-Verlag.

13. Diday, E. (1987). Canonical analysis from the automatic classification point of view. Control and Cybernetics, Vol. 15, n° 2-3.

14. Lance, G.C., Williams, W.T. (1967). A general theory of classification sorting. Computer Journal 9(10) and 10(3).

15. Jambu, M., Lebeaux, M.O. (1978). Classification automatique pour l'analyse des données. Dunod.

16. Lerman, J.C. (1980). Classification et analyse ordinale des données. Dunod.

17. Meunier, Ph., Rasson, J.P., Diday, E. (1986). Une méthode de sélection typologique de variables. Fourth Intern. Symp. Data Analysis and Informatics. North Holland.

18. Mc Queen, J.B. (1967). Some methods for classification and analysis of multivariate observations. Proc. of the 5th Berkeley Symp. on Methematical Statistical and Probability, vol. 1, pp. 281-297.

19. Sneath PHA, Sokhal, R.R. (1973). Numerical Taxonomy. Freeman.

SIMULTANEOUS CLUSTERING OF CASES AND VARIANCES

Yasuo Ohashi

University Hospital Computer Center
University of Tokyo Hospital
7-3-1, Hongo, Bunkyo-ku, Tokyo, JAPAN

I. INTRODUCTION

It is our usual experience that the results of clustering cases depend, often largely, on the variables used for analysis and, *vice verse*, those of clustering variables depend on the cases used, whether those results are obtained by the application of automated classification techniques such as *k*-means method or by the visual inspection of the output of ordination techniques such as principal component analysis. So, it is also an usual practice for consultants of data analysis to recommend their clients to reanalize the data by deleting or subsetting cases and/or variables in order to verify the "stability" and "generality" of the results of the analysis.

One possible remedy for the problems in selecting cases and/or variables to be used in classification is to classify all cases and all variables simultaneously, as Hartigan points out (8). Although a number of block clustering methods (7) (8), one of which is implemented in BMDP as P3M(2), are proposed for the simultaneous clustering, it seems to the author that they are *ad hoc* and an approach

ISBN 0-12-215485-1

which has some correspondence with well-established data analysis techniques is worth developing.

In this paper, it is assumed that data is given in the form of a cases × variables data matrix where measurements are taken on continuous scale. Existing models for approximating a rectangular data matrix are summarized and (simultaneous) clustering is characterized as a method of increasing the goodness of fit of each model or decreasing the number of parameters of a well-fitted model through two operations named "localization" and "merging", respectively.

The author should confess that this paper discusses only a conceptual framework. Some important practical problems including development of efficient algorithms, determination of the number of clusters, empirical comparison with other approaches and interpretability of the results are little discussed in this paper, and the object of application of the proposed approach is limited to a relatively small data. The author would also appreciate if he is informed of important research works on simultaneous clustering he overlooks.

II. MODELS AND THEIR FITTING

A. Approximation of a Rectangular Data Matrix

Let $\mathbf{X} = (X_{\alpha i})$ be a data matrix where $\alpha = 1,\ldots,A$ and $i = 1,\ldots,I$ refer to cases and variables, respectively. A variety of models are proposed for approximating $\mathbf{X}$ by $\hat{\mathbf{X}} = (\hat{X}_{\alpha i})$ as follows:

Constant model: $$\hat{X}_{\alpha i} = c \qquad \text{(MO)}$$

Case (row) effect model: $\hat{X}_{\alpha i} = t_\alpha$ (MC)

Variable (column) effect model: $\hat{X}_{\alpha i} = u_i$ (MV)

Additive model: $\hat{X}_{\alpha i} = t_\alpha + u_i$ (MA)

$(\Sigma t_\alpha = 0 \text{ or } \Sigma u_i = 0)$

Multiplicative model: $\hat{X}_{\alpha i} = dp_\alpha q_i$ (MM)

$(\Sigma p_\alpha{}^2 = 1,\ \Sigma q_i{}^2 = 1)$

Mixture model: $\hat{X}_{\alpha i} = t_\alpha + dp_\alpha q_i$ (MXC)

$(\Sigma p_\alpha{}^2 = 1,\ \Sigma q_i{}^2 = 1,\ \Sigma q_i = 0)$

$\hat{X}_{\alpha i} = u_i + dp_\alpha q_i$ (MXV)

$(\Sigma p_\alpha{}^2 = 1,\ \Sigma q_i{}^2 = 1,\ \Sigma p_\alpha = 0)$

$\hat{X}_{\alpha i} = t_\alpha + u_i + dp_\alpha q_i$ (MXA)

$(\Sigma t_\alpha = 0 \text{ or } \Sigma u_i = 0,\ \Sigma p_\alpha{}^2 = 1,\ \Sigma q_i{}^2 = 1,\ \Sigma p_\alpha = 0,\ \Sigma q_i = 0)$

Saturated model: $\hat{X}_{\alpha i} = v_{\alpha i}$ (MS)

where restricted conditions in parentheses are necessary for identifiability of each parameter value.

We adopt the following squared norm as a measure of goodness of fit of each model:

$$\|X-\hat{X}\|^2_{\mathbf{W},\mathbf{V}} = \sum_\alpha \sum_\beta \sum_i \sum_j w^{\alpha\beta} v^{ij} (X_{\alpha i} - \hat{X}_{\alpha i})(X_{\beta j} - \hat{X}_{\beta j}) \quad (1)$$

where $\mathbf{W} = (w^{\alpha\beta})$ and $\mathbf{V} = (v^{ij})$ are metric matrices which accommodate intra-cluster correlation structure or give weights to cases and variables, respectively. If each case is independent and considered to have to give the same weight to analysis, $w^{\alpha\beta}$ should be equal to

$$\delta^{\alpha\beta} = 1 (\text{when } \alpha = \beta),\ 0\ (\text{otherwise}). \quad (2)$$

It is well known that k-means method assumes implicitly a metric matrix of (2) and the resulting clusters tend to be spherically shaped and "natural" but long-shaped clusters often fail to be detected (4). Although there are some attempts for estimating intra-cluster variable correlation ((1) and ACECLUS in SAS (16)), in the following we adopt the usual squared Euclidean norm (the residual sum of squares; RSS):

$$\|X-X\|^2_{I,I} = \sum_{\alpha}\sum_{i} (X_{\alpha i}-\hat{X}_{\alpha i})^2 \tag{1'}$$

as a measure of goodness of fit.

The models (MO)-(MA) are classical ANOVA models. The model (MM) was first introduced by Fisher and Mackenzie instead of an additive model (MA) and applied to the data of manurial response of potato varieties (5). A generalization of (MM):

$$\hat{X}_{\alpha i} = \sum_{k=1}^{K} d_k p_{k\alpha} q_{ki} \tag{MSV}$$

was studied by Eckart and Young (3) and Householder and Young (9) and is known as the Eckart-Young decomposition in quantitative psychology. It is well known that the least squares (LS) solution which minimizes RSS is given by the largest K singular values and the corresponding eigenvectors of the raw data matrix $\mathbf{X}$. The ordination technique based on the solution of K=2 is called biplot.

Mandel (12) and Gollob (6), independently and possibly influenced by the work of Fisher and Mackenzie, developed a generalization of (MXA):

$$\hat{X}_{\alpha i} = t_{\alpha}+u_{i}+ \sum_{k=1}^{K} d_k p_{k\alpha} q_{ki}. \tag{MMA}$$

Generalizations of (MXC) and (MXV) are also possible.

Statistical inference theory based on (MMA) or (MXA) under the normality assumption is summarized in Krishnaiah and Yochmowitz (11).

For each model, the LS solution of an additive part is easily obtained as the usual ANOVA solution for a two-way table without replication, and that of a multiplicative part is calculated from the singular value decomposition of the residual matrix whose element is given by ($X_{\alpha i}$-($\hat{X}_{\alpha i}$ *by the additive part*)).

Okamoto (14) called the singular value decomposition (principal component analysis) based on (MXA), (MSV), (MXC) and (MXV) as *M* method, *N* method, *R* method and Q method, respectively.

When enough goodness of fit is not attained with a simpler model, we usually proceed to a more complex one with more parameters and, *vice verse*, when enough fit is attained, we proceed to a more parsimonious model for the easiness of interpretation. But it is not always easy to interpret complicated models such as (MXA) and (MMA) and the difference of degrees of freedom between models are large when A or *I* are large. (See Fig. 1 for the degrees of freedom of RSS of each model.)

In the next section, simultaneous clustering is characterized giving an intermediate continuum, through two operations defined as follows, between (MO) and (MXA) (or (MS)), which would be more interpretable than the complex mixture model with more than one multiplicative term.

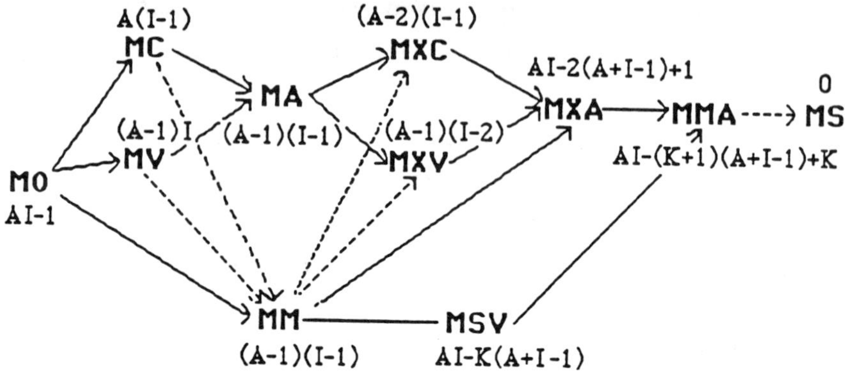

Fig. 1 Hierarchy of models and degrees of freedom of RSS

B. Two Operations Generating Simultaneous Clustering

In this paper, we confine ourselves to a relatively simple simultaneous clustering where clustering structure of cases does not depend on variables as shown in Fig. 2a and exclude a "mosaic" block structure as shown in Fig. 2b.

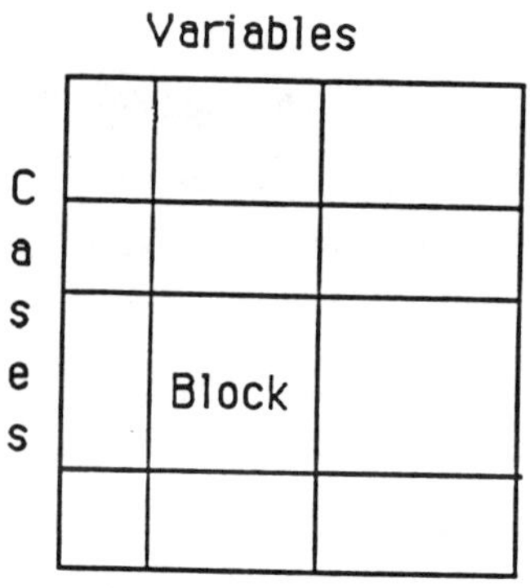

Fig. 2a
"Simple" structure

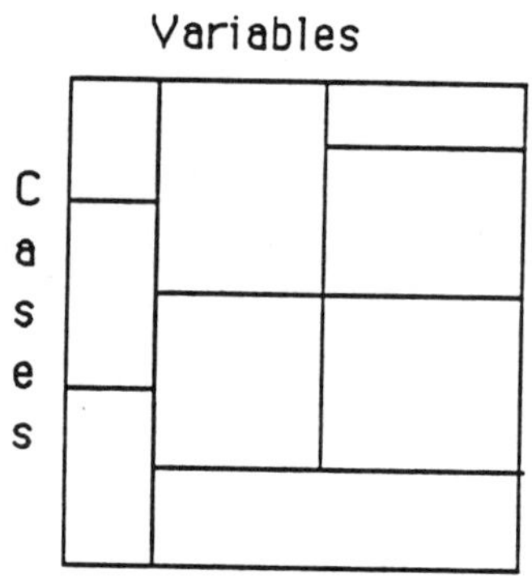

Fig. 2b
Complicated "mosaic" structure

Let C(•) be a function which gives the label of cluster to which each case or variable belongs and define $\alpha* = C(\alpha)$ and $i* = C(i)$. The label $\alpha*$ varies from 1 to L and $i*$ varies from 1 to M where L and M are the number of clusters of cases and variables, respectively. We denote the parameters c, t_α, u_i, d, p_α, q_i and $\upsilon_{\alpha i}$ in (MO)-(MS) representatively by Z. We define that Z is global as to cases or variables when Z does not include suffix α or i, respectively and define Z is local otherwise.

<u>Localization</u>. This operation is defined as "localizing" a global parameters so that the value varies among clusters. We denote the localized parameters as to cases, variables and both by $Z|_{\alpha*}$, $Z|_{i*}$ and $Z|_{\alpha*i*}$, respectively. For example, models $c|_{\alpha*}$, $c|_{i*}$ and $c|_{\alpha*i*}$ have L, M and LM parameters, respectively.

<u>Merging</u>. This operation is defined as equating the value of local parameters within each cluster. We denote the merging operator and the merged parameters as to cases, variables and both by

$$Z_{\alpha*} = (Z_\alpha)_{\cdot\alpha*},\ Z_{i*} = (Z_i)_{\cdot i*},\ Z_{\alpha*i*} = (Z_{\alpha i})_{\cdot\alpha*i*},$$

respectively.

It is easy to show the following rules hold:

$$(Z)|_{\alpha*i*}=(Z|_{\alpha*})|_{i*}=(Z|_{i*})|_{\alpha*},\ (Z_{\alpha i})_{\cdot\alpha*i*}=(Z_{\alpha*i})_{\cdot i*}=(Z_{\alpha i*})_{\cdot\alpha*} \tag{3}$$

$$(Z_1+Z_2)|_{X*} = Z_1|_{X*}+Z_2|_{X*},\ (Z_1+Z_2)_{\cdot X*} = Z_{1\cdot X*} + Z_{2\cdot X*} \tag{4}$$

$$(Z_1\cdot Z_2)|_{X*} = Z_1|_{X*}\cdot Z_2|_{X*},\ (Z_1\cdot Z_2)_{\cdot X*} = Z_{1\cdot X*}\cdot Z_{2\cdot X*} \tag{5}$$

$$Z_X|_{X*} = Z_X,\ Z_{\cdot Y*} = Z \text{ if } Z \text{ does not include suffix } Y \tag{6}$$

where $X*$ is $\alpha*$, $i*$ or $\alpha*i*$ and Y is α or i.

Restrictions to parameters for identifiability are

modified as follows:

localization: $\Sigma Z_Y = \text{const.} \rightarrow \Sigma Z_{Y|X*} = \text{const.}$ (7)

merging: $\Sigma Z_X = \text{const.} \rightarrow \Sigma N(X*)Z_{X*} = \text{const.}$ (8)

where $N(\cdot)$ is a function which gives the size of each cluster.

Estimation of parameters under an localized model is easily carried out by treating each block (defined by clustering) as a whole matrix and applying the LS calculation as described before independently to other blocks. For a merged model, estimation can be carried out as follows:

•For (MO)-(MV) and the additive parts of (MXC) and (MXV), each parameter is estimated as the corresponding mean of merged cases or variables.

•For (MA) and the additive part of (MXA), $t_{\alpha*}$ and u_{i*} are estimated by the application of the ANOVA method for a two-way table with different replications.

•For the multiplicative part, a whole matrix is replaced by a $L \times M$ matrix whose element is given

$$(N(\alpha*))^{\frac{1}{2}}(N(i*))^{\frac{1}{2}}(\tilde{X}_{\alpha*i*} - (\hat{X}_{\alpha*i*} \text{ by the additive part}))$$

where $\tilde{X}_{\alpha*i*}$ is a mean of the block, that is

$$\tilde{X}_{\alpha*i*} = \sum_{C(\alpha)=\alpha*} \sum_{C(i)=i*} X_{\alpha i}/(N(\alpha*)N(i*)).$$

Singular value decomposition is applied to the resulting matrix to yield the parameters d, $\tilde{p}_{\alpha*}$ and $\tilde{q}_{i*}$ and, finally $p_{\alpha*}$ and q_{i*} are obtained by

$$p_{\alpha*} = \tilde{p}_{\alpha*}/(N(\alpha*)^{\frac{1}{2}}, \quad q_{i*} = \tilde{q}_{i*}/N(i*)^{\frac{1}{2}}.$$

C. Model generation and relationship to existing methods

Intermediate models are generated by applying a localization operation to parsimonious models and, *vice verse*, applying a merging operation to nonparsimonious models, taking account of the rules (3)-(6). One model can be derived by more than one different path, for example,

$$v_{\alpha*i*} = c|_{\alpha*i*}, \tag{MHA}$$

$$v_{\alpha*i} = u_i|_{\alpha*}, \tag{MKM}$$

$$v_{\alpha i*} = t_\alpha|_{i*}.$$

It is worth noting some models which are bases of existing (simultaneous) clustering methods are derived by applying these two operations to a simple model. Hartigan's block clustering and BMDP3M can be interpreted as fitting the above (MHA) model. Sarle assumes the same model and develops an iterative computer program for finding a nonhierarchical optimal solution (15). SAS VARCLUS (16) implicitly assumes the model

$$\hat{X}_{\alpha i} = u_i + d|_{i*} p_\alpha|_{i*} q_i \tag{MVC}$$

which is derived from (MXV) and a localization as to variables.

III. STRATEGIES OF DATA ANALYSIS

As usual clustering techniques of cases or variables

based on an optimizing criterion, two strategies are possible for simultaneous clustering proposed here; that is, a hierarchical approach and non-hierarchical one. And, moreover, the former is dichotomized into an agglomerative one and a divisive one. As regards an non-hierachical approach, technical difficulties such as determination of the number of clusters, remedies of avoiding local optimum solutions and development of rapid algorithms are multiplied in simultaneous clustering. At present, it seems to the author that the hierarchical approach with an interactive computer program is feasible and practical solution for exploring a data matrix which is possibly cluster-structured unless we have certain information on the number of clusters.

Merging operations which start from (MS), (MA) or (MM) are appropriate for an agglomerative approach which yields more parsimonious models, and localization operations are appropriate for an divisive one which seeks homogeneous blocks in which a simple model such as (MO), (MC) or (MV) fits well locally. The starting models for an agglomerative approach and the local models for an divisive approach should be selected, problem by problem, based on physical interpretability and goodness of fit of global models.

In the following, examples of strategies are briefly discussed using three data types. The detailed results of analysis will appear on another report and be available upon request to the author.

<u>Applicants × ratings</u> Kendall, Stuart and Ord gives an example ((10) p. 358) where 48 applicants are evaluated from score 0 to 10 on 15 items. The objective of analysis is grouping applicants together whose scoring patterns are similar and, at the same time, grouping items together which take similar values on each applicant. The model $\upsilon_{\alpha * i *}$ =

$c|_{\alpha*i*}$ is appropriate for this objective. Starting from the saturated model $v_{\alpha i}$, merging of two cases or two variables which yields the minimum increase of RSS can be carried out successively just as Ward's method. Mean square defined as

(Increase of RSS/difference of degrees of freedom)

may be used in order to determine whether cases or variables should be merged at some stage of agglomeration.

<u>Individual × measurements of sizes of various parts</u> The famous Fisher's Iris data is this example. Here, individuals come from a mixture of several "natural" clusters and, at least conceptually, we can take as many measurements as we want from each individual, however, we can assume the existence of hypothetical factors (in Iris data, "size" of petal and "size" of sepal) and each variable has high correlation with one of these factors. Therefore, adding measurements (variables) does not always contribute to the separation of "natural" clusters, and to make matters worse, adding measurements which have high correlation with one factor comes to neglect other factors possibly with high discriminating power.

One approach for finding natural clusters while taking account of such structure of variables is to estimate intra-cluster correlation among variables, as is mentioned in the Introduction. Another possible approach from the point of simultaneous clustering is to fit a model:

$$\hat{X}_{\alpha i} = u_{i|\alpha*} + d|_{\alpha*i*} p_{\alpha|i*} q_{i|\alpha*}.$$

This model is a localization of (MVC) as to cases and assumes that individuals in each cluster have one dimensional structure in each cluster of variables. When the numbers of clusters of cases and variables are fixed, the LS solution can be sought by alternating the minimization step as to cases (while fixing the cluster structure of variables) and that as to variables (while fixing the cluster structure of cases). The algorithm for the former step is developed by modifying the algorithm of k-means method and the latter step is carried out by modifying the VARCLUS algorithm.

Varieties × manurial treatments The data of Fisher and Mackenzie mentioned earlier, to which the multiplicative model (MM) fits better than the additive model (MA) and the additive model after log-transformation (13) and, at the same time, gives a physically meaningful interpretation, is an example of this type of data. Merging operations starting from a well-fitted model give us more parimonious interpretation of response pattern and necessary calculation steps are straightforward if we confine ourselves to the hierarchical clustering. Whether cases or variables should be merged at some stage of agglomeration is determined by taking account of degrees of freedom as in the first example. Some sort of "stopping rules" may be invented by comparing the increase of RSS and RSS of the initial model.

REFERENCES

1. Art, D. et al. (1982) Utilitas Mathematica, 21A, 75-79.
2. BMDP Statistical Software Manual (1985) University of California Press.
3. Eckart, C. and Young, G. (1939) Bull. Amer. Math. Soc., 45, 118-121.
4. Everitt, B. (1979) Biometrics, 35, 169-181.
5. Fisher, R.A. and Mackenzie, W.A. (1923) J. Agricultural Sci., 13, 311-320.
6. Gollob, H.F. (1968) Psychometrika, 33, 73-115.

7. Hartigan, J.A. (1975) Clustering Algorithms, Wiley.

8. Hartigan, J.A. (1982) "Classification" IN: Encyclopedia of Statistical Sciences (Kotz, S. and Johnson, N.L. eds.), vol. 2, 1-10, Wiley.

9. Householder, A.S. and Young, G. (1938) Amer. Math. Monthly, 45, 165-171.

10. Kendall, M.G. et al. (1983) The Advanced Theory of Statistics, vol. 3, fourth ed., Griffin.

11. Krishnaiah, P.R. and Yochmowith, M.G. (1980) IN: Handbook of Statistics (Krishnaiah, P.R. ed.), vol. 1, 973-994, North-Holland.

12. Mandel, J. (1969) J. Res. National Bureau of Stand., B, 73B, 309-328.

13. Ohashi, Y. (1982) (in Japanese) Kokyuroku, 526, 13-45.

14. Okamoto, M. (1972) J. Japan Statist. Soc., 2, 63-69.

15. Sarle, W. (1982) Proc. 7th SAS User's Group International, 651-653.

16. SAS User's Guide: Statistics, Version 5 Ed. (1985). SAS Inc.

TECHNIQUES OF APPROXIMATION FOR BUILDING TWO TREE STRUCTURES

M. Roux

Centre d'études phytosociologiques et écologiques (CNRS)
B.P. 5051
34033 MONTPELLIER CEDEX, FRANCE

I.- INTRODUCTION

The object of this paper is twofold. First we state two new algorithms, one for building a hierarchic tree clustering and the other providing with a non-hierarchic tree representation. Both algorithms start with a distance or a dissimilarity matrix. The second purpose is to compare the results with those of classical algorithms such as average link agglomerative method and minimum spanning tree.

Among clustering methods the most popular are certainly the so-called "sequential, agglomerative, hierarchic, non-overlapping" techniques (SAHN according to Sneath and Sokal 1973). We want to present here a quite different algorithmic approach which is neither agglomerative nor divisive.

It is well-known that any scaled hierarchic tree may

ISBN 0-12-215485-1

be associated an ultrametric distance (Jardine and Sibson 1971, Benzecri 1973). Our first algorithm works up directly the distance matrix, modifying step by step its values in order to fulfill the ultrametric inequality. The second one is designed to build up an additive tree metric out of the given dissimilarities. It works in an analagous way, striving to meet with the four-points condition ; this condition is the characteristic property of additive (non-hierarchic) trees, in which the distance between two vertices is measured as the sum of the edge lengths along the path joining them (Buneman 1971 pp 387-395, Dobson 1974).

In a previous conference (Roux 1985) we explained the construction of the additive tree without any demonstration. To day the first algorithm is shown in details and its convergence is demonstrated, while the second one is only outlined. Then a real life example is presented to which the Minimum spanning tree construction is applied together with the average link clustering. The results are compared to those of the two previous algorithms.

II.- CONTINUOUS APPROACH OF AN ULTRAMETRIC

A.- The algorithm.

The ultrametric inequality stands as follows for any triplet i, i', i" of distincts objects :

$$d(i,i'') \leq \mathrm{Max}\ (d(i,i'),\ d(i',i''))$$

where d(i,i') is the distance between objects i and i'. It

is known that whenever this inequality holds, all triangles are isosceles, the equal sides being longer than the third one. This gives the basis of our algorithm the principle of which is to consider all the triplets, one at a time. For each and ever triplet we record the distance modifications which would be necessary to get the two longest sides equal to eachother, their new common length being the arithmetic mean of their actual lengths.

Once all triangles are examinated, each dissimilarity is modified by adding the mean value of all modifications required for it. Up to now the dissimilarities are not ultrametric, but the whole set of triangles is repeatedly considered in the same way until the sum of the absolute values of all modifications becomes small enough.

B.- Convergence of the algorithm

We need some notations :

I = set of objects under study (OTUs)
P = set of pairs of distincts objects
p = (i,i') = a pair of such objects
d = the given dissimilarity
d_K = state of the dissimilirity at step k
T = set of triplets of distinct objects
t = (i,i',i") one of these triplets
P(t) = the set of three pairs of elements of t

We call "step" a complete scanning of all triplets made of different elements of I. The algorithm deals with a set of partial ultrametrics d_k^t defined after the values of

d_k on triplet t. Let Dmin(t) denote the value of the smallest dissimilarity in t ; let Dmed(t) be the median one and Dmax(t) be the largest one.

If (i,i') is not included in t then $d_k^t(i,i') = 0$

If $d_k(i,i') = \mathrm{Dmin}(t)$ then $d_k^t(i,i') = d_k(i,i')$

If $d_k(i,i') = \mathrm{Dmed}(t)$ or $d_k(i,i') = \mathrm{Dmax}(t)$ then
$d_k^t(i,i') = (\mathrm{Dmax}(t) + \mathrm{Dmed}(t))/2$

With these notations there is another way to state the successives values of the dissimilarities as the arithmetic mean of the partial ultrametrics :

$$d_{k+1}(i,i') = (1/N)\Sigma_{t \varepsilon T} \; d_k^t(i,i')$$

where N = Card(I) - 2 is the number of triplets including the pair (i,i').

The demonstration is conducted in three parts ; first the positive quantity

$$R_k = \Sigma_{(i,i')\varepsilon P} d_k^2(i,i')$$

is decreasing as k tends to infinity. So there exists a limit R_*. Next the distance between two consecutive states of the dissimilarities is shown to tend to zero. The last point is to demonstrate that, for any p, the distance between state k and state k+p of the dissimilarities tends to zero with k growing up to infinity, so that the sequence of dissimilarities meets with Cauchy's criterion, hence it has a limit d_*.

Demonstration, part 1 :

Let us look at R_{k+1}

$$R_{k+1} = \Sigma_{(i,i')\varepsilon P} d^2_{k+1}(i,i')$$

$$R_{k+1} = \Sigma_{(i,i')\varepsilon P}((1/N)\Sigma_{t\varepsilon T}\ d_k^t(i,i'))^2$$

The second sum has only N non-zero terms. It is then required to square the mean of N values which is less or equal to the mean of the squared values :

$$R_{k+1} \leqq (1/N)\Sigma_{(i,i')\varepsilon P}\Sigma_{t\varepsilon T}\ (d_k^t(i,i'))^2$$

$$R_{k+1} \leqq (1/N)\Sigma_{t\varepsilon T}\Sigma_{(i,i')\varepsilon P}\ (d_k^t\ (i,i'))^2$$

As partial ultrametrics d_k^t are null out of triplet t, the second sum is only made of three terms, the values of which are :

Dmin(t), (Dmed(t)+Dmax(t))/2 , (Dmed(t)+(Dmax(t))/2

Adding the squares of these values provides with a result less or equal to the sum of squares of the initial values :

$$\text{Dmin}\ (t) + 2((\text{Dmed}(t)+\text{Dmax}(t))/2)^2 \leqq \text{Dmin}^2(t) + \text{Dmed}^2(t) + \text{Dmax}^2(t)$$

because of the same property used previously : the square of the mean is less or equal the mean of the squared values. And finally

$$R_{k+1} \leqq (1/N)\Sigma_{t \varepsilon T} \Sigma_{(i,i') \varepsilon P(t)} d_k^2 (i,i')$$

In this double sum each pair (i,i') appears exactly N times, so that it yields :

$$N \Sigma_{(i,i') \varepsilon P} d_k^2 (i,i')$$

Hence

$$R_{k+1} \leqq R_k$$

Since the successive values of R are decreasing, they do have a limit R_*.

Demonstration, part 2

Let us consider the criterion

$$D(k,k+1) = \Sigma_{(i,i') \varepsilon P} (d_{k+1}(i,i') - d_k(i,i'))^2$$

$$D(k,k+1) = \Sigma_{(i,i') \varepsilon P} d_{k+1}^2 (i,i') + \Sigma_{(i,i') \varepsilon P} d_k^2 (i,i') - 2 \Sigma_{(i,i') \varepsilon P} d_{k+1} (i,i').d_k(i,i')$$

In this expression the first sum is equal to R_{k+1} while the second one is equal to R_k. Let us examine the last one :

$$S = \Sigma_{(i,i') \varepsilon P} d_{k+1} (i,i').d_k(i,i')$$

where we replace $d_{k+1}(i,i')$ by its definition.

$$S = \Sigma_{(i,i')\varepsilon P}(1/N)\Sigma_{t\varepsilon T}\ d_k^{\ t}(i,i').d_k(i,i')$$
$$S = (1/N)\Sigma_{t\varepsilon T}\ \Sigma_{(i,i')\varepsilon P(t)}\ d_k^{\ t}(i,i').d_k(i,i')$$

The sum over P may be restricted to P(t) because $d_k^{\ t}(i,i')$ is null outside the triplet t. Now let us look at some particular triplet t and call i,i',i" its elements ; we suppose they meet the following inequalities :

$$d_k(i,i') \leqq d_k(i,i") \leqq d_k(i',i")$$

Then applying the algorithm leads to :

$$d_k^{\ t}(i,i') = d_k(i,i')$$
$$d_k^{\ t}(i,i") = (\ d_k(i,i") + d_k(i',i")\ )/2$$
$$d_k^{\ t}(i',i") = (\ d_k(i,i") + d_k(i',i")\ )/2$$

Transfering these values in our last sum gives :

$$\Sigma d_k^t(i,i').d_k(i,i') = (d_k^{\ t}(i,i'))^2 + (d_k(i,i") + d_k(i',i"))^2/2$$

$$= (d_k^{\ t}(i,i'))^2 + (d_k^{\ t}(i,i"))^2 + (d_k^{\ t}(i',i"))^2$$

$$= \Sigma_{(i,i')\varepsilon P(t)}\ (d_k^{\ t}(i,i'))^2$$

$$= \Sigma_{(i,i')\varepsilon\ P}(d_k^{\ t}(i,i'))^2$$

And finally :

$$S = \Sigma_{(i,i')\varepsilon\ P}(1/N)\Sigma_{t\varepsilon T}\quad (d_k^{\ t}(i,i'))^2$$

The term after the first sum is still the mean of a sum of

squares, hence this quantity is greater or equal to the square of the mean value :

$$S \geqq \Sigma_{(i,i')\varepsilon P}\left((1/N)\Sigma_{t \varepsilon T} d_k^{\,t}(i,i') \right)^2$$

$$S \geqq \Sigma_{(i,i')\varepsilon P} d^2_{k+1}(i,i')$$

Replacing this value in the expression of D yields :

$$0 \leqq D(k,k+1) \leqq R_{k+1} + R_k - 2\,R_{k+1} = R_k - R_{k+1}$$

As R_k and R_{k+1} have the same limit their difference tends to zero as k tends to infinity.

Demonstration, part 3.

To show the convergence one can write the successive inequalities :

$$D(k,k+1) \leqq R_k - R_{k+1}$$
$$D(k+1,k+2) \leqq R_{k+1} - R_{k+2}$$

...

$$D(k+p-1,\ k+p) \leqq R_{k+p-1} - R_{k+p}$$

Summing all these inequalities, term by term leads to :

$$D(k+p-1,\ k+p) + \ldots + D(k,\ k+1) \leqq R_k - R_{k+p}$$

and, owing to the triangular inequality on D :

$$D(k+p,\ k) \leqq R_k - R_{k+p}$$

For any fixed p the difference $R_k - R_{k+p}$ tends to zero, hence the sequence of d_k meets with Cauchy's criterion and converges.

III.- CONTINUOUS APPROACH OF AN ADDITIVE TREE METRIC

A.- Description of the algorithm

The characteristic inequality of an additive tree is the following one :

$$d(i,j) + d(k,l) \leqq \text{Max}(\ d(i,k) + d(j,l)\ ,\ d(i,l) + d(j,k)\ )$$

where i, j, k, l, are any four objects of the set I. It is known (Patrinos and Hakimi 1972) that if this inequality, called the four points condition, holds then, among the three sums of dissimilarities involved above, the two largest ones are equal.

For convenience we use some new notations :

Q is the set of all quadruplets q = (i, j, k, l) made of distincts elements of I.

We call Smin(q), Smed(q) and Smax(q) the smallest, the median and the largest sum of dissimilarities, respectively, involved in the four points condition.

At the beginning of the algorithm we set d = d, the given dissimilarity. Then, at each step n, the whole set Q of quadruplets is examinated in an analogous way to that of previous algorithm, defining for each quadruplet q a partial additive tree metric (ATM) d_n^q , the values of which are defined as follows .

Let C(q) = (Smax(q) - Smed(q))/4 ; it is the correction of the dissimilarities due to quadruplet q.

If $d_n(i,i')$ contributes to Smin(q) then $d_n^q(i,i') = d_n(i,i')$

If $d_n(i,i')$ contributes to Smed(q) then $d_n^q(i,i') = d_n(i,i') + C(q)$

If $d_n(i,i')$ contributes to Smax(q) then $d_n^q(i,i') = d_n(i,i') - C(q)$

If i or i' does not belong to q then $d_n^q(i,i') = 0$.

Once all quadruplets are examinated and the $d_n^q(i,i')$ evaluated the new values of d are set up :

$$d_{n+1}(i,i') = (1/N)\Sigma_{q \varepsilon Q}\ d_n^q(i,i')$$

where N is the number of quadruplets containing the pair (i,i'), that is :

$$N = (\text{Card } I - 2)*(\text{Card } I - 3)/2$$

The process is stopped when the distance D(k, k+1) between two successive states of d becomes small enough.

B.- Basis of the convergence.

The proof of the convergence is carried out in a similar fashion as for the ultrametric algorithm. First the quantity

$$R_n = \Sigma_{(i,i')\varepsilon P} d_n^2 (i,i')$$

is shown to be ever decreasing, hence, as a positive function, it has a limit R_*. This is done using the same principle as previously, that is the squared mean-value of a set of real numbers is less or equal to the mean of the squared numbers.

This principle is used anew to demonstrate that the distance D(n,n+1) between to consecutive states of the dissimilarities tends to zero when k tends to infinity :

$$0 \leqq D(n,n+1) \leqq R_n - R_{n+1}$$

Then the demonstration is achieved, as for the ultrametric algorithm, with the fact that, for any positive integer p :

$$D(n,n+p) \leqq R_n - R_{n+p}$$

so D(n,n+p) tends to zero with n growing up ; the sequence d is thus a Cauchy's sequence, which ensures its convergence.

IV.- AN EXAMPLE IN ECOLOGY

This example is extracted from Mr. P.Bach's thesis (Bach, 1985) studying the associations of fishes in some

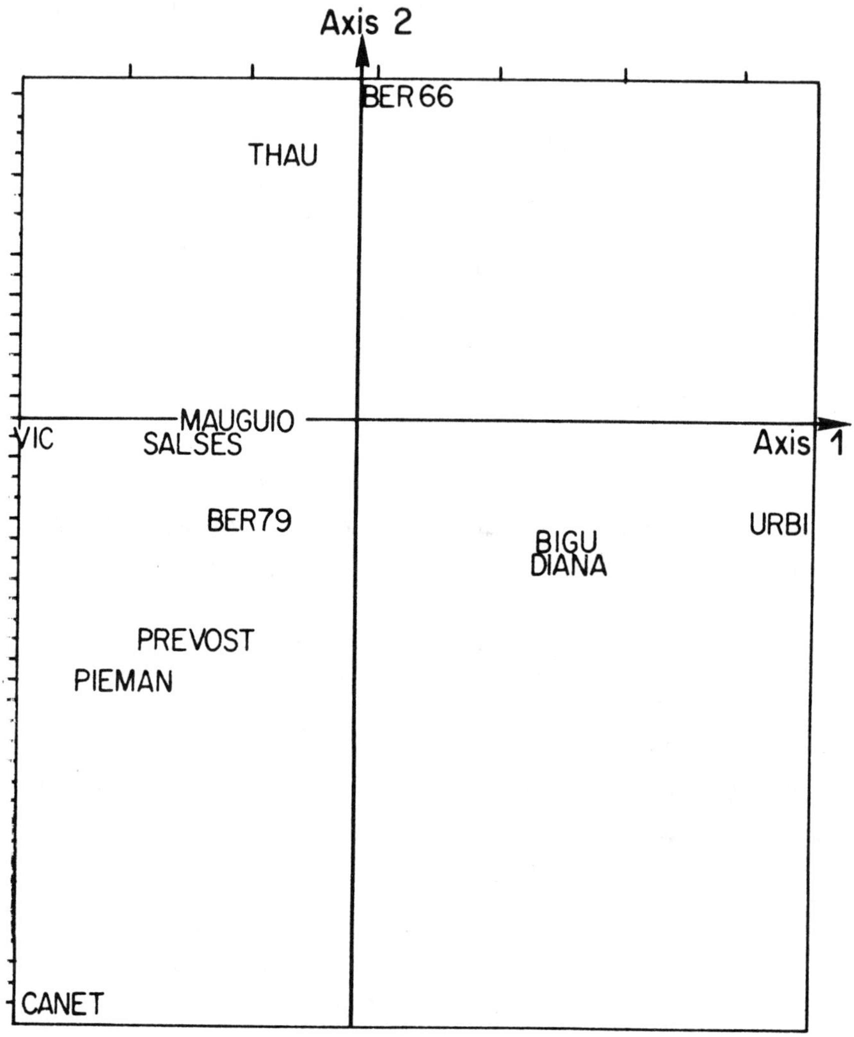

Fig. 1.- Factor analysis of correspondences, Axis 1 and 2. Mediterranean ponds, raw data. First axis accounts for 25 % and second axis for 17.7 % of the overall inertia.

mediterranean coastal ponds. The raw data are made of 0-1 values in a rectangular table the lines of which are species while the column are 12 ponds. If a species is present in a certain pond then the corresponding value is 1, otherwise it is 0. The starting point for ours clusterings is the set of the Jaccard's coefficients of dissimilarity among these 12 ponds, given in table 1 (magnified 1000 times).

	BIGU	URBI	DIANA	BER66	BER79	PIEMA
URBIna	348					
DIANA	362	23				
BER66	517	484	492			
BER79	538	617	625	558		
PIEMAnson	667	756	761	755	577	
PREVOst	692	659	667	673	462	524
VIC	596	679	661	526	553	563
MAUGUio	532	519	528	418	389	600
THAU	500	517	525	403	511	702
CANET	639	761	766	759	593	474
SALSEs	511	585	593	455	526	618

	PREVO	VIC	MAUGU	THAU	CANET
VIC	559				
MAUGUio	514	366			
THAU	667	537	392		
CANET	476	618	649	735	
SALSEs	529	375	300	431	588

Table 1.-Jaccard's dissimilarities between 12 mediterranean coastal ponds.

Several methods of data analysis have been used, the

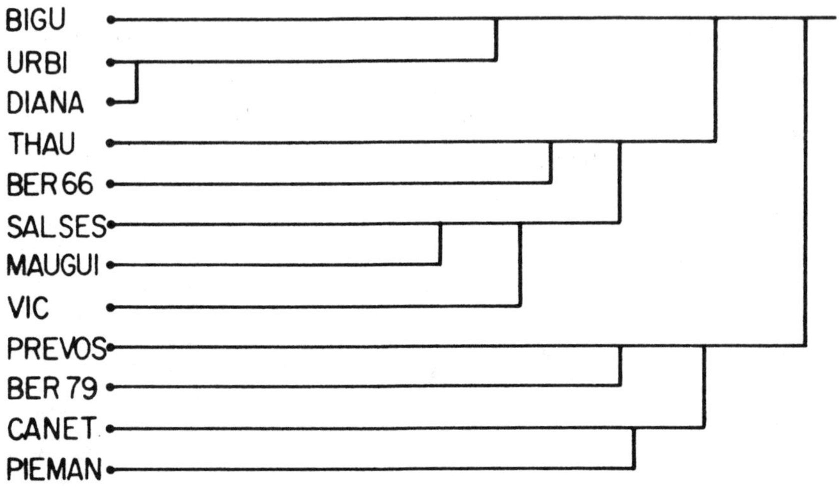

Fig. 2.- Average link agglomerative hierarchy.

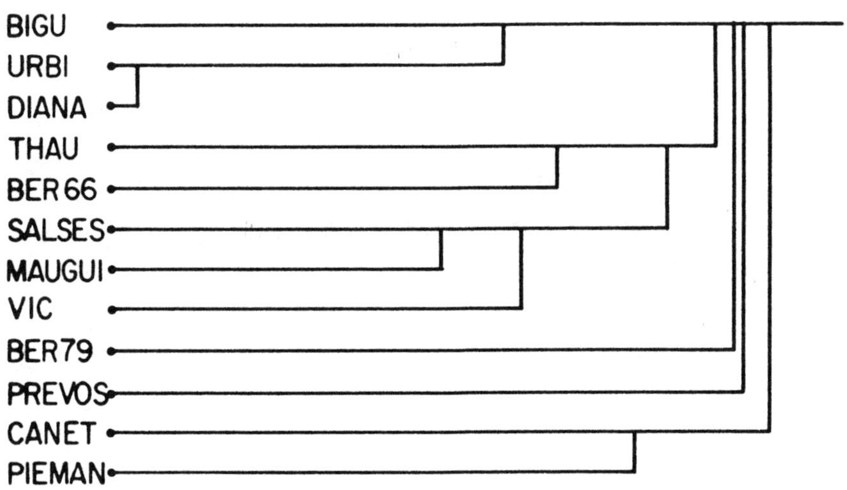

Fig. 3.- Continuous approach of an ultrametric.

results of which are displayed in the subsequent figures. Nevertheless, whatever be the method, these figures show up the same results :

- there is a first group of three ponds, Biguglia, Urbina and Diana, well isolated from the others ; this is not surprising since they are the only ones located in Corsica island, while the remaining ponds are continental.

- a second group made of five ponds : Thau, BER66 (short for Berre pond investigated in 1966), Salses, Mauguio, Vic, more heterogeneous than the previous one, is still visible.

- the four remaining ponds, though even more heterogeneous, should be aggregated in one group, according to correspondence analysis as well as to the weighted average linkage cluster method (In this group BER79 denotes Berre pond investigated in 1979).

In order to compare these results five criteria are computed, three of them being metric, the two others being based upon rank comparisons. In the subsequent formulas d denotes the given dissimilarity while d_* represents the result of some algorithm.

Maximum divergence

$$\text{Max. div.} = \text{Max Abs } (d(i,i') - d_*(i,i'))$$

where "Abs" means "absolute value", and the maximum is taken over the set P of pairs of distinct objects (i,i').

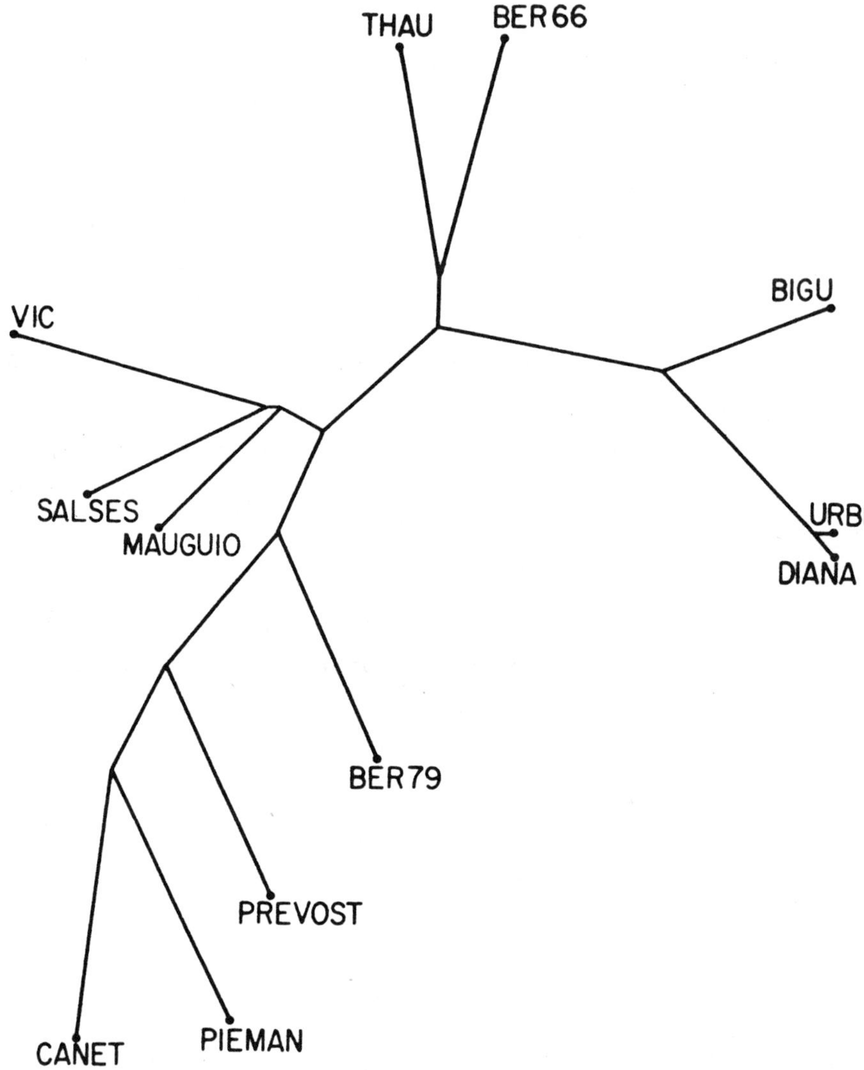

Fig. 4.- Mediterranean pond. Additive tree resulting from the Continuous approach algorithm.

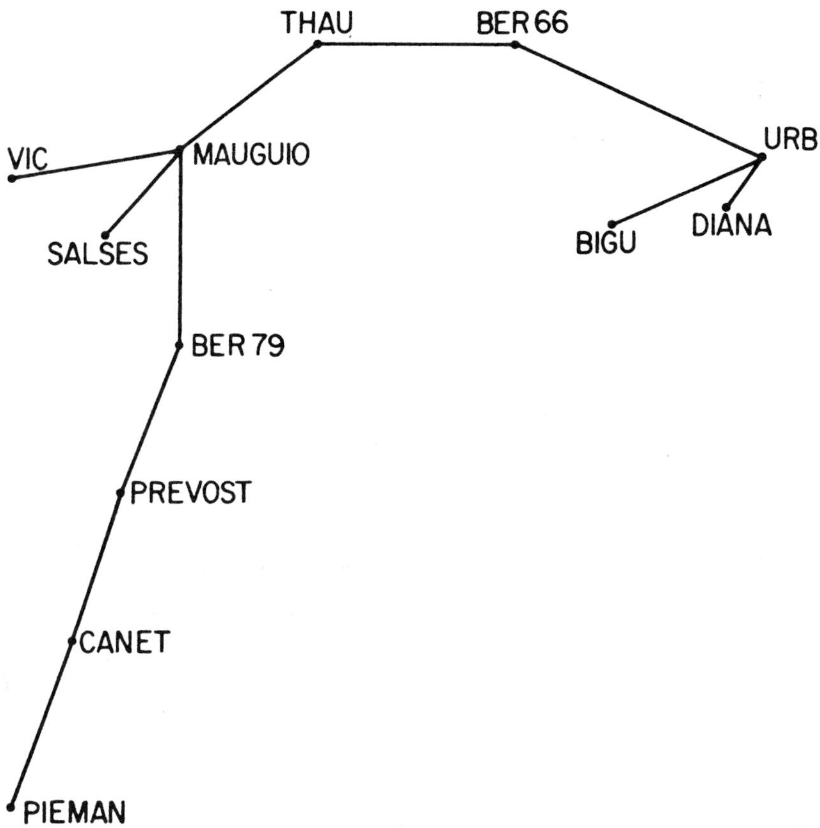

Fig. 5.- Mediterranean ponds, Minimum spanning tree.

Mean absolute divergence (N is the number of pairs in P).

$$\text{Mean div.} = (1/N)\Sigma_{(i,i')\varepsilon P}\text{Abs}(\ d(i,i') - d_*(i,i')\ )$$

Mean square divergence ("Sqr" means "square root")

$$\text{Msq. div.} = \text{Sqr}(\ (1/N)\Sigma_{(i,i')\varepsilon P}(\ d(i,i') - d_*(i,i'))^2)$$

Matching and unmatching inequalities

For all quadruplets (i,j,k,l) of distincts objects the two following quantities are evaluated :

$$d(i,j) - d(k,l)$$
$$d_*(i,j) - d_*(k,l)$$

If they have the same sign then they are considered as matching, otherwise they are unmatching. If one of them, or both, are null the quadruplet is discarded.

	Weighted Average Link	Continuous Approach Ultrametric	Continuous Approach Add. tree	Minimum Spanning Tree
Max. div.	242.7	179.1	88	283
Mean div.	54.8	58.2	28.1	106.3
Msq. div.	72.2	73.0	35.8	138.3
Match. ineq.	1324	1403	1945	1256
Unmatch. in.	194	237	191	406

Table 2.- Five criteria for comparing original data to the results of some tree structuring algorithms.

In the table 2 the distance associated to the Minimum spanning tree is obtained as the length of the longest edge on the path joining two points. That is to say this distance is nothing but the subdominant ultrametric (Jardine and Sibson 1971), which may also be obtain from the single link agglomerative cluster method.

Two main features are obvious in this table : the Continuous approach of an ultrametric yields almost as good a result as the average link cluster method ; the additive tree structure fits better to the data than the hierarchic ones. These results confirm the previous observations (Roux 1985) made on artificial and real life data.

V.- CONCLUSION

We have designed two new algorithms to set up an ultrametric and an additive tree metric out of a dissimilarity matrix. Though they do not have any clear advantage over usual methods, neither in computing time nor in optimizing some criterion, they show a quite new approach of these problems. It is hoped they will suggest new ideas and new improvements in clustering and tree representations.

REFERENCES

1. Bach P. 1985.- La pèche dans l'étang de Thau . Thèse de 3 ème cycle. U.S.T.L. Montpellier, France.
2. Benzecri 1973.- L'analyse des données . 2. La taxinomie

Dunod. Paris.

3. Buneman P. 1971.- The recovery of trees from measures of dissimilarity. In "Mathematics in the archeological and historical sciences" (F.R. Hodson, D.G. Kendall and P. Tautu Eds), Edinburgh University Press.

4. Dobson J. 1974.- Unrooted trees for numerical taxonomy. Journal of applied probability , 11, pp 32-42.

5. Jardine N. and Sibson R. 1971.- Mathematical taxonomy . 286 p, Wiley, London and New-York.

6. Patrinos A.N. and Hakimi S.I. 1972.- The distance matrix of a graph and its tree realization. Quaterly of applied mathematics , 30, pp 255-269.

7. Roux M. 1985.- Représentation d'une distance par un arbre aux arètes additives. Fourth international symposium on Data analysis and Informatics . I.N.R.I.A. Versailles, France, pp 3-15.

8. Sneath P.H.A. and Sokal R.R. 1973.- Numerical taxonomy 573p. Freeman and Co. San Francisco.

A HIERARCHICAL CLUSTERING METHOD FOR DISSIMILARITY MATRICES WITH INDIVIDUAL DIFFERENCES

Tadashi Imaizumi

St. Paul's(Rikkyo) University
3, Nishi-Ikebukuro, Toshima-ku, Tokyo, JAPAN

I. INTRODUCTION

In recent years, the varieties of clustering methods have been used to find groupings of n objects $O = \{o_i : 1 \leq i \leq n\}$ from the given data sets. The objects are allocated to the groups such that the objects within groups are more "similar" in some sense than objects between groups.

The methods for hierarchical clustering are possible to visualize the result of clustering by a tree diagram. Several methods have been proposed (1). Those techniques operate on a matrix $\Delta = (\delta_{ij})$ of dissimilarity of n objects:

$$\delta_{ij} = \delta_{ji} \geq \delta_{ii} = 0, \qquad \text{for any } (i, j). \tag{1}$$

A tree diagram can be used to construct a new dissimilarity matrix between objects. Though these methods are basically descriptive methods, those procedures can be be treated as one of fitting a rooted tree structure using some optimization algorithm as Hartigan suggested (2).

ISBN 0-12-215485-1

And the hierarchical clustering process can be described as a process to fit a rooted tree structure model to each of sub-matrices of dissimilarity matrix. Because this process can be halted by stopping when a inter-cluster differences are all greater that d_0, where d_0 is an arbitrary thershold level as Mardia, Kent and Bibby described (3).

We try to generalize this clustering process mentioned to the case that dissimilarity matrix is given from each of individual subjects or other data sources. And we assume the individual differences model in which each of a rooted tree structures and inter-cluster differences are assumed to be the same for all subjects, but the weights assigned to them are not assumed to be identical for all subjects.

II. MODEL

A. Model for dissimilarity matrix

Let $g = g(d_0)$ denote the numbers of clusters:

$$\sum_{i=1}^{g} C_i = 0. \tag{2}$$

Then we assume the model:

$$\Delta = \begin{bmatrix} \Delta_{11} & \Delta_{12} & \cdots & \Delta_{1g} \\ \cdot & \cdot & & \cdot \\ \Delta_{g1} & \Delta_{g2} & \cdots & \Delta_{gg} \end{bmatrix} \simeq D = \begin{bmatrix} V_{11} & B_{12} & \cdots & B_{1g} \\ \cdot & \cdot & & \cdot \\ B_{g1} & B_{g2} & \cdots & V_{gg} \end{bmatrix}, \tag{3}$$

where matrix V_{ii} satisfies a rooted tree structure model for each cluster C_i and matrix B_{ij} is a constant matrix:

$$B_{ji} = B_{ij} = b_{ij}J, \tag{4}$$

where b_{ij} takes some positive value.

B. Individual differences model

Individual differences model is formulated as follows: Let Δ^h denote dissimilarity matrix of individual h (h = 1, 2,..., N), then we assume the model:

$$\Delta^h \simeq D^h = \begin{bmatrix} w_{11}^h V_{11} & w_{12}^h B_{12} & \cdots & w_{1g}^h B_{1g} \\ \cdot & \cdot & & \cdot \\ w_{g1}^h B_{g1} & w_{g2}^h B_{g2} & \cdots & w_{gg}^h V_{gg} \end{bmatrix} \quad (5)$$

where g × g matrix $W^h = (w_{ij}^h)$ is a weight matrix of individual h. This individual differences model has an indeterminacy in the decomposition of Δ_{ii} in terms of w_{ii}^h and V_{ii}; that is, if $\Delta_{ii} = w_{ii}^h V_{ii} = w_{ii}^{h*} V_{ii}^*$, then $w_{ii}^h = a_i w_{ii}^{h*}$ for some positive constant a_i.

C. The number of parameters in the model

It is of interest to show the number of parameters in the proposed models. We see that the number of free parameters M[m(V),g] is defined by

$$M[m(V),g] = m(V) + g(g - 1)/2 + N[g + g(g - 1) / 2], \quad (6)$$

where $m(V) = \sum_{i=1}^{g} m(V_{ii})$ denotes the number of free parameters in V_{ii}. Since the structure of matrix V_{ii} is not well-defined, we can not specify the value of the m(V). And no goodness of fit test can be carried.

III. METHOD

A. Thershold level and the number of clusters

The value of the threshold d_0 is arbitrary one. This value and the number of clusters g will be functionally related. And We specify g instead of d_0 in this paper.

B. Choice of loss function

When the dissimilarity matrices $\{\Delta^h: h = 1, 2, \ldots, N\}$ are given, we can fit the proposed model under the condition $\{C_i: \sum_{i=1}^{g} C_i = 0\}$. And we must define a fitting function to choose $\{C_i\}$, $\{V_{ii}\}$, $\{B_{ij}\}$, and $\{W^h\}$. We choose $\{\hat{C}_i\}$, $\{\hat{V}_{ii}\}$, $\{\hat{B}_{ij}\}$, and $\{\hat{w}^h_{ij}\}$ which minimize a simple least squares criterion:

$$L(h,g) = \sum_{h=1}^{N} tr(\Delta^h - D^h)'(\Delta^h - D^h) \qquad (7)$$

C. The choice of model parameters

Let $\bar{\Delta} = (\bar{\delta}_{kl})$ denote the mean dissimilarity matrix:

$$\bar{\delta}_{kl} = \sum_{h=1}^{N} \delta^h_{kl} / N, \qquad \text{for any } (k, l). \qquad (8)$$

We define the average of pairwise dissimilarities between cluster C_i and C_j, $\tilde{\delta}_{ij}$, by

$$\tilde{\delta}_{ij} = \sum_{o_k \in C_i} \sum_{o_l \in C_j} \bar{\delta}_{kl} / (n_i \times n_j), \qquad (9)$$

where n_i is the number of objects in the cluster C_i.

The value of $\hat{b}_{ij}$ which minimizes L(g) = L(1,g) is $\tilde{\delta}_{ij}$. By the indeterminacy in the decomposition of Δ^h_{ii} for any i, we can use $\tilde{\delta}_{ij}$ as the $\hat{b}_{ij}$ which minimize L(h,g).

In other hand, it is difficult to obtain $\hat{V}_{ii}$ which minimizes L(g). And we employ the method of the average linkage between merged groups as one approximation solution (1). The elements of matrix V_{ii} are defined by the average of pairwise dissimilarities between two sub clusters which are merged. We denote this matrix by $\tilde{V}_{ii}$. It is more complex to obtain V_{ii} in the model (5) than that in the model (3). And we apply the procedure mentioned above to the mean dissimilarity matrix $\bar{\Delta}$.

It is easy to obtain $\hat{W}^h$ which minimizes $tr(\Delta^h - \hat{D}^h)'(\Delta^h - \hat{D}^h)$ for given g, $\{\tilde{V}_{ii}\}$, and $\{\hat{B}_{ij}\}$. It is estimated by

$$\hat{w}^h_{ij} = \begin{cases} tr(\Delta^{h\prime}_{ii}\tilde{V}_{ii}) \,/\, tr(\tilde{V}'_{ii}\tilde{V}_{ii}), & \text{for } i = j, \\ tr(\Delta^{h\prime}_{ij}\hat{B}_{ij}) \,/\, tr(\hat{B}'_{ij}\hat{B}_{ij}), & \text{for } i \neq j. \end{cases} \quad (10)$$

IV. ALGORITHM

The minimization of the loss functions L(h,g) can be facilitated by proceeding in two stages. First, we minimize L(h,g) over $\{V_{ii}\}$, $\{B_{ij}\}$ and $\{W^h_{ij}\}$ for fixed $\{C_i : \sum_{i=1}^{g} C_i = 0\}$, and second, we minimize over C_i . The first minimization is carried out numerically, and the second minimization must be done by some combinatorial procedure.

A. Initial allocation

We allocate n objects to g clusters by applying the average linkage between merged groups method to the matrix $\bar{\Delta}$.

B. Fitting the proposed model

This fitting process consists of the following steps for a given C_i .

1. We fit our model (3) to the matrix $\bar{\Delta}$, and obtain $\{\tilde{V}_{ii}\}$ and $\{\hat{B}_{ij}\}$.
2. The weight matrix $\hat{W}^h$ is obtained by using the equation (10), and we evaluate L(h,g)

C. Re-allocation

We allocate any object o_k in cluster C_i to another cluster C_j when the number of members of this cluster is greater than 1. And we try to fit model under this new allocation. When the value of the L(h,g) is smaller than that of the original allocation, we move o_k to the cluster C_j.

We repeat from the step B to the step C until no improvement occurs

V. EXAMPLE

We gathered dissimilarity matrices of 10 nations from 5 subjects. The nations listed were: Japan, U.S.A., U.S.S.R., China, India, Brazil, Czechoslovakia, France, Congo, and Egypt. Each subject rated the degree of dissimilarity between two nations on 8 point category scale (0 = Most similar, 7 = Most dissimilar). We fitted the proposed model (5) to these data matrices. The values of the loss function for the various number of clusters are shown in Table 1.

We choose g = 4 by inspecting the values in Table 1. And the allocation and a matrix D are shown in Table 2. There are four clusters, {Japan, China}, {U.S.A., France}, {U.S.S.R., Czechoslovakia}, and {India, Brazil, Congo, Egypt}, that might be described as "Far East", "Economically developed", "Pro-communist", and "Underdeveloped". The weight matrices for a given D is shown in Table 3. In table 3, the columns are indices of clusters. And the estimated values are only comparable between subjects.

Table 1 The values of the loss function L(h,g)

g	2	3	4	5	6
L(h,g)	310.977	274.966	233.708	215.250	193.000
L(h,g) / df*	1.438	1.388	1.291	1.345	1.429

* df is defined by df = N × n(n − 1) / 2 − M[0,g]
This quantity is an over-estimated value of N.D.F.

Table 2 The allocation and a matrix D

Japan	1									
China	1	2.2								
U.S.A.	2	4.7	4.7							
France	2	4.7	4.7	3.4						
U.S.S.R.	3	5.2	5.2	5.0	5.0					
Czechoslovakia	3	5.2	5.2	5.0	5.0	2.2				
India	4	5.2	5.2	5.8	5.8	5.6	5.6			
Brazil	4	5.2	5.2	5.8	5.8	5.6	5.6	4.6		
Congo	4	5.2	5.2	5.8	5.8	5.6	5.6	3.0	4.6	
Egypt	4	5.2	5.2	5.8	5.8	5.6	5.6	3.4	4.6	3.4

Table 3 Weight vectors

Subject	11	22	33	44	12	13	14	23	24	34
1	0.91	0.88	1.82	0.92	0.85	1.01	0.86	0.80	0.91	0.83
2	1.36	0.88	0.91	1.12	0.80	1.06	1.12	1.40	1.10	1.17
3	1.36	1.77	0.91	1.42	0.96	1.06	1.10	1.15	1.08	1.19
4	0.91	0.88	0.46	0.74	1.44	1.20	1.27	0.85	1.13	1.01
5	0.46	0.59	0.91	0.78	0.96	0.68	0.62	0.80	0.78	0.79

REFERENCES

1. Anderberg, M.R. (1973). Cluster Analysis for Applications, Academic Press.
2. Hartigan, J.A. (1967). Representation of similarity matrices by trees, Journal of American Statistical Association, 62, 1140 - 1158.
3. Mardia, K.V., Kent, J.T. and Bibby, J.M. (1979). Multivariate Analysis, Academic Press.

APPLICATIONS OF MULTIVARIATE AND CLADISTIC DATA ANALYSES FOR THE CLASSIFICATION OF DERMAPTEROUS INSECTS

Seiroku Sakai

Institute of Biology and Life Sciences
Daito Bunka University
No.560 Iwadono-Nagasaka,Higashi-Matsuyama,
Saitama,Japan 355

I. INTRODUCTION

In biological classification-data analysis,firstly,it is very important how select original data of marked taxonomic features.If the original data matrices are made by inadequate way,the multivariate analyses are not significant. It is very difficult to pick up effective taxonomic features from numerous taxonomic information but the experienced taxonomist has rich ability in oder to make good original data matrices of any special taxa group by his intuitive and accumulated brain information on any special taxa group.

I would like to present about the applications of multivariate and cladistic data analyses for the classification on Dermapterous insects.

II. DERMAPTERAN FOSSILS

A probable ancestor of the present Dermaptera,*Protodiplatys fortis* Martynov(1925) was found from the Jurassic period in the Mesozoic era of South Kazakstan(Turkestan). The fossil was observed the primitive segmented cerci. 40 Dermapteran fossils were found from Karatau near Caspian Sea,Zhejiang,China and Florissant,Colorado.The fossils of more than 25 species were found from Tertiary period such as upper Ecocene,lower Oligocene and Miocene epochs.

RECENT DEVELOPMENTS IN
CLUSTERING AND DATA ANALYSIS

ISBN 0-12-215485-1

Popham(1969) proposed that Dermaptera were evolving throughout Gondwanaland as it disintegrated towards the end of the Mesozoic era.Bey-Bienko(1936) proposed that there are two centers of the origin of Dermaptera,namely,the Eastern and Western of the Angaran Continent in the Jurassic period.

III. DERMAPTERAN BIOGEOGRAPHY

Dermaptera was found about 1850 species around the world. Sakai(1971,1977) attempted to analyse the distributional pattern of all families by multivariate analyses involving PCA and clustering analyses.The dominant fauna occur more in southern hemisphere than the northern.The highest potencies of the fauna are in Oriental and Neotropical regions.The higher potencies are in Ethiopean and Australian regions.

20 environmental variables concerning with temperature, precipitation,altitude,latitude andlongitude were used for multivariate analyses of Spongiphoridae(Labiidae).According to PCA analysis,each genus of 10 subfamilies has got a different pattern of distribution.The present distribution seems to demonstrate the characteristic differentiation pattern from the dispersion-origin of the Oriental region in Jurassic period based upon the multivariate score distribution.The present author speculated 3 dispersal routes from the results.

IV. DERMAPTERAN MULTIVARIATE MORPHOMETRICS

Kevan(1985) pointed out the value of non-traditional physical and chemical characteristics including numerical computer analyses,laser holography,electrophoresis and the other chemical analyses as applied to Orthopteroid insects. The present author understands that modern integrated taxonomy is a very broad subject encroaching on conception,theory and practice of many scientific techniques and disciplines although such non-traditional analyses and results do not provide final taxonomic answers as pointed out Kevan(1985).As a result,the more accurate identification is advanced by such integrated taxonomy as well as advance in traditional systematics. Sakai(1971-1984) attempted to cluster the species level of all the Dermaptera by using external morphological characters including forceps and to use several taxonomic distances and correlation.The individual variation of the external characters is relatively larger than the genitalia on Dermaptera.

There are few satisfactory external characters including forceps,pygidium and telson but the systematics on Dermaptera is largely based on male genitalia. In the common seaside earwig,*Anisolabis maritima* (Bonelli),The present author demonstrated smaller values of the fluctuating variation of male genitalia than the other external morphological characters using PCA and cluster analyses. These studies deal with 18 variables including length of body,length of head,width of head,length of pronotum,width of pronotum, length of mesonotum, maximum width of mesonotum, length of metanotum,length of abdomen,maximum width of abdomen, length of forceps, length of paramere,length of median membrane of paramere,length of inner membrane of paramere,length of penis lobe, and width of tip part of penis lobe, width of middle part of penis lobe, and width of hinge part of penis lobe.The original data matrices of the respective quantitative characters were key punched and processed on FACOM M-140-6 system electronic computer.

54 individuals collected at 35 different localities in all Japan were used for PCA and cluster analyses.

The score distribution of genital quantitative characters in PCA is shown remarkably aggregation of score values in three dimentional space like a cloud but the score distribution of the external quantitative characters is scattered in three dimensional space on the contrary.The male genitalia may be recognized as important character which have small fluctuating variation than the other external characters by morphometric analyses. The study selected 6 different cluster analyses. Methods used group average,Ward method and weighted pair group method.The distances used are Euclid and Mahalanobis distance. There is no tendency of biogeographical cline in respective dendrogram by 6 cluster analyses.

56 individuals of the Japanese striped earwig,*Labidura riparia japonica* (de Haan) collected at 30 different localities in all Japan are used for PCA and cluster analyses.

Similar multivariate analyses of the Japanese striped earwig indicated similar tendency as the Common seaside earwig. In general,taxonomist like to consider the recognitions of the form,shape and coloration with concerning to phylogeny and evolution but physicist and chemist like to consider them with concerning to any numerical unit.The present author seems to need the non-traditional consideration to biological form and coloration from the points of view of physical and chemical aspects. Sakai(1976) attempted a topological and stereological analyses for the form of the Dermapteran forceps. As pointed out by de Hoff and Rhines(1968),maximum calipers diameter,Nassenstein's diameter,equivalent area diameter, stereological graph of the rose of the number of intersection in radial axis and omega degree of orientation of the form are useful for the practical analysis of the form in insects as well as allometric analysis. Sakai and Roomi(1977) and Sakai (1979) analysed the forms of the forceps of the above two earwigs. PCA score distribution showed to be relatively scattered in 4 regions of Z_1 and Z_2 axes.The both species are not exhibit an inclination for geographical cline variation.In multivariate analyses in Forficulidae,13 divisible different types of the paramere forms of male genitalia were recognized.The

PCA score distribution indicated to be relatively scattered in 4 axes in 2 dimensions.

V. DERMAPTERAN PHYLOGENETIC AND CLADISTIC INFORMATION

It would need to check important morphological character from the point of view of Hennigian phylogenetic reconstruction-line and transformed cladistics-line.

Sakai(1984) presented,under the preliminary analysis, Hennigian argumentation scheme, Wagner network by Farris(1970) and procladogram by Camin and Sokal(1965) on thr family-level. The 7 character states are as follows: A1:Forceps segmented in nymph:0(Plesiomorphic state),Forceps not segmented in nymph:1 (Apomorphic state);B2:Virga without reniform vesicle:0,with reniform vesicle:1;C3:Paired penis lobes directed forwards:0 directed backwards:1,Single penis lobe directed backward in paired penis:2,Single penis lobe directed backward:3(Apomorphic state);D4:Paired functional penis lobes:0,Single functional penis lobe:1;E5:Second tarsal segment simple:0,produced below 3rd,but not expanded laterally:1,produced below 3rd and expanded laterally:2(Apomorphic state);F6:Parameres without epimerite:0,Parameres with epimerite:1;G7:last tergite not produced posteriorly to form an anal process:0,produced posteriorly to form an anal process:1. The monophyletic scheme by modified method from Hennig(1966) almost agree with the empirical classification of Dermaptera. Wagner's network by using Manhattan matrix computed that 6 families branched away at the point of HTU 1(Hypothetical Taxonomic Unit) except Diplatyidae and Spongiphoridae after achieved parsimony.The results need to more consideration. The phylogenetic reconstruction by modified method from Camin and Sokal(1965) presented and the X-axis was introduced the evolutionary step concept. The reconstruction may be almost similar to the present empirical or intuitive classification. However the more revision of system need to joint with critical use of contemporary integrated taxonomic techniques under the consideration that convergence is commonplace.

VI. DERMAPTERAN PHYSICAL TAXONOMY AND FUTURE CLASSIFICATION

Physical taxonomic informations such as Moiré picture, scanning electron microscopical photographs,Omnicon image analyzer measurement and laser hologram wre used in order to understand three dimensional pattern.In future,taxonomists will be easy to examine any taxonomic information from Video-image print system,optical disc plate system and stereo virtual image of hologram mount of specimens preserved in any foreign Museum or institution. The author will show several examples of this aspects.We,taxonomists would need to collect any taxonomic image information including the features since Linnaeus-1758 in oder to make such Video or laser disc plates.We should return to original description and type specimens on traditional classification from non-traditional techniques on the image storage aspects.

REFERENCES

1. Bey-Bienko,G.Y.(1936) Faune de l'URSS,Dermapteres, Acad. Sci.URSS,Moscou & Leningrad.
2. Camin,J.H. and Sokal,R.R.(1965). A method for deducing branching sequences in phylogeny.Evolution,19,311-326.
3. De Hoff,R.T.and F.N.Rhines(1968).Quantitative microscopy, McGraw Hill Publ.Co.
4. Farris,J.S.(1970).Methods for computing Wagner trees. Systematic Zool.,19,83-92.
5. Hennig,W.(1966). Phylogenetic systematics, Univ.Illinois

Press.Urbana.

6. Kevan,D.K.McE.(1985,-1981).New trends in systematics:Value of non-traditional physical and chemical characteristics as applied to Orthopteroid insects,Proc.3rd Triennial Mtg. ,Pan Amer.Acrid.Soc.,5-12.

7. Popham,E.J. and Manly,B.F.(1969). Geographical distribution of the Dermaptera and the continental drift hypothesis, Nature, 222(5197),981-982.

8. Sakai,S.(1970 - 1985). Dermapterorum Catalogus Praeliminaris,I-XVIII,Tokyo. 4939 pp.

9. Sakai,S.(1984). Integrated taxonomy on Dermaptera I,Numerical taxonomic information on male genitalia. 17th International congress of Entomology,Abstract,1-16.

COMPARING RELATIONAL VARIABLES ACCORDING TO LIKELIHOOD OF THE LINKS CLASSIFICATION METHOD

I.C. Lerman
IRISA
Campus de Beaulieu, 35042 Rennes Cédex, FRANCE

I. INTRODUCTION ; REPRESENTATION OF RELATIONAL VARIABLES

Most of the Data Analysis methods in current use involve geometrical representation of the data table : Objects x Variables (OxV). However the descriptive variables as they appear, are rarely numerical variables. For several reasons it is of importance to respect faithfully the mathematical nature of the descriptive scale behind each variable. Our aim in this paper is to give some general principles and to mention some recent development of an approach of relationnal data analysis, using hierarchical classification scheme. This method -on which we have been working for about twenty years- is based on the likelihood of the observed relationships between variables.

Indeed, the most general structure for clustering problem consists of a family V of relations (of any type) on a set O of objects. Each relation corresponds -in a very general sense- to a descriptive variable. If p=card(V) and n=card(O), we will denote

$$V=\{w^j/1\leqslant j\leqslant p\} \tag{1}$$

$$O=\{o_i/1\leqslant i\leqslant n\}. \tag{2}$$

Generally O represents a sample from a large population that we want to study.

Faced with this data we have to deal with two complementary and dual problems. The former is of "understanding" and the la-

ISBN 0-12-215485-1

ter is of "management".

The hierarchical cluster organization of the variable family into "significant" classes and subclasses -provided by the classification tree level Statistics ("global" and "local")- contributes in a relevant way to the first problem (7),(8). On the other hand the same type of decomposition of the object set is showed to be a powerful tool of management and information retrieval (13).

We mainly consider in this paper -very oriented towards our contribution- the classification problem of the set V of relational descriptive variables and then, specially the elaboration of an association coefficient between two relational variables taken in the family V. We will suppose -as it is often the case- to be able to consider the two variables as defining the same type of relation on the set O of objects.

The problem of establishing proximity structure on the set O of objects described by the family V of relational variables of any type, is extensively studied in (12).

On the other hand, these past few years, we have performed several algorithms of hierarchical classification concerning large data sets and using the likelihood of the maximal link criterion that we have invented (1970)(5),(7).

The logical idea of mutual comparison between variables consists of set theoretic representation of the relation defined by a variable. This representation may be situated at the level of subset set of O, or of OxO, or even of (OxO)x(OxO). More precisely, the level of this representation is the set of subsets of

O: if the variable is "logical attribute" (i.e. 0-1 variable), or "quantitative variable" or "row" (resp. "column") of contingency table.

OxO: if the variable is "nominal qualitative", or "ordinal qualitative" (totally or partially), or "weighted graph" on O,

(OxO)x(OxO): if the variable is "preordonnance" (i.e. total or partial preorder on OxO) or "weighted graph" on OxO.

More precisely let us give some typical situations :

- A "logical attribute" $\underline{a}$ can be represented by the subset O(a) of the objects which possess the feature associated to the attribute. A "quantitative variable" $\underline{v}$ is considered as defining a "weighting" on the set O of objects : $\xi_i = v(o_i)$ is the weight associated to the object o_i.

- A "nominal qualitative" variable $\underline{c}$ defines on O a partition which can be represented at the level of

$$O^{[2]}=\{(x,y)/x\in O,\ y\in O,\ x\neq y\} \tag{3}$$

by the subset of the object pairs that it joins. Namely, if we denote by $\{O_j/1\leqslant j\leqslant k\}$ the partition classes, the representation of c and then of π is

$$R(\pi)=\sum_j\{O_j^{[2]}/1\leqslant j\leqslant k\}.\ \text{(set sum)} \tag{4}$$

In fact the representation can be restricted at the level of the set $O^{\{2\}}$ of unordered object pairs.

- An "ordinal qualitative" variable $\underline{d}$ defines a total preorder ω on O. If we denote by $(O_j/1\leqslant j\leqslant k)$ its ordered classes, $\underline{d}$ can be represented -at the level of $O^{[2]}$- by

$$R(\omega)=\sum_j\{O_j x O_h/1\leqslant j\leqslant h\leqslant k\}.\ \text{(set sum)} \tag{5}$$

- A "preordonnance" variable $\underline{e}$ defines a total preorder on OxO, or on $O^{[2]}$, or on $O^{\{2\}}$. Let us -without loss of generality- consider the last case of $O^{\{2\}}$. The formal representation of the variable $\underline{e}$ is at the level of $O^{\{2\}}xO^{\{2\}}$. In fact we represent it for simplicity and coherence reasons by a ranking function on $O^{\{2\}}$. More precisely, if we denote by $(H_q/1\leqslant q\leqslant Q)$

the ordered sequence of preorder classes and if we indicate by ℓ_q the cardinality of $H_q[\sum\{\ell_q/1\leqslant q\leqslant Q\}=L=n(n-1)/2]$, we will assign to each pair $\{x,y\}$ belonging to H_q, the following rank:

$$r(x,y)=\sum\{\ell_p/1\leqslant p\leqslant(q-1)\}+(\ell_q+1)/2. \quad (6)$$

By this way, we always have got the relation :

$$\sum\{r(x,y)/\{x,y\}\in O^{\{2\}}\}=L(L+1)/2. \quad (7)$$

We cannot -without destroying the real structure defined- make poorer the representation level of a relational qualitative variable. But we can make richer this level of representation. Thus, we can look at qualitative nominal variable as definning a "preordonnance" on O ; that is to say and more precisely, total preorder on $O^{[2]}$ with two classes, that we denote by $S(\pi)$ (subset of the pairs of which the two components are separated by the partition π) and $R(\pi$ [cf.(4) above]. More precisely :

$$S(\pi)=\sum\{O_j xO_h/1\leqslant j\neq h\leqslant k\} \text{ and } S(\pi)<R(\pi) \quad (8)(9)$$

To realize all the richness of the set theoretic representation point of view, we shall consider below different possible levels of representation of a logical descriptive attribute $\underline{a}$:

- Subset $O(a)$ of O (cf. above).

- Partition $\pi(a)$ of which the only classe which includes more than one single object, is $O(a)$:

$$\pi(a)=\{O(a),\{x\}/x\in O-O(a)\}. \quad (10)$$

- Total preorder $\omega(a)$ with two classes $O(\bar{a})=[O-O(a)]$ and $O(a)$, where we set $O(\bar{a})<O(a)$.

- The partition $\pi(a)$ -as we have just mentionned it above- can be represented by a preordonnance on O [cf.(8) and (9) above].

- Let $\underline{a}$ denote the logical attribute opposite to a. We consider the following ordinal relation which represents resemblance perception on the set of modality pairs :

$$(a,\bar{a})\sim(\bar{a},a)<(\bar{a},\bar{a})<(a,a). \qquad (11)$$

This relation determines a preordonnance with three classes on O :

$$[O(a)xO(\bar{a})] \cup [O(\bar{a})xO(a)]<[O(a)]^{[2]}<[O(a)]^{[2]}. \qquad (12)$$

We hope to nearly lead experimental analysis on the influence of representation level of descriptive variables. But we have some tendency to beleive that the higher the level is, the more stable the results are. This is the reason why we give great importance to preordonnance coding of qualitative variables.

II. COMPARING RELATIONAL VARIABLES

For comparing two qualitative relational variables, we have established the following general diagram :

$$(\alpha,\beta)\in AxB \longrightarrow [R(\alpha),R(\beta)]\in \Omega x \Omega$$

$$s(\alpha,\beta)=card[R(\alpha)\cap R(\beta)]$$

hypothesis of non-link (or independence) taking into account the cardinality characteristics of α and β.

$$S=card[R(\alpha^*)\cap R(\beta^*)]$$

$$Q(\alpha,\beta)=[s-E(S)]/\sqrt{var(S)}$$

Figure 1

In this diagram, α and β are the two relations on O determined by the two variables to be compared. A(resp.B) is the set of all possible structures of the same nature as α(resp.β). R(α) [resp.R(β)] is the set theoretic representation of α [resp.β]. R(α) [resp.R(β)] is a subset of O or OxO or (OxO)(OxO) (cf. above). Ω is the set of representation subsets. s is called the "raw" index. α^* and β^* are two independent relational variables, respectively associated to α and β, according to hypothesis of non-link ("h.n.l.") which takes into account cardinality characteristics of α and β. S is the "random raw index", the mathematical expectation and variance of which being denoned by E(S) and var(S). Q(α,β) is the "standardized" index.

To make this completely clear diagram, let us consider the comparison of two qualitative ordinal variables. α and β are two total preorders (i.e. partitions with ranked classes) on O. We denote by t(α) [resp.t(β)] the type of α(resp.β) ; that is to say, the ordered sequence of the class cardinals. Then A (resp.B) is the set of all total preorders on O having the type t(α) [resp.t(β)]. R(α) [resp.R(β)] corresponds to R(ω) [cf. (5) of §I] and consists of the object couples (x,y) such as x strictly precedes y for α(resp.β). Ω can be defined as the set of all subsets of $O^{[2]}$, representing total preorders on O.

This diagram leads to well known coefficients. Thus K. Pearson's coefficient $\rho_1(a,b)$ (16) between logical attributes (a and b) is obtained from one particular form N_1 of the hypothesis of non-link, by the means of the following relation:

$$Q_1(a,b)=\sqrt{(n-1)}\ \rho_1(a,b). \tag{1}$$

In fact we showed that there are exactly three fundamental forms of the "h.n.l." for comparing two logical attributes (7).

In different general situations of comparing relational variables and <u>whatever is the representation level</u>, we have established an analogous relation to (1), namely :

$$Q(\alpha,\beta)=\sqrt{n}\ \rho(\alpha,\beta), \quad (2)$$

where $\rho(\alpha,\beta)$ is a pure coefficient, the limit of which being independent of n(11).

The M.G. Kendall's τ coefficient (3) between two ranking functions r and s -defining two total and strict orders on O- can also be obtained from the diagram, but by replacing the denominator of Q(r,s). More precisely :

$$\tau(r,s)=(s(r,s)-E[S(r^*,s^*)])/\max\{s(r,s)-E[S(r^*,s^*)]\}. \quad (3)$$

Some researchers like L. Hubert (2) consider association coefficient between two qualitative variables α and β as necessarily having the same form as (3). Namely

$$\tau(\alpha,\beta)=(s(\alpha,\beta)-E[S(\alpha^*,\beta^*)])/\max\{s(\alpha,\beta)-E[S(\alpha^*,\beta^*)]\}. \quad (4)$$

On the other hand, they consider that the statistic $Q(\alpha,\beta)$ should be devoted to testing independence hypothesis between and β. But we showed the non relevance of the tests of independence hypotheses between descriptive variables in Data Analysis (11). On the other hand, the maximization problem of $s(\alpha,\beta)$ is a very difficult one. Relative to the latter, we have very recently obtained an exact bound by the means of a recursive algorithmic approach, in the case of comparing two partitions(14). In this research report we show that generally, the exact value of the denominator of (4) is remplaced by a poor bound which does not take any account of the exact nature and the constraints of the compared structures α and β on O. But $\rho(\alpha,\beta)$

[cf.(2) above] considers intimately the comparison of the two structures. We are now trying a new type of association coefficient directly deduced from $Q(\alpha,\beta)$ by the means of the following formulae :

$$R(\alpha,\beta)=Q(\alpha,\beta)/\sqrt{Q(\alpha,\alpha)Q(\beta,\beta)}. \tag{5}$$

We have -in the framework of the general diagram (cf.figure 1)- clearly set up the problem of partial comparison of two qualitative relational variables α and β, neutralizing the influence of a third variable γ(9),(10).

We present below two classification trees that we obtain by the likelihood of the links method on the same set of qualitative ordinal variables. For the first, a total association coefficient has been used. But the second tree has been built from a partial association coefficient. The sample determining the set O corresponds to the students of a given graduate class of Polytechnic School (Palaiseau-France) [n=card(O)=198]. The variables are defined from questions of ordinal evaluation. The number of modalities per variable is not the same for all the variables ; its is around 4 to 5. The questions concern the following taught topics : "Hilbertian Analysis", "Distribution Theory", "Probability", "Quantum Mecanics", "Mecanics", "Chemistry" and "Informatics". Three dimensions are treated by the variables : "interest", "work" and "difficulty". In both trees the dimension of "difficulty" does not appear. But, as we have mentionned it above, to build the second tree, we neutralized the effect of the difficulty variable in establishing the association coefficient between two variables concerning "interest" or "work" aspects. This neutralization is done with respect to the topic or to the couple of topics covered by the two associated variables. It appears clearly that the higher is the place of Calculus in the topic, the lower is the association degree between "interest" and "work" for the same field (e.g. "Hilbertian Analysis", "Distribution Theory", "Probability").

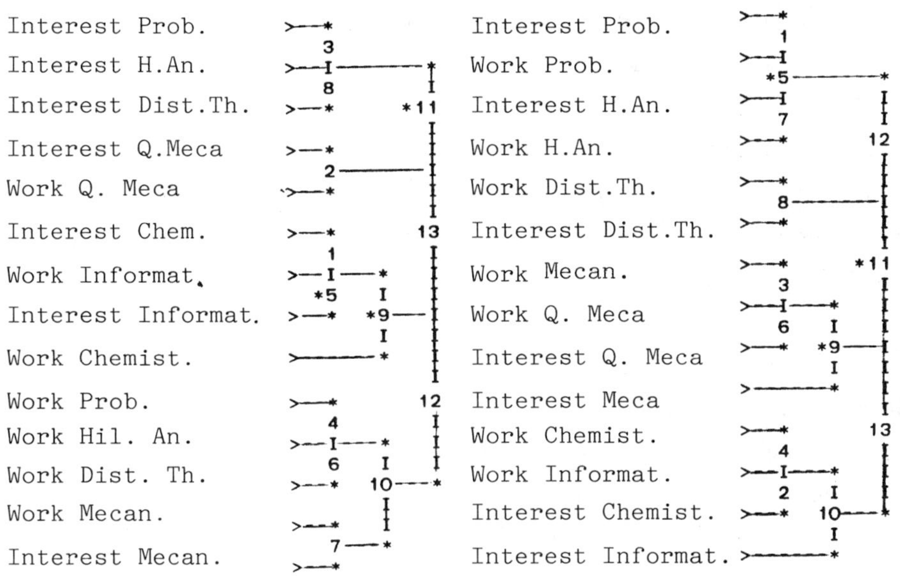

2.1. Total association 2.2. Partial association

Figure 2

Both preceding trees are in fact reduced to their levels where "significant nodes" appear. These nodes are denoted by stars. Our method of dynamical interpretation of the sequence of the tree levels, which leads to the recognition of the most "significant levels" and of the "significant nodes" is in fact based upon association coefficients conceived according to the preceding diagram. These coefficients that we call "global" and "local" level statistics, concern the comparison between the last partition to have emerged -at a given level- and the "preordonnance" associated to the similarity measure S on the set E to be classified. That preordonnance -denoted by $\omega(E)$- is defined as follows :

$$[\forall(p,q)\in F\times F],\ p\leq q \iff S(p)\leq S(q), \tag{6}$$

where $F=P_2(E)$ is the unordered pair set of elements of E. The representation set for comparing the two relational variables on E [$\omega(E)$ and a partition on E, $\pi(E)$] is FxF.

The analysis of the structural form of the obtained criteria lead us to propose a clear solution for comparing two preordonnance variables on the set O (7),(8). Let us consider the case where a preordonnance $\omega(O)$ is defined by a total preorder on $O^{[2]}=\{(o_i,o_j)/1\leq i\neq j\leq n\}$. Then we introduce the representation of $\omega(O)$ by a ranking function on $O^{[2]}$, the principle of which being given in the preceding paragraph [cf.(6) and (7), §I, where the ranking function was defined on $O^{[2]}$]. If we denote by k and h the two ranking functions representing the two preordonnances $\omega(O)$ and $\bar{\omega}(O)$ to be compared, the raw index will take the following form :

$$s[\omega(O),\bar{\omega}(O)]=\sum\{h(i,j)k(i,j)/(i,j)\in I^{[2]}]. \tag{7}$$

The random raw index can be written according to one of two one side forms :

$$S[\omega^*(O),\bar{\omega}(O)]=\sum\{h[\sigma(i),\sigma(j)]k(i,j)/(i,j)\in I^{[2]}\}$$

and

$$S[\omega(O),\bar{\omega}^*(O)]=\sum\{h(i,j)k[\tau(i),\tau(j)]/(i,j)\in I^{[2]}\}, \tag{8}$$

or according to the two side form :

$$S[\omega^*(O),\bar{\omega}^*(O)]=\sum\{h[\sigma(i),\sigma(j)]k[\tau(i),\tau(j)]/(i,j)\in I^{[2]}\}, \tag{9}$$

where σ and τ are two independent random permutations, taken in the set G_n of all permutations, provided by an uniform probability measure.

Then, we can obtain the standardized coefficient $Q[\omega(O),\bar{\omega}(O)]$ and this one, which is directly deduced :

$$R[\omega(O),\bar{\omega}(O)]=Q[\omega(O),\bar{\omega}(O)]/\sqrt{Q[\omega(O),\omega(O)]Q[\omega(O),\omega(O)]} \qquad (10)$$

In fact, all the association coefficients that we have elaborated in the diagram framework can technically be obtained from a general association coefficient between two "weighting" variables on $O^{[2]}$. This last is analogous to the above one, but we have to replace $\{h(i,j)/(i,j)\in I^{[2]}\}$[resp.$\{k(i,j)/(i,j)\in I^{[2]}\}$] (2),(4),(6),(7).

The first to have considered a permutational distribution of a statistic like $S(\xi^*,\eta)$ was -to our knowledge- H.E. Daniels (1). The aim of Daniels was to study the covariation between Kendall's τ and Spearman's ρ. The first clear calculation of the variance of $S(\xi^*,\eta)$ is considered in the Mantel paper (15). Mantel dit not mention Daniels' work and he developed his calculation with regressional goal. We ignored Mantel's paper when we have considered after Lecalvé's treatment (very inspired from Daniels'calculation type), clear combinatorial generalization of our association coefficients between relational qualitative variables. We have obtained very clear formal expressions of the moments of $S(\xi^*,\eta)$ distribution.

After defining the following general coefficient :

$$Q(\xi,\eta)=(s(\xi,\eta)-E[S(\xi^*,\eta^*)])/\sqrt{var[S(\xi^*,\eta^*)},\qquad (11)$$

where ξ and η are two "weightings" on $O^{[2]}$ (cf. above), the question arises about the real interest of the combinatorial approach -which led to the diagram (cf. Figure 1)- where the manipulated mathematical objects are directly : subsets, partitions, total preorders, preordonnances,... In fact all the relevancy of this last approach is preserved, because it is only by this way that we recognize all the combinatorial phenomenons associated to the total or partial comparisons between qualitative relational variables. This approach is a powerful conceptual tool for discovering new association coefficients concerning new structural situations in data analysis (17)(18).

REFERENCES

1. Daniels, H.E. (1944). The relation between measures of correlation in the universe of sample permutations, Biometrika, vol.33, 129-135.
2. Hubert, L. (1983). Inference procedures for the evaluation and comparison of proximity matrices, in Numerical Taxonomy, Ed. J. Felsenstein, NATO ASI Series, Springer Verlag.
3. Kendall, M.G. (1970), Rank correlation methods, Charles Griffin, fourth edition (first edition in 1948).
4. Lecalvé, G. (1976), Un indice de similarité pour des variables de type quelconque, Stat. et Anal. des Données, 01-02, 39-47.
5. Lerman, I.C. (1970), Sur l'analyse des données préalable à une classification automatique (proposition d'une nouvelle mesure de similarité), Rev. Math. & Sc. Hum., 8è année, n°32.
6. Lerman, I.C. (1976), Formal analysis of a general notion of proximity between variables, Congrès Européen des Statisticiens, Grenoble, published by North Holland in 1977.
7. Lerman, I.C. (1981), Classification et analyse ordinale des données, Dunod, Paris.
8. Lerman, I.C. (1983a), Sur la signification des classes issues d'une classification automatique des données, in Numerical Taxonomy, Ed. J. Felsenstein, NATO ASI Series, Springer Verlag.

9. Lerman, I.C. (1983b), Indices d'association partielle entre variables qualitatives nominales, R.A.I.R.O., vol.17 n°3, 7-46.
10. Lerman, I.C. (1983c), Indices d'association partielle entre variables qualitatives ordinales, Publ. I.S.U.P., XXVIII, fasc. 1,2, 7-46.
11. Lerman, I.C. (1984), Justification et validité statistique d'une échelle 0,1 de fréquence mathématique pour une structure de proximité sur un ensemble de variables observées, Publ. I.S.U.P., XXIX, fasc. 3-4, 27-57.
12. Lerman, I.C. (1987), Construction d'un indice de similarité entre objets décrits par des variables d'un type quelconque. Application au problème du consensus en classification, Rev. Stat. Appl. Vol. XXXV(2), 39-60.
13. Lerman, I.C. (1985), Organisation et consultation d'une banque de "petites annonces" à partir d'une méthode de classification hiérarchique en parallèle, in Data Analysis and Informatics IV, to be published by North Holland.
14. Lerman, I.C. et Peter, Ph. (1986), Structure maximale pour la somme des carrés d'une contingence aux marges fixées. Une solution algorithmique programmée, P.I. n°318, Octobre 1986, 90 pages.
15. Mantel, N. (1967), The detection of disease clustering and a generalized regression 'approach', Cancer Research, vol. 27, n°2, 209-220.
16. Pearson, K. (1904), Mathematical contributions to the theory of evolution XIII. On the theory of contingency and its relation to association and normal correlation, Draper's Company Research Memoirs, Biometric Series 1. 35 pp.
17. Lerman, I.C., Gras R. et Rostam H., (1981), 'Elaboration et évaluation d'un indice d'implication pour des données binaires', Rev. Math. & Sc. Hum., 19è année, n° 74, 5-35,

n°75, 5-47.

18. Lerman I.C. (1984), 'Analyse classificatoire d'une correspondance multiple, typologie et régression', in Data Analysis and Informatics III, Eds E. Diday & al., North Holland, 193-221.

ROLE OF COMPUTER GRAPHICS IN INTERPRETATION OF CLUSTERING RESULTS

Noboru Ohsumi

The Institute of Statistical Mathematics
4-6-7, Minami-Azabu, Minato-ku, Tokyo, JAPAN

I. INTRODUCTION

The utility of computer graphics has now won the recognition of many researchers specializing in data analysis, and a variety of graphical representation methods have been proposed by them. What data analysts expect most from graphical representation is to obtain a firm grasp of the intrinsic structure of collected data by visualizing their features.

The hardware environment bears very closely upon free use of graphical representation as a means of data analysis. In this connection, it deserves attention that the functional improvement of microcomputers and workstations, realized recently in upgraded graphics functions as well as increased performance in peripheral equipment, has provided a wider freedom of graphical representation and greater ease of graphics equipment operation.

These graphical representation methods have some common features: (1) representing the features of multivariate data, (2) visual inspection of the relationship among variables, (3) observing data distribution, (4) projections based on data transformation and features selection, and (5) enhancement of intrinsic features in data.

RECENT DEVELOPMENTS IN
CLUSTERING AND DATA ANALYSIS

ISBN 0-12-215485-1

II. GRAPHICAL REPRESENTATION IN AUTOMATIC CLASSIFICATION

Graphical representation is also applied for visualizing the results of automatic classification. While these can be represented on a computer graphics display with little difficulty, the interpretation of displayed graphs and figures is not as easy as the observation of histograms and scatter plots and tends to be influenced by subjective judgement. It is for this reason that computer graphics leaves something to be desired despite the increasingly important role it is playing in automatic classification.

In most cases, these representations of classification results must be considered the preliminary stage to developing more suitable graphical representation methods for grasping the meanings of classification. At the same time, many of clustering methods are heuristic in nature, and importance is attached to the comparison between classification methods as well as the comparison of classification results. Thus, what the analyst expects from classification can be summarized as follows.

1. Immediate visual inspection of classification results.
2. Fast recognition of what is being classified in what way.
3. Easy identification of the class to which an individual belongs.
4. Simplified recognition of the variables contributing to classification in multivariate data.
5. Rapid understanding of the differences or similarities between clusters by visual inspection.

However, in fulfilling these requirements with computer graphical methods, a broad overview and consistent approach is imperative. The assumptions, algorithms, and procedures underlying a particular computerized representation must be carefully examined to assure compatibility and consistency with other graphical methods used by the analyst in his analysis. Failure to maintain such consistency in both individual modules and, especially, in integrated packages will tend to magnify the subjective element inherent in visual interpretation of graphical representations. Hardware consistency is also required. The hardware environment must be constructed systematically to optimize the capabilities of the software system.

In addition to consistency, the ideal system should have flexibility, allowing the analyst to select from a wide variety of graphical representation methods and conventional multivariate methods and to do so rapidly and easily.

For the past several years, we have been working on the development of such an integrated, consistent hardware-software system for use in performing automatic classification and in visualizing and interpreting the results.

III. CONCEPT OF COMPUTER GRAPHICS SYSTEM IN AUTOMATIC CLASSIFICATION

Because of the phenomenal development in computer technologies in recent years, it is now possible to meet all standard requirements for data analysis using terminals connected to large-scale computers or microcomputers. However, in order to meet the demands of data analysts as described above, development of a computer graphics system designed specifically for the use of experts is much desired. A project started some years ago provided an opportunity for us to embark on the development of such a computer system in 1982. We started with a minicomputer because the powerful microcomputers seen today were not available at that time. The system has since been refined and upgraded by stages and is now a powerful system using graphics. It has been given the name Automatic Classification Techniques for Interpretation, Visualization and Evaluation, or simply, the ACTIVE workstation.

A. System Design Policy and Features

The ACTIVE workstation has been built in stages. At first, a minicomputer with intelligent color graphics display was introduced as an intelligent terminal connected to a large-scale host system. Next, the peripheral equipment were strengthened by introducing color XY-plotters, tablet digitizers, color video cameras, video recorders, image recorders, etc. Recently, professional workstations, graphics server processor and high-resolution color display were added, and as completed, the system consists mainly of the workstation. Figure 1 shows the configuration of the ACTIVE workstation. The ACTIVE workstation has the following features.

1. The workstations and minicomputer can be used independently in the system.
2. Both serve as intelligent terminals to the host system.
3. Each of the special microprocessors built into the color graphics displays is available for image analysis and provides intelligent functions.
4. A channel controller links the various components to each other and to a host system, allowing free access to any part of the system.
5. Files can be freely transferred between subsystems.
6. Additional microcomputers can be linked to the workstation

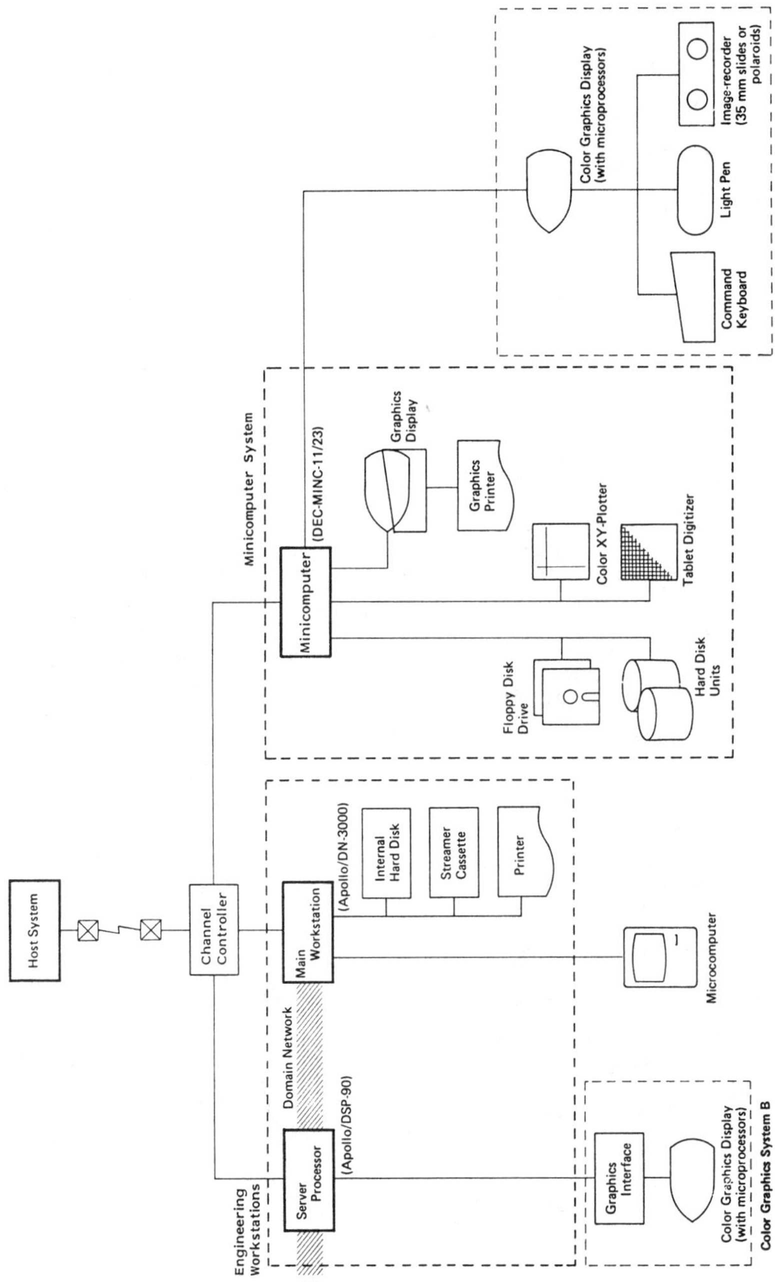

Figure 1 Configuration of ACTIVE Workstation

to serve as new terminals (e.g., Macintosh, NEC-9800 series).

What is notable about the system design is that all its components are organically interconnected with the workstation installed as the system core, permitting real-time access to any subsystem at any time according to the needs of the data analysis.

B. Software Environment

The system currently provides a well-integrated hardware environment, but its software environment requires further sophistication. The system software developed thus far includes: (1) integrated set of data analysis techniques, (2) management software for graphical devices, (3) interface management software, and (4) the handlers and associated software for peripheral devices. The integrated set of data analysis techniques includes the hierarchical classification methods, k-means method, classification methods such as the MST (Minimum Spanning Tree) and multivariate AID methods, principal components analysis, correspondence analysis, simple descriptive statistical methods, etc.

Graphics functions are available for monochrome bitmap displays and color graphics displays. The two color graphics displays are intelligent and have a number of graphics and mathematical primitives included in removable ROM cartridges. The handlers developed to use these ROMs realize plotting and color handling of almost all kinds. Although kinetic display is rather difficult by virtue of the system configuration, it can nevertheless be realized by adequate application of the graphics server processor. Other system components include tablets for linework information input and color XY-plotters for graphics output.

The ACTIVE workstation is controlled from the main workstation via a channel controller which allows any of the subsystems to be connected directly, or indirectly, to any other or to the host system. Analysis data and results, both statistical and graphics, can be freely transferred among the subsystems. Finally, each of the subsystems can be used independently as a standalone system.

IV. CASE STUDIES

In the following sections, some of major studies made using the ACTIVE workstation will be introduced. Although many of these studies have already been reported at some conferences, we consider it worthwhile to review them here as examples of the effective application of computer graphics and automatic classification methods. This is because each of them serves as evidence of our contention that the data analysis that we hope for can never be realized by a mere aggregation of graphical representation techniques.

A. Areal Clustering System and Its Applications

The areal clustering system of the ACTIVE workstation is an integrated set of classification techniques developed for areal partition using multivariate observations that have locate coordinates on the plane (5), (6). These techniques can be broadly classified into two phases according to the features of the data handled. One is NTAP (Numerical Techniques for Areal Partition) and the other is IMAGE (Image Generator of Colored Maps). The preceivable application for NTAP include classification using LANDSAT data and areal partition using land-use data. NTAP is extensively used in the analysis of data from remotely-sensed measurements.

The objective of IMAGE is to generate image information covering the entire area from discrete and irregularly distributed multivariate observations on the plane (accordingly, on the objective area) for visual presentation of the general features of such partial data and for drawing out the characteristics which are hidden in them. From the partially located data on the screen, IMAGE generates a visual map covering the whole screen and is aimed at stimulating the observer's cognitive powers. As an example of IMAGE applications of the past, we can cite the case where it was used in the analysis of attitude survey data of urban dwellers for the purpose of generating various "colored attitude maps" from the survey results of urban environment assessment. The distribution of attitude data in the survey area, obtained from respondent answers to any specific questions, can be displayed as such colored attitude maps. With a simple glance at these maps, the observer can acquire a firm grasp of respondent opinions to the question, and can also compare the attitude survey data with other kinds of area information, such as amount of greenery. Furthermore, if several questions are asked, the data obtained from respondents can be converted to composite scores by correspondence analysis and principal components analysis and displayed as colored maps. It is also possible to overlay the colored maps of individual questions and to generate composite color maps by image processing.

B. Use of Color Imaging in Automatic Classification

There are many ways of representing multivariate data using computer graphics. New, colored representation methods have been proposed because the raster graphics display has been made much easier to use in recent years (10). Color image handling is one of the features of our system, and it is possible to develop programs making fuller use of various color handling functions. Development of such programs is made easy by the powerful graphics software and the integrated set of color model transformation programs that we have developed (8). To present the results of various analyses, including automatic classification, immediately on the color display, it is necessary to use a color handling interface program for linking the data analysis techniques and the color monitor. Special attention should be paid to the color models used to simplify the user interface. In other words, the program must be developed using the RGB (Red, Green, and Blue) model for hardware control and the HLS or HSV (Hue, Saturation, and Lightness or Value) models which are closer to the human color perceptional models, for the analyst. If development of this program is to be completed in a short time, it is essential that the color handling system be applied using color models, based on the set of transformation programs between color models. A program for color imaging of automatic classification results is generated using basic software consisting of these programs and the graphics service routines provided as part of the standard system software (9).

Unlike the case where color assignment is made using a color look-up table and a limited number of colors, the program thus generated presents the following features, and problems.

1. Changes in color are controlled on the basis of the three color elements (hue, saturation, and lightness).
2. Accordingly, each color bears a certain meaning and corresponds to a certain tendency of the classification results (colors are not just painted on).

3. However, perfect color handling is still very difficult because of numerous problems remaining in color science. Hence, program development must be planned with a clear understanding of the differences between the color image on the computer graphics display and the human color perception.
4. Peripheral equipment must be used with a clear understanding of the differences between the color system adopted for printing and reproducing color image operations and that adopted for the computer graphics.

Colored Pattern Matrix and Color Patterns

The greater part of the quantitative data handled in data analysis is made up of multivariate data matrices consisting of individuals by variables. The objective here is to represent the features of such a matrix as a color pattern. The color model transformation programs play an important role in the following process required for features representation, although the process itself is quite simple.

(S1) Individuals of the given multivariate data matrix X are classified by a suitable method and reordered.

(S2) Variables are also reordered by a suitable method using, for example, principal component loadings.

(S3) The doubly reordered matrix and each of its elements is converted to color information using the HLS model. The new matrix prepared is taken as "colored pattern matrix" Y, and each of its converted elements is transmitted to the color monitor through the RGB color model to represent the matrix Y as a "color pattern." Next, the lightness is linearly rescaled to reflect the dispersion of distribution of each variable of the multivariate data. The hue and saturation are determined using the component loadings derived from the principal components analysis of the original data matrix X.

Coloring Principal Component Scores

Since the vector corresponding to each individual of colored pattern matrix Y is expressed by means of the three color elements, it is called a "color vector." A plot is prepared that shows the arrangement of scores obtained for any two components derived from the matrix X, and the color vector is placed on the coordinates showing the principal score of each individual. In this way, each of the multivariate observations can be observed as a "color belt" within the reduced dimensions. This method offers the following advantages: (1) ambiguity of principal components scores can be removed, and the meaning of the scores can be shown as color, (2) patterns of clustering can be enhanced in color, (3) discrepancy among clusters and variable features cause by outliers can be grasped with ease in color, and (4) variables characterizing individuals can be compared as changes in color.

The most outstanding advantage of this method is that the features of the original data matrix X can be observed directly as colored images. Dendrograms, scatter diagrams and distribution of principal components scores can naturally observed in time with the obtained color patterns (see the experimental results described below).

C. Classification of Large Data Sets and Its Representation

The main memory size of the workstations and minicomputer of the ACTIVE workstation is not very large. For this reason, considerable difficulty was involved in classifying large-scale data sets (of at least 5,000 to 10,000 cases) or to represent classification results quickly on the display. However, since workstations and minicomputer can be used as terminals of a large capacity host system, it is certainly possible to classify large-scale data sets using the statistical systems implemented in the host system (SPAD, GENSTAT, SAS, MINTS). The results of such clustering can be transmitted to workstations and minicomputer by file transfer and represented on the display after editing and data modification. Nevertheless, there is no doubt that interactive workstations and minicomputer are more suitable to increasingly detailed and sophisticated data analysis while watching the classification process step by step. Moreover, various ideas are required in order to apply large-scale data clustering to microcomputers with small memory sizes, and it is useful to watch the classification process by graphical representation because the algorithm is often heuristic in nature.

The objective here is not to directly display the distribution of large-scale data. The graphical representation is applied as a means of presenting the skeleton of the data structure using the classification results. In other words, it is applied for the purpose of representing the intrinsic data structure or the fundamental data features on the display. The process actually adopted for classification is as follows.

(S1) A partial data set is generated by sampling a given data set at random in a certain proportion.

(S2) Sampled data is subjected to the initial classification. At this time, the number of clusters is designated to be as large as possible to generate a large number of groups. The distance table is not used because it calls for the use of a large matrix. The k-means method is used in a hierarchical manner to obtain the tree structure.

(S3) The classification thereby completed is displayed by tree representation as the "preliminary classification." Next, a tree showing cluster features is produced using the fractal recursivity described below.

(S4) All individuals in the data set subjected to the initial classification are classified consecutively using the tree obtained in (S3) as a decision tree (i.e., each individual is assigned consecutively to some cluster in the tree by binary decision rule).

(S5) After all individuals is assigned, "reclassification" is executed using the centroid vector of each cluster at the terminal end of the tree. The reclassification made in this step is the fine adjustment of the relationship between each cluster and the individuals belonging to each cluster. At this time, the tree structure is reconstructed by making use of the reciprocal nearest neighbor (this reconstruction is called "refinement using back-tracking").

(S6) The fractal tree structure obtained by reclassification is displayed and observed again, and is compared with that obtained by the preliminary classification.

(S7) A comparison of any two clusters in the tree is observed by displaying it as a type of Lissajous figure.

Method of Tree Structure Representation

The tree structure obtained in (S5) and (S6) above differs slightly from ordinary dendrograms. To generate this tree representation, attention is directed to the fractal technique of tree models. The fractal plotting is intended to make the plotted tree model look like a natural tree as much as possible. Here, attention is focused on its recursivity and parameter control, with ingenuity exercised so that the tree structure can reflect the data structure to an extent and the computer program can be simplified for faster image generation.

In general, when generating a tree structure, the changes in the tree form are controlled by specifying the following parameters: (1) exterior angle between the limbs, (2) ratio among limb lengths (3) number of limbs, and (4) limb diameter, decline in limb diameter toward the treetop, limb bending, etc.

The tree structure can be made as complex as desired by varying the values of these parameters. Since branching of each limb of a cluster is equivalent to the decomposition of the sum of squares, the following rules are established: (1) the between-cluster sum of squares corresponds to the angle between the limbs, (2) the within-cluster sum of squares (or within-cluster variance) is applied to the length of the limbs, (3) limbs are allocated to the left and right regularly according to the values of the within-cluster sum of squares. This is related to the bend of the limbs.

Graphical Representation of Relationship between Clusters

The tree structure has a drawback to intuitively grasping the similarity between the clusters. Hence, it is necessary to develop a method of observing the relationship between clusters. Since multiple individuals are contained in each cluster and all are multivariate data, the following method was devised.

(S1) Mean vectors of clusters are compared, and the similarity among them is displayed by graphical representation.

(S2) Andrews' functional plotting, a graphical representation method for multivariate observations, is taken and projected to produce a two-dimensional plot.

The p-dimensional mean vector of a cluster is expressed as $\underset{\sim}{x}$, and that of another cluster as $\underset{\sim}{y}$. At this time, the two polynominals obtained by Andrews' method are expressed as follows:

$$f(\underset{\sim}{x}\,;\,t) = x_1 / \sqrt{2} + x_2 \sin t + x_3 \cos t + x_4 \sin 2t + \ldots$$

$$f(\underset{\sim}{y}\,;\,t) = y_1 / \sqrt{2} + y_2 \sin t + y_3 \cos t + y_4 \sin 2t + \ldots$$

where, $-\pi \leqq t \leqq \pi$.

These two functions are projected within the two-dimensional space to obtain the trajectory of the point $(f(\underset{\sim}{x};t),\ f(\underset{\sim}{y};t))$. The figure plotted by the trajectory clearly shows the similarity between the two clusters, and is a type of Lissajous figure. It is possible to compare mean vectors of clusters by drawing a large number of one-dimensional Andrews' plots. However, observation of the relationship between the two clusters by two-dimensional plotting at each stage of the tree structure in sequence is displayed as the changes in cluster structure, and consequently produces quite different information to that obtained from an overall observation of the tree. Further discussion will be required to explore the mathematical significance of the figures obtained by this method and the relationship between the Lissajous patterns and the data structure. It is generally thought, however, that Lissajous figure can be used as an effective tool of visual comparison of two sets of multivariate observations. It can also be used for graphical representation of similarity or dissimilarity among the individuals.

D. Some Experimental Results

Below are the experimental results obtained using artificial data sets for the purpose of examining the effectiveness of the methods discussed in B. and C. above. The artificial data was prepared by combining five groups of three-dimensional normal random numbers (see reference 2). The number of individuals is 5,000; Figure 2 shows the scatter diagram.

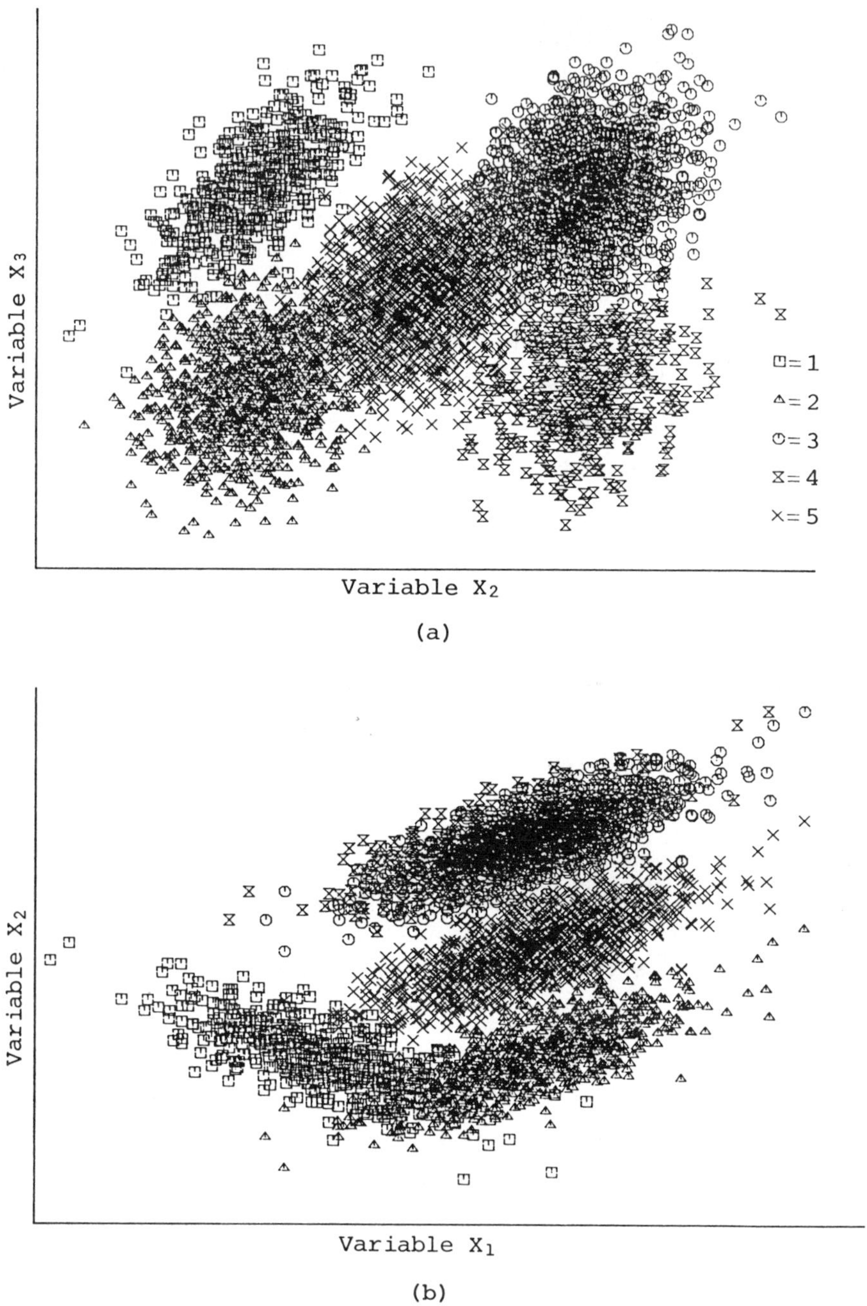

Figure 2 Five thousand observations produced by generating three-dimensional normal random numbers

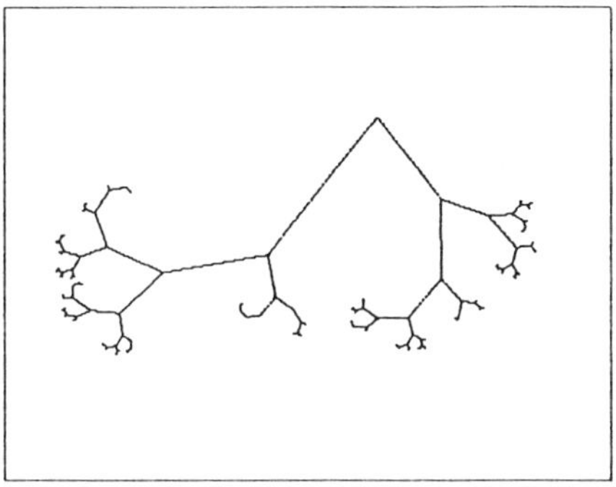

(a) Result of preliminary classification

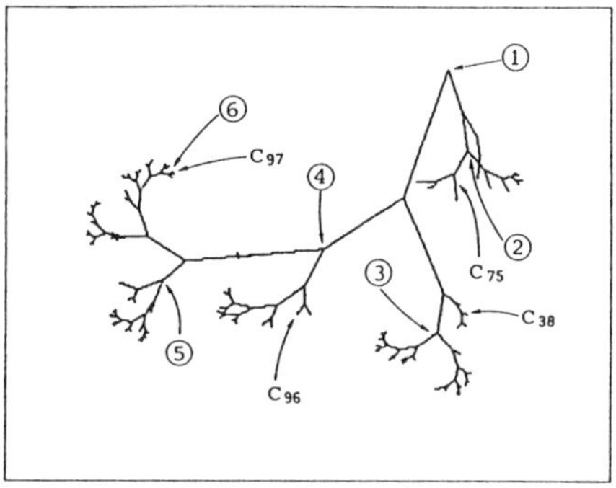

(b) Result of reclassification

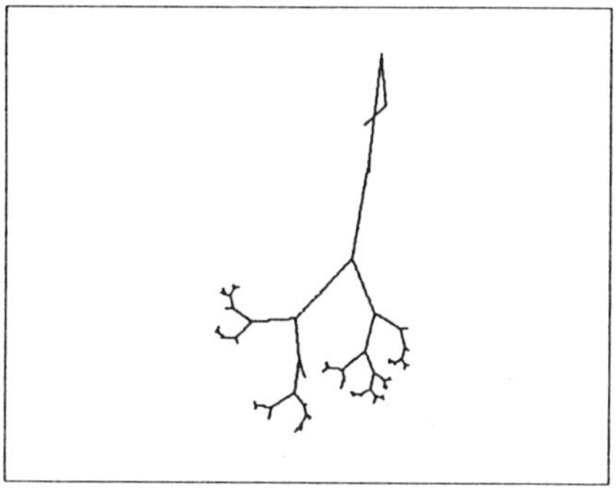

(c) Another result of reclassification

Figure 3 Tree structures presenting the results of classification

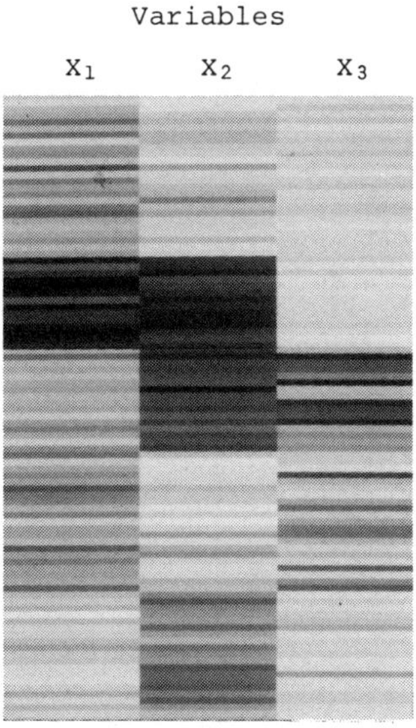

Figure 4 Example of colored pattern matrix

Result 1

Figure 3 (a) and (b) show the tree structures presenting the results of preliminary classification and reclassification of the artificial data. It can be seen that the features of the original data are fairly well exposed by the tree structure, as indicated by the fairly large limb angles. In other words, the outline of the large given data set is enhanced by the tree. For the purpose of comparison, a certain group of data (the group shown in the upper left corner of Figure 2-(a)) was sampled from the given data for similar tree representation shown in Figure 3-(c). Unlike the previous results, the homogeneity of data can be clearly observed from the small angle of limb branching.

Result 2

The feature of clusters obtained in the analysis can be also observed as the color pattern described before (see Figure 4). Although the pattern is printed here as a monochrome copy, it shows the features of each cluster quite well. Compared with the results of Result 1, the dispersion in each cluster is represented more clearly as changes of color in the variables, especially, variable x_2.

Result 3

Comparison among the clusters was performed using Lissajous figures. Figure 5 shows a comparison of two clusters numbered ① to ⑥ , respectively, in Figure 3-(b). Since three-dimensional data was used in this case, the generated curves are rather simple. If data with more dimensions is used, however, complex, yet still useful figures will be generated. The symmetry, inclination, location of

Figure 5 Visualizing the relationships among the clusters

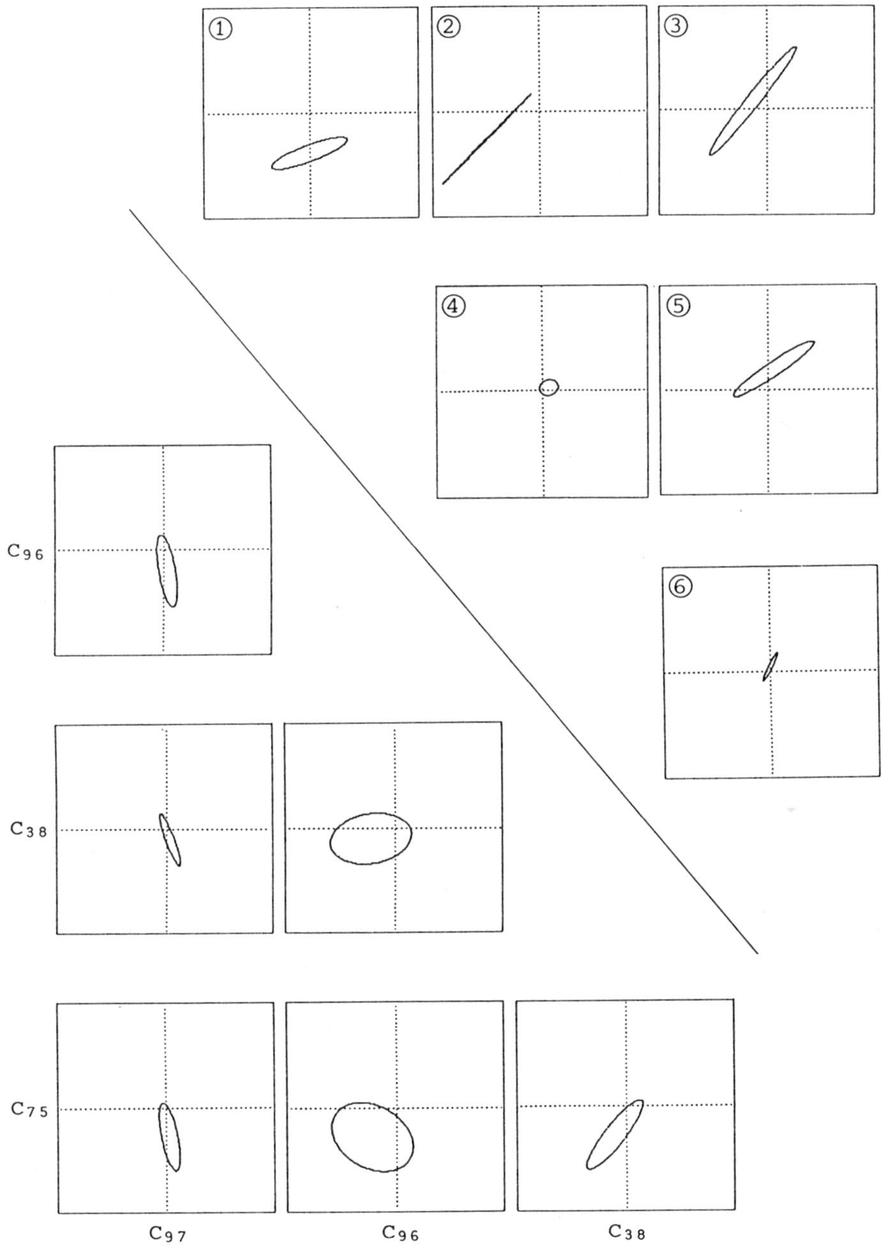

Figure 6 Pair comparison between two clusters

center, convexity of the figures shown in this analysis represent the similarity and difference among the clusters. If we take four clusters from the figure (C_{38}, C_{75}, C_{96} and C_{97}) and draw a pair comparison of their mean vectors as a Lissajous figure, the result is shown in Figure 6. These are quite useful at showing the relationships among the clusters. This is equivalent to plotting the distance table for the clusters. Thus Lissajous figures aid in visualizing cluster relationships which are insufficiently clear in the tree structures.

V. CONCLUSION

There are none who argue against the high utility of graphical representation for quick visual inspection of complex data structures. At the moment, however, reluctance is generally felt in using the information obtained by graphical representation as a direct clue for making decisions in data analysis. The graphical representation methods are constructed rather as a means of grasping intuitive information that can be used in opening up the way to higher sophisticated model analyses. It is precisely for this reason that great importance is attached to the building of a well-integrated computer system environment which is suitable for graphical data analysis. Though small in scale, the ACTIVE workstation introduced in this paper has shown its efficiency as an experimental system. The following improvements are planned for further functional upgrading: (1) development of a programming language with easier-to-use graphics and of object-oriented tools with sufficient extensibility to smoothly link graphics functions using classification results with the algorithms of graphical representation, and (2) raising the system to the level of an expert system.

ACKNOWLEDGEMENTS

The author is very indebted to Miss M. Koiso who assisted us in the development of several useful computer programs related with section C. in part IV.

REFERENCES

1. Everitt, B. (1978): Graphical Techniques for Multivariate Data, Heinemann Educational Books.
2. Matusita, K. and Ohsumi, N. (1980): A criterion for choosing the number of clusters in cluster analysis, in "Recent Developments in Statistical Inference and Data Analysis," (K. Matusita, ed.) 203-213, North-Holland.
3. Morineau, A. and Lebart, L. (1986): Specific clustering algorithms for large data sets and implementation in SPAD software, in "Classification as a Tool of Research," (W. Gaul and M. Schader eds.), North-Holland.
4. Murtagh, F. (1985): Multidimensional Clustering Algorithms, COMPSTAT Lecture 4, Phisica-Verlag.
5. Ohsumi, N. and Mizuno, K. (1984): Microcomputer graphics clustering techniques for urban environmental evaluations, Proceedings of the XII International Biometric Conference, 148-156.
6. Ohsumi, N. (1984): Practical techniques for areal clustering, in "Data analysis and Informatics III," (E. Diday and others eds.), North-Holland.
7. Ohsumi, N. and Mizuno, K. (1985): Areal clustering system and its application to human ecological problems

(in Japanese), The Proceedings of the Institute of Statistical Mathematics, 33, 2, 176-198.

8. Ohsumi, N. (1986): Transformation programs between color models for computer graphics (in Japanese), The Proceedings of the Institute of Statistical Mathematics, 34, 1, 37-57.

9. Ohsumi, N. (1986): Practical use of color imaging in automatic classification, COMPSTAT-86: Proceedings in Computational Statistics 1986, (F. De Antoni and others eds.), 489-497, Physica-Verlag.

10. Wegman, E. J. and DePriest, D. J. eds. (1986): Statistical Image Processing and Graphics, Marcel Dekker, Inc.

CLASSIFICATION OF FRUIT AND VEGETABLE VARIETIES BY CHEMICAL ANALYSIS OF FRAGRANCE SUBSTANCES

Shigeru Suzuki

National Institute of Agrobiological Resources
Yatabe, Tsukuba, Ibaraki-ken, 305 JAPAN

I. INTRODUCTION

In distinguishing fruit and vegetable varieties, fragrance is sometimes one of the very important characteristics. It is generally evaluated by the panel as sensory scores. The evaluation is of course reproducible and reliable by the experienced panel, but in other cases rather fragile and subjective. So, more solid and objective evaluation methods are explored.

As a result of recent development in chemical analysis techniques, distribution and identification of chemical materials in fragrance substance became not difficult to study by Gas-Chromatograph(GC) and Mass-Spectrograph(MS) analysis.

Identification and analysis of fragrance substances by GC data in foods have been reported(1), but quantitative assay of those in crop varieties is still unexploited(2).

Helped by the multivariate statistical analysis, relationship between sensory ratings and peak values of some fragrance substances in GC chart were examined and classification of the crop varieties was attempted.

ISBN 0-12-215485-1

II. MATERIALS AND METHODS

Varieties of these crops examined are listed in Fig. 1 and 2: twenty-one varieties of cultivated melon composed of 4 groups (Makuwa, Winter melon, Netted melon and Carotin type) and 21 varieties of grapes composed of 4 groups (Foxy, Muscat, Fragranceless and Specific type). They are grouped by the kinds of fragrance, and sensory ratings by the panel are shown for grapes (Fig. 2).

Edible portions of ripe fruits were crushed and outcoming fragrance substances were subjected to the chemical analysis. Head Space gas of melon varieties, replicated 3 times, and successively distilled and extracted samples from grape varieties by the improved Dean and Stark apparatus, replicated twice are used for the GC analyses.

Observed important peaks appeared in the GC chart were numbered and represented by the relative proportion of the occupied area in the chart. They were then chemically identified by GC-MS analysis, and some dubious materials including instrumental contamination were discarded. Using these reliable selected peak values, discriminant analysis was carried out intending to establish a criterion separating groups by the objective and reproducible chemical analysis data. Discriminant analysis for two groups(DIS) and that for multiple groups (DISM) were applied using CMAP at the Computing Centre for Research in Agriculture, Forestry and Fisheries(3).

III. EXPERIMENTAL RESULTS

Twelve peak values obtained over 63 samples of 21 melon varieties by the Head Space gas analysis were computed to get discriminant functions by DIS. Any pair of two groups out of the four was clearly separated. Miss-classification probability was extremely low throughout, and non of the samples showed higher fitting probabilities for any other assigned group than their own.

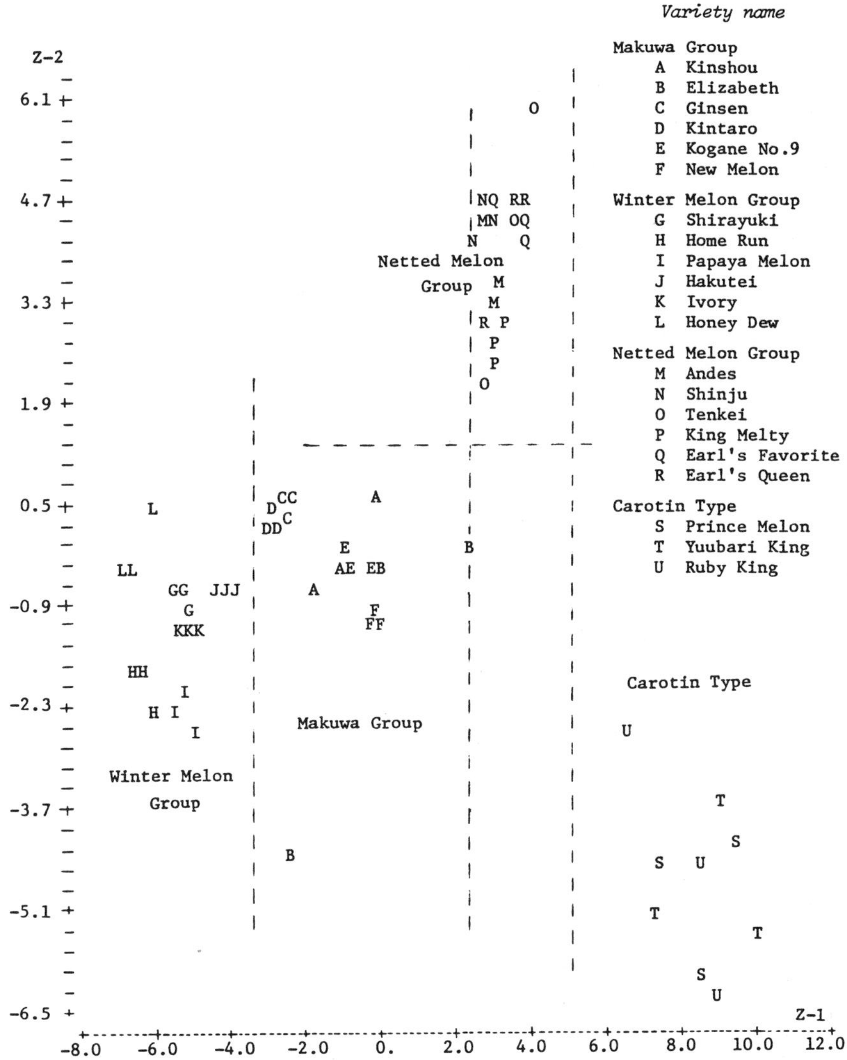

Fig. 1 Discrimination of melon varieties by canonical axes derived from GC analysis of fragrance substances.

Using DISM, all samples were completely assigned into 4 original groups. Sixty-three samples were plotted on the Z-1 / Z-2 plain in Fig 1.

For grape varieties rerationship between sensory ratings of fragrance strength and peak values in the GC chart was examined. Eighteen reliable peaks from the chart of distilled and extracted GC analysis were selected for the analysis. Thirty-two samples, twice replicated 16 varieties(upper 16 of those listed in Fig.2) where strength scores were available, were subjected to multiple regression analysis of the sensory score on 18 peak values. Stepwise regression method resulted in a formula consist of 7 selected peaks to get multiple correlation coefficient R = 0.975**.

DIS of 21 varieties of grapes, replicated twice, using 18 peak values revealed a good separation of any pair among the 4 groups in general. With minor exceptions like Catawba, Buffalo or Red Millennium, all samples yielded high probabilities to be settled in their own groups. Even for these exceptions, probabilities were always highest for their own groups, although the values were rather small.

DISM showed almost complete separation of Foxy and Muscat groups from the other 2 as displayed in Fig. 2.

IV. DISCUSSION

Fragrance is very important for fruit and vegetable crops as the quality character. Kind and strength of fragrance are sometimes claimed to characterize new varieties. But fragrance has long been beyond objective measurement, and left for the experienced panel for evaluation. Development of chemical analysis and advanced statistical techniques realized a hope to establish objective and reliable evaluation.

Variety name		*Score*
Foxy Group		
A	Catawba	9
B	Adiron-dack	7
C	Red Port	7
D	Campbell's Early	5
E	Portland	5
F	Niagara	5
G	Kyohou	3
H	Pionnier	3
Muscat Group		
I	Muscat-Hamburg	7
J	Neo Muscat	5
K	Italia	3
Fragranceless Group		
L	Koshuu	1
M	Delaware	1
N	Muscat Berry A	1
O	Riesling	1
P	Suntory Blanc	1
Specific Type		
Q	Buffalo	
R	Cabernet Sauvignon	
S	Red Millennium	
T	Mellow	
U	New York Muscat	

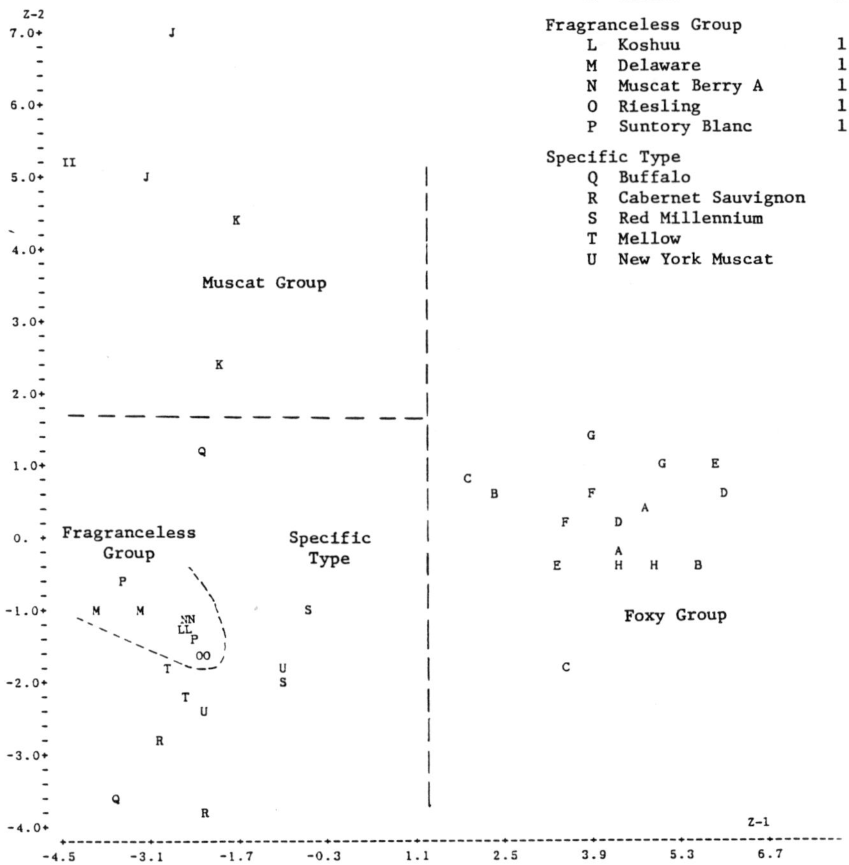

Fig. 2 Discrimination of grape varieties by canonical axes derived from GC analysis of fragrance substances.

Melon varieties were reasonably classified into groups by DIS. Every samples fell into originating groups following to the highest-probability rule, although a few were at delicate balance. According to Fig. 1, DISM gave a good presentation on the feasibility of grouping. Four groups were almost separable only on the Z-1 axis, with some overups for Makuwa and Netted Melon. While, they are clearly classified on the Z-2 axis. Possibility of melon variety classification using GC data is thus advocated.

Experiment of grape varieties showed a close relationship between the sensory score in strength of fragrance and stepwise selected GC peak values. This may lend hope in establishing more reliable formulas to estimate fragrance strength on a practical scale in future.

Foxy and Muscat groups have apparently different genetic background, which is well reflected in Fig. 2. Replicates of the same variety are generally coming closely together reasonably on the figure, with minor exceptions.

Fragile fragrance substances have been identified and quantified through GC analysis, and they were shown to be effective in variety classification. Further refinement is needed to establish better standards for quality evaluation in wider range of crop varieties.

REFERENCES

1. Kept, T.R., Stoltz, L.P. and Knavel, D.E.(1972). J. Agricultural Food and Chemistry. 20(2). 196-198

2. Shimizu, J. (1982). Agricultural and Biological Chemistry. 46(9). 2265-2274.

3. Agricultural Research Council Secretariat.(1982) CMAP. Computing Centre for RAFF Users' Manual 13. 82-84

Section 3

Scaling Method and Correspondence Analysis from the Viewpoint of Practical Approach

ASSESSING THE NUMBER OF AXES THAT SHOULD BE CONSIDERED IN CORRESPONDENCE ANALYSIS

Yves Escoufier

ENSA-INRA-USTL
Unité de Biométrie
9, place Pierre Viala
34060 MONTPELLIER CEDEX
(France)

I. INTRODUCTION

When performing a correspondence analysis, statisticians most frequently have to cope with the following dilemma - choose either a simple interpretation based on the first two or three axes obtained, or a more subtle interpretation (obliging them to handle and view more information) based on a great number of axes.

One could wish to find in the inferential statistic approaches, efficient processes to objectivize the choice of the number of axes. The dramatic limitations of available tools are reviewed in Section IV.

The problem can then be solved to through more exploratory approaches based either on least squares approximation properties of the solutions, or on cross validation techniques.

Sections II and III of this article introduce those approaches.

ISBN 0-12-215485-1

II. CORRESPONDENCE ANALYSIS AS AN APPROXIMATION OF THE DATA MATRIX

A. Let P be a table of frequencies the size of which is IxJ, with $\sum_{i=1}^{I} \sum_{j=1}^{J} P_{ij} = 1$.

Δ_I and Δ_J are diagonal matrices defined by marginal weights of the rows and columns of P respectively.

$$\Delta_I = \begin{pmatrix} P_{1.} & & \\ & \ddots & \\ & & P_{I.} \end{pmatrix} \qquad \Delta_J = \begin{pmatrix} P_{.1} & & \\ & \ddots & \\ & & P_{.J} \end{pmatrix}$$

Consider the triplets $(\lambda_\alpha, \psi_\alpha, \phi_\alpha)$ defined for $\alpha = 1, \ldots, N = \min(I,J)$ by

$$\Delta_I^{-1} P \Delta_J^{-1} P' \psi_\alpha = \lambda_\alpha \psi_\alpha \quad \text{with } \psi'_\alpha \Delta_I \psi_\alpha = 1 \tag{1}$$

$$\Delta_J^{-1} P' \Delta_I^{-1} P \phi_\alpha = \lambda_\alpha \phi_\alpha \quad \text{with } \phi'_\alpha \Delta_J \phi_\alpha = 1 \tag{2}$$

$$\lambda_1 \geqslant \lambda_2 \geqslant \ldots > \lambda_N \tag{3}$$

For $\lambda_\alpha \neq 0$, the so called transition formulas are easily deduced from 1) and 2)

$$\phi_\alpha = \frac{\Delta_J^{-1} P' \psi_\alpha}{\sqrt{\lambda_\alpha}} \tag{4}$$

$$\psi_\alpha = \frac{\Delta_I^{-1} P \phi_\alpha}{\sqrt{\lambda_\alpha}} \tag{5}$$

Therefore $\sqrt{\lambda_\alpha}\, \phi_{\alpha j}$ is the barycenter of the $\psi_{\alpha i}$'s bearing the weights $\frac{P_{ij}}{P_{i.}}$, while $\sqrt{\lambda_\alpha}\, \psi_{\alpha i}$ is the barycenter of the $\phi_{\alpha j}$ bearing the weights $\frac{P_{ij}}{P_{.j}}$.

Geometric considerations (Lebart (1977), p. 58) lead then to the conclusion that for any α, $\lambda_\alpha \leqslant 1$.

Note $\underset{\sim}{1}_I$ as the vector of R^I, all the components of which are equal to the unit, and $\underset{\sim}{1}_J$ as the vector equivalent to R^J. It is easy to check that :

$$\Delta_I^{-1} \; P \; \Delta_J^{-1} \; P' \; \underset{\sim}{1}_I \; = \; 1 \times \underset{\sim}{1}_I \tag{6}$$

$$\Delta_J^{-1} \; P' \; \Delta_I^{-1} \; P \; \underset{\sim}{1}_J \; = \; 1 \times \underset{\sim}{1}_J \tag{7}$$

The triplet $(1, \underset{\sim}{1}_I, \underset{\sim}{1}_J)$ is therefore one of the triplets $(\lambda_\alpha, \psi_\alpha, \phi_\alpha)$, which leads, when isolating it, to renumber the remaining triplets from 1 to N-1. These notations being implemented, any presentation of Correspondence Analysis gives the two following formulas :

Data reconstitution formula

$$\forall (i,j) \in IxJ \qquad P_{ij} = P_{i.} P_{.j} (1 + \sum_{\alpha=1}^{N-1} \sqrt{\lambda_\alpha} \; \psi_{\alpha i} \; \phi_{\alpha j}) \tag{8}$$

Sum of latent root significance

$$\sum_{i=1}^{I} \sum_{j=1}^{J} \frac{(P_{ij} - P_{i.} P_{.j})^2}{P_{i.} \; P_{.j}} = \sum_{\alpha=1}^{N-1} \lambda_\alpha \tag{9}$$

B. When seen in the context of an approximation of a matrix by a matrix of lower rank, these two formulas seen to be particular cases of more general formula.

In fact consider $A = \Delta_I^{-1} \; P \; \Delta_J^{-1}$ and define N matrices A_k by :

$$A_0 = \underset{\sim}{1}_I \; \underset{\sim}{1}'_j$$

$$\forall k = 1, \ldots, N-1 \quad A_k = \underset{\sim}{1}_I \; \underset{\sim}{1}'_J + \sum_{\alpha=1}^{k} \sqrt{\lambda_\alpha} \; \psi_\alpha \; \phi'_\alpha \tag{10}$$

It was shown (Escoufier and Junca (1986)) that A_k is the best approximation of order k+1 for A and very accurately that :

$$\mathrm{Tr}((A-A_k)\ \Delta_J (A-A_k)'\ \Delta_I) = \sum_{\alpha=k+1}^{N-1} \lambda_\alpha \tag{11}$$

which also can be written for $k \geqslant 1$

$$\sum_{i=1}^{I} \sum_{j=1}^{J} \left(\frac{P_{ij}}{P_{i.}\ P_{.j}} - (1 + \sum_{\alpha=1}^{k} \sqrt{\lambda_\alpha}\ \psi_{\alpha i}\ \phi_{\alpha j})\right)^2 P_{i.}\ P_{.j} = \sum_{\alpha=k+1}^{N-1} \lambda_\alpha \tag{12}$$

For A_0, the following is obtained

$$\sum_{i=1}^{I} \sum_{j=1}^{J} \left(\frac{P_{ij}}{P_{i.}\ P_{.j}} - 1\right)^2 P_{i.}\ P_{.j} = \sum_{\alpha=1}^{N-1} \lambda_\alpha \tag{12'}$$

Noticing that $A - A_k = \sum_{\alpha=k+1}^{N-1} \sqrt{\lambda_\alpha}\ \psi_\alpha\ \phi'_\alpha$, formula 11 can be rewritten in order to show the factorizations of the total error made in the approximation $\sum_{\alpha=k+1}^{N-1} \lambda_\alpha$, according to rows, columns or even cells (i,j) of array P. The following is obtained :

$$\sum_{\alpha=k+1}^{N-1} \lambda_\alpha = \mathrm{Tr}((A-A_k)\ \Delta_J (A-A_k)'\ \Delta_I)$$

$$= \sum_{i=1}^{I} \sum_{j=1}^{J} P_{i.}\ P_{.j} \left(\sum_{\alpha=k+1}^{N-1} \sqrt{\lambda_\alpha}\ \psi_{\alpha i}\ \phi_{\alpha j}\right)^2$$

Noticing that for $\alpha \neq \beta$ $\sum_{i=1}^{I} P_{i.}\psi_{\alpha i}\psi_{\beta i} = \sum_{j=1}^{J} P_{.j}\phi_{\alpha j}\ \phi_{\beta j} = 0$ the following is obtained :

$$\sum_{\alpha=k+1}^{N-1} \lambda_\alpha = \sum_{\alpha=k+1}^{N-1} \sum_{i=1}^{I} \sum_{j=1}^{J} P_{i.} P_{.j}\ \lambda_\alpha (\psi_{\alpha i})^2\ (\phi_{\alpha j})^2 \tag{13}$$

It can be deduced that as long as the triplet $(\lambda_\alpha, \psi_\alpha, \phi_\alpha)$ is not used in the approximation, an error λ_α is made, which can be attributed to it.

The error can be factorized by the rows

$$\lambda_\alpha = \sum_{i=1}^{I} P_{i.} \left[\sum_{j=1}^{J} P_{.j} \lambda_\alpha\ (\psi_{\alpha i})^2\ (\phi_{\alpha j})^2 \right]$$

$$= \sum_{i=1}^{I} P_{i.} \lambda_\alpha (\psi_{\alpha i})^2$$

Similarly, it can be factorized by columns

$$\lambda_\alpha = \sum_{j=1}^{J} P_{.j} \lambda_\alpha (\phi_{\alpha j})^2$$

and also by cells :

$$\lambda_\alpha = \sum_{i=1}^{I} \sum_{j=1}^{J} P_{i.} P_{.j} \lambda_\alpha (\psi_{\alpha i})^2 (\phi_{\alpha j})^2$$

C. These results can be used to assess the quality of the approximations made to the order k+1 in a finer way than for the simple consideration of the quantity $\sum_{\alpha=k+1}^{N-1} \lambda_\alpha$.

When constructing the matrix of the item $\sum_{\alpha=k+1}^{N-1} P_{i.} P_{.j} \lambda_\alpha (\psi_{\alpha i})^2 (\phi_{\alpha j})^2$ as a contribution from the cell (i,j) of P to the approximation error $\sum_{\alpha=k+1}^{N-1} \lambda_\alpha$, the sum of terms of the ith row is equal to $\sum_{\alpha=k+1}^{N-1} \lambda_\alpha (\psi_{\alpha i})^2 P_{i.}$, as a contribution from the ith row of P to the approximation error.

Similarly, the sum of the items of the jth column is equal to $\sum_{\alpha=k+1}^{N-1} \lambda_\alpha (\phi_{\alpha j})^2 P_{.j}$, as a contribution from the jth column of P to the approximation error.

A simple visualization of the matrix makes it possible to see whether factorizations of the total error show an identifiable structure or not. When this error is both small and unstructured, the number of axes used is therefore sufficient.

It should be noted that small $P_{i.} P_{.j}$ values can hide very high real residuals $\sum_{\alpha=k+1}^{N-1} \lambda_\alpha (\psi_{\alpha i})^2 (\phi_{\alpha j})^2$. These residuals are negligible from a methodological point of view since they are linked to very low weightings.

III. DETERMINING THE NUMBER OF AXES TO BE RETAINED BY CROSS VALIDATION

A. The calculation of item $(A_k)_{i,j}$ as carried out in formula (10) requires all items A, and more especially A_{ij}. It may seem highly favorable to use item A_{ij} in order to estimate this item itself. The principle of cross validation processes consists precisely in trying to avoid that situation. Approximation $(A_k)_{i,j}$ shall be computed without considering A_{ij}. A poor reconstruction of A_{ij}'s by $(A_k)_{i,j}$'s will show that the model underlying the approximation of A by A_k lets too great a part of the data variability escape : the specificity of a cell A_{ij} escapes the approximation $(A_k)_{i,j}$ that can be deduced from other items of A.

B. Omitting item A_{ij} could be considered with references to the processes used when data are missing (Greenacre (1984)).

In order to allow for the objective of cross validation this algorithm shall be used for each cell of matrix A, i.e., IxJ times. Assuming that r iterations of the algorithm are necessary (or deliberately chosen) for each cell of matrix A, r x I x J analyses should then be performed.

A less costly process is preferred. It was introduced within a Principal Component Analysis of a triplet (X, Q, D) by Holmes-Junca (1985). Adapting it to the Correspondence Analysis must allow for the fact that in Correspondence Analysis, matrices Q and D depend on data X.

Consider table $P^{(i)}$ obtained when removing row i from P. The size of this table is (I-1) x J. Correspondence Analysis of this table will produce $N^{(i)} = \min((I-1), J)$ triplets $(\lambda_\alpha^{(i)}, \psi_\alpha^{(i)}, \phi_\alpha^{(i)})$.

Similarly column j can be removed from table P in order to obtain a table $P^{(j)}$. The correspondence analysis of this table will produce $N^{(j)} = \min(I, (J-1))$ triplets $(\lambda_\alpha^{(j)}, \psi_\alpha^{(j)}, \phi_\alpha^{(j)})$.

Let $N^- = \min(N^{(i)}, N^{(j)})$. For $k = 1, \ldots, N^- - 1$ the following is then defined

$$(A_k)_{ij} = 1 + \sum_{\alpha=1}^{k} \sqrt{\sqrt{\lambda_\alpha^{(i)} \lambda_\alpha^{(j)}}\, \psi_{\alpha i}^{(j)} \phi_{\alpha j}^{(i)}}$$

The quality of the reconstruction can be determined in calculating the matrix of the following terms

$$(A_{ij} - (A_k)_{ij})^2 \, P_{i.} \, P_{.j}$$

IV. TESTS ON THE SUM OF THE NON-RETAINED LATENT ROOTS

A. Basing the determination of the number of axes to be retained on a testing process, first implies considering that the observations performed consist in a sample representative of a greater population. Let E be the table, I x J, of numbers from which table P of frequencies was calculated. $E_{i.}$, $E_{.j}$ and $E_{..}$ will be the total numbers for the line i, the column j and the entire table, respectively. It is usual practice to rewrite formula (9) as

$$\sum_{i=1}^{I} \sum_{j=1}^{J} (E_{ij} - E_{i.} E_{.j})^2 \,/\, (E_{i.} E_{.j}/E_{..}) = (\sum_{\alpha=1}^{N-1} \lambda_\alpha) \times E_{..}$$

The following conventional statement is thus obtained. Under the Independence Hypothesis of rows and columns of E, $(\sum_{\alpha=1}^{N-1} \lambda_\alpha) \times E_{..}$ respects the law of chi-square with (I-1) x (J-1) degrees of freedom.

In fact, this result comes down to checking that none of the latent roots observed is significantly different from zero.

Used strictly, this result should lead not to perform a correspondence analysis when the independence hypothesis is not rejected.

In fact, it is often interesting to explore deviations from independence model even if those deviations are not important enough to be judged significant by the test. Symmetrically the chi-square may be significant without the correspondence analysis being the method adapted for an explanation of the dependence structure. These results follow from the independence of latent roots of a Wishart matrix with regard to their sum. This point has ben detailed by Lebart (1976) and Lebart et al. (1977).

B. For the problem under discussion, the necessary result would be to check that, from a certain index k, latent roots are not significantly different from zero. In this respect several solutions were proposed. Relying on O'Neill's results (1978), Greenacre's (1984) and Goodman's (1986) articles have shown that all these proposals were erroneous and produced too optimistic results.

Let Π be the probability table in the population under consideration, and state

$$\Pi_{ij} = \Pi_{i.}\ \Pi_{.j}\ (1 + \sum_{\alpha=1}^{N-1} \sqrt{\mu_\alpha}\ \xi_{\alpha i}\ \eta_{\alpha j})$$

where triplets $(\mu_\alpha, \xi_\alpha, \eta_\alpha)$ are defined for Π by analogy with triplets $(\lambda_\alpha, \psi_\alpha, \phi_\alpha)$ linked to P through formulas (1), (2) and (3).

Let $k < N-1$

For any $s > k$, any $t > k$ and any $\alpha \leqslant k$, set down

$$E_c(s, t, \alpha) = \sum_{i=1}^{I} \xi_{si}\ \xi_{ti}\ \xi_{\alpha i}\ \Pi_{i.}$$

Similarly, for any u > k, any v > k and any $\alpha \leqslant k$ set down

$$E_e(u, v, \alpha) = \sum_{j=1}^{J} \eta_{uj} \eta_{vj} \eta_{\alpha j} \Pi_{.j}$$

O'Neill has shown that with the hypothesis

$\mu_\alpha = 0$ for any $\alpha > k$

quantify $\left(\sum_{\alpha=k+1}^{N-1} \lambda_\alpha \right) \times E_{..}$ is distributed as the sum of N-k squares of normal dependant variables and that a chi-square is obtained with (I-1-k) x (J-1-k) degrees of freedom if, and only if

for any u,v,s,t > k $\quad \sum_{\alpha=1}^{k} \sqrt{\mu_\alpha}\, E_e(u,v,\alpha)\, E_c(s,t,\alpha) = 0$

Most frequently, this purely theoretical condition shows above all that any testing process dealing with the last latent roots will be questionable.

C. Conclusion

At the time being, Correspondence Analysis users are not provided with very efficient testing tools for retaining a number of axes to be studied. This lack should give rise to works on that subject. Experiments were made to use Bootstrap's and Jackknife's techniques which have not brought yet a precise and easy-to-implement tool to users.

REFERENCES

1. Escoufier, Y. and Junca S. (1986). Least squares approximation of frequencies or their logarithms. Inst. Stat. Rev. Vol. 54, n° 3, pp. 279-283.

2. Goodman, L. (1986). Some Useful Extensions of the Usual Correspondence Analysis Approach and the Usual Log-Linear Models Approach in the Analysis of Contingency Tables. Int. Stat. Rev. Vol. 54, n° 3, pp. 243-270.

3. Greenacre, M.J. (1984). Theory and Applications of Correspondence Analysis. Academic Press, Inc.

4. Holmes-Junca, S. (1985). Outils Informatiques pour l'Evaluation de la Pertinence d'un Résultat en Analyse des Données. Thèse de 3ème cycle. U.S.T.L. Montpellier.

5. Lebart, L. (1976). The Significance of Eigenvalues Issued from Correspondence Analysis. In "Proceedings in Computational Statistics" (Compstat), pp. 38-45. Physica-Verlag, Vienna.

6. Lebart, L., Morineau, A., Tabard, N. (1977). Techniques de la Description Statistique : Méthodes et Logiciels pour l'Analyse des Grands Tableaux. Dunod. Paris.

7. O'Neill, M.E. (1978). Distributional Expansions for Canonical Correlations from Contingency Tables. J.R.S.S. B. 40, n° 3, pp. 303-312.

SINGULAR VALUE DECOMPOSITION OF MULTIARRAY DATA AND ITS APPLICATIONS

Tadashi Yoshizawa

Yamanashi University
4-3-11, Takeda, Kofu, Japan

I. INTRODUCTION

Iwatsubo (3) proposed a quantification method for three-way discrete data which was derived from the maximization criterion of the so-called correlation coefficient of the third order. Yoshizawa (7) showed that Iwatsubo's method could be considered to be a kind of decomposition of the interaction of the highest order defined by Lancaster (5) in the case of multiple contingency tables, and therefore, is an extension of Hayashi's quantification method of the third type or correspondence analysis.

Yoshizawa (8) further extended the concept of singular value decomposition of matrices to general multiway data. The decomposed configuration given by the author has two distinguishing characteristics: "nesting" and "interpretability similar to that of the ANOVA model". In addition, an extended reciprocal averaging algorithm was formulated to decompose multiway data.

Difficulties in analyzing multiarray data, in other words, multiway data or n-mode data, lie in the complexity of the configurations and in the interpretation of structures. Tucker (6) described three methods to deal with the estimation

ISBN 0-12-215485-1

of the parameters in the three-mode principal component models. One of this models, which Kroonenberg (4) called the Tucker-3 model, is fundamental in three-mode data analysis. Kroonenberg broadly reviewed three-mode principal component analysis, and briefly outlined n-mode extensions in section 3.9, of his excellent monograph titled, "Three Mode Principal Component Analysis".

In this paper, the theory of singular value decomposition of multiarray data will be roughly sketched. The configuration of the extended singular value decomposition is demonstrated, in comparison with the Tucker-3 model for three way data. Next, the criteria and algorithms to derive the decomposition, the nested configuration, and its interpretation will be described. Finally, some applications will be shown.

II. SINGULAR VALUE DECOMPOSITION

A. Case of matrix

It is well known that any real l x m matrix $A = \{A_{ij}\}$ may be expressed by singular value decomposition as follows:

$$A_{ij} = \sum_{p=1}^{r} c_p g_{ip} h_{jp}, \quad i=1,\ldots,s, \; j=1,\ldots,t \qquad (1)$$

where c_p's are singular values, $\{g_{ip}\}$ and $\{h_{jp}\}$ are columnwise orthonormal eigenvectors of AA' and A'A respectively, and r is the rank of A.

The above quantities satisfy the following equations:

$$c_p = \sum_{i,j} A_{ij} g_{ip} h_{jp} \tag{2}$$

$$\left.\begin{aligned} \sum_j A_{ij} h_{jp} &= c_p g_{ip} \\ \sum_i A_{ij} g_{ip} &= c_p h_{jp} \end{aligned}\right\} \tag{3}$$

In addition, they are the solutions to the following problems:

(a) Maximize

$$\sum_{i,j} A_{ij} g_{ip} h_{jp} \tag{4}$$

for a fixed p, subject to the orthonormal conditions of $\{g_{iq}\}$ and $\{h_{jq}\}$ for $q=1,2,\ldots,p$, and

(b) Minimize

$$\sum_{i,j} (A_{ij} - \sum_{q=1}^{p} c_q g_{iq} h_{jq})^2 \tag{5}$$

subject to the orthonormal conditions of $\{g_{iq}\}$ and $\{h_{iq}\}$.

Equations (3) are the determinantal equations of problems (a) and (b) and are called "reciprocal averaging" as a numerical process by Hill (2). After the first and largest singular value, c_1, and the corresponding $\{g_{i1}\}$ and $\{h_{j1}\}$ were obtained, the second components c_2, $\{g_{i2}\}$ and $\{h_{j2}\}$ can be solved using the same determinantal equations (3). That is, the sequence of solutions

$$\sum_{q=1}^{p} c_q g_{iq} h_{jq}, \text{ for } p=1,2,\ldots,r \tag{6}$$

is nested.

B. Tucker's model

Tucker (6) proposed three methods to deal with the estimation of the parameters in the so called three-mode principal component models. One of his models, the Tucker-3 model, is basic to three-mode data analysis.

Following Kroonenberg's notation with some modification, the Tucker-3 model can be formulated as the factorization of the three-mode data matrix $A = \{A_{ijk}\}$ such that

$$A_{ijk} = \sum_{p=1}^{s} \sum_{q=1}^{t} \sum_{r=1}^{u} c_{pqr} g_{ip} h_{jq} e_{kr} \tag{7}$$

for $i=1,\ldots,l$; $j=1,\ldots,m$; $k=1,\ldots,n$. The coefficients g_{ip}, h_{jq}, and e_{kr} are the entries of the component matrices $G(l \times s)$, $H(m \times t)$, and $E(n \times u)$; l, m, n are the number of elements, and s, t, u are the number of components of the first, second and third mode, respectively. G, H, and E are assumed to be columnwise orthonormal real matrices with the number of rows larger than or equal to the number of columns. The c_{pqr} are the elements of the three-mode core matrix $C(s \times t \times u)$.

If we are interested in exactly decomposing A into all of its components, Tucker's methods would suffice to provide a solution for decomposition, as Kroonenberg noted (see Bahadur (1) also). It is easily understood that any columnwise orthonormal matrices G, H and E satisfy the equation (7) when s=l, t=m, u=n.

Moreover, Kroonenberg noted in section 4.5 of his book that the configurations of three-mode principal component analysis in its alternating least squares formulation are in general not nested.

C. Singular value decomposition of three-way data

Extending problem (a) and (b) above, consider the following problems:

(c) Maximize

$$\phi = \sum_i \sum_j \sum_k A_{ijk} g_i h_j e_k \tag{8}$$

subject to the normal condition, such that

$$\sum_i g_i^2 = \sum_j h_j^2 = \sum_k e_k^2 = 1, \text{ and} \tag{9}$$

(d) Minimize

$$Q = \sum_i \sum_j \sum_k (A_{ijk} - c g_i h_j e_k)^2 \tag{10}$$

subject to the normal conditions of $\{g_i\}$, $\{h_j\}$ and $\{e_k\}$.

The determinantal equations of the two problems are the same and are easily obtained as follows:

$$\left.\begin{aligned} \sum_{i,j} A_{ijk} h_j e_k &= c g_i \\ \sum_{k,i} A_{ijk} e_k g_i &= c h_j \\ \sum_{i,j} A_{ijk} g_i h_j &= c e_k \end{aligned}\right\} \tag{11}$$

These equations may be regarded as the direct extension of (3). Let the solutions of problem (c) be c_{111}, g_{i1}, h_{j1}, and e_{k1}. Then these quantities are least square estimates for the Tucker-3 model when s = t = u.

Unfortunately, we cannot nest the equations (11) above to obtain the second components c_{222}, g_{i2}, h_{j2}, and e_{k2} simply by substituting A_{ijk} with

$$A_{ijk} - c_{111}g_{i1}h_{j1}e_{k1}.$$

Using c_{111}, g_{i1}, h_{j1}, and e_{k1} obtained above, let the following quantities be defined:

$$\left.\begin{aligned} A_{ij*} &= \sum_k A_{ijk}e_{k1}; \\ A_{i*k} &= \sum_j A_{ijk}h_{j1}; \\ A_{*jk} &= \sum_i A_{ijk}g_{i1}; \end{aligned}\right\} \quad (12)$$

$$B_{ijk} = A_{ijk} - A_{ij*}e_{k1} - A_{i*k}h_{j1} - A_{*jk}g_{i1} + 2c_{111}g_{i1}h_{j1}e_{k1} \quad (13)$$

If we solve (11), replacing A_{ijk} with B_{ijk}, we can obtain the second components c_{222}, g_{i2}, h_{j2}, and e_{k2}, which are orthonormal with g_{i1}, h_{j1}, and e_{k1}, i.e.,

$$\sum g_{i2}^2 = \sum h_{j2}^2 = \sum e_{k2}^2 = 1 \quad (14)$$

$$\sum g_{i1}g_{i2} = \sum h_{j1}h_{j2} = \sum e_{k1}e_{k2} = 0 \quad (15)$$

Furthermore, if we define C_{ijk} as equation (13) using B_{ijk} and the second components associated with B, we obtain the third components c_{333}, g_{i3}, h_{j3}, and e_{k3}.

Example 1. Suppose that l=m=n=2. Then, B_{ijk} defined by equation (13) may be expressed as follows:.

$$B_{ijk} = c_{222}g_{i2}h_{j2}e_{k2}, \text{ and} \tag{16}$$

A_{ijk} is decomposed as

$$\begin{aligned} A_{ijk} &= c_{111}g_{i1}h_{j1}e_{k1} + c_{122}g_{i1}h_{j2}e_{k2} \\ &+ c_{212}g_{i2}h_{j1}e_{k2} + c_{221}g_{i2}h_{j2}e_{k1} \\ &+ c_{222}g_{i2}h_{j2}e_{k2}. \end{aligned} \tag{17}$$

we may observe that the three components c_{112}, c_{121}, and c_{211} vanish.

III. ALGORITHM AND CRITERIA

A. Nested algorithm

In Yoshizawa (9), Kronecker product notation has been used to describe the theory of singular value decomposition of multiarray data. To avoid the complicated notation here, let $A = \{A_{ij...k}\}$ denote a multiarray of dimension l x m x ...x n, and let $\{g_{ip}\}$, $\{h_{jq}\},\ldots,\{e_{kr}\}$ denote the vectors which correspond to each index of the multiarray. An algorithm to obtain a singular value decomposition of multiarray data is defined as follows:

Nested algorithm for singular value decomposition

Given a multiarray data $A_{ij...k}$, set

$$A^{(1)}_{ij...k} = A_{ij...k}$$

. Repeat the following for p=1, 2,... until

$A^{(p)}_{ij\ldots k} = 0$ (for all i,j,...,k)

. Compute the largest c and the corresponding and normalized vectors $\{g_i\}$, $\{h_j\},\ldots,\{e_k\}$ which satisfy the following determinantal equations:

$$\left.\begin{array}{l} \sum\limits_{\hat{}j\ldots k} A_{ij\ldots k}\,\hat{}\,h_j\ldots e_k = cg_i \\ \sum\limits_{i\hat{}\ldots k} A_{ij\ldots k}\,g_i\,\hat{}\ldots e_k = ch_j \\ \quad\cdots\cdots\cdots\cdots\cdots \\ \sum\limits_{ij\ldots\hat{}} A_{ij\ldots k}\,g_i h_j\ldots\hat{} = ce_k \end{array}\right\} \qquad (18)$$

where ^ means that the element pointed to is deleted.

. Compute the following quantities:

$$A^{(p+1)}_{ij\ldots k} = \sum_{i'j'\ldots k'} A^{(p)}_{i'j'\ldots k'} \times (\delta_{i'i} - g_{i'}g_i)(\delta_{j'j} - h_{j'}h_j) \times\ldots\times (\delta_{k'k} - e_{k'}e_k) \qquad (19)$$

where δ denotes Kronecker's delta.

It is easily shown that the number of repetitions in this algorithm is finite and is equal to or less than the minimum of {1,m,...,n}.

Now let c_p and $\{g_{ip}\}$, $\{h_{jp}\},\ldots,\{e_{kp}\}$ denote the solutions of equation (18) at the p-th step of the algorithm. It may then be proved that each of $G = \{g_{ip}\}$, $H = \{h_{jp}\},\ldots,E = \{e_{kp}\}$ is a columnwise orthonormal matrix.

B. Criteria and reciprocal averaging

The equations (18) are the determinantal equations for the following problem, which is a simple extension of problem (c):

(e) Maximize

$$\phi = \sum_{i,j,\ldots,k} A_{ij\ldots k} g_i h_j \ldots e_k \tag{20}$$

subject to the normal conditions

$$\sum_i g_i^2 = \sum_j h_j^2 = \ldots = \sum_k e_k^2 = 1 \tag{21}$$

The determinantal equations (18) may be regarded as an extension of reciprocal averaging process by Hill (2) and may be solved numerically by a simple repetition method.

IV. NESTED CONFIGURATION AND INTERPRETATION

A. Generalized concept of interaction

Given a three-way data A_{ijk} of dimension l x m x n and any vectors $\{g_i\}$, $\{h_j\}$, $\{e_k\}$ of dimension l, m, n respectively, let the following quantities be defined as in the ANOVA model:

$$\begin{aligned}
b(0) &= \sum A_{ijk} g_i h_j e_k \\
b(1)_i &= \sum A_{ijk} h_j e_k \\
b(2)_j &= \sum A_{ijk} g_i e_k \\
b(3)_k &= \sum A_{ijk} g_i h_j \\
b(12)_{ij} &= \sum A_{ijk} e_k \\
b(23)_{jk} &= \sum A_{ijk} g_i \\
b(13)_{ik} &= \sum A_{ijk} h_j
\end{aligned}$$

Grand mean term

$$B(0)_{ijk} = b(0)\ g_i h_j e_k$$

Main effect

$$B(1)_{ijk} = b(1)_i h_j e_k - B(0)_{ijk}$$
$$B(2)_{ijk} = b(2)_j g_i e_k - B(0)_{ijk}$$
$$B(3)_{ijk} = b(3)_k g_i h_j - B(0)_{ijk}$$

Interaction of first order (two-factor effect)

$$B(12)_{ijk} = b(12)_{ij} e_k - B(1)_{ijk} - B(2)_{ijk} - B(0)_{ijk}$$
$$B(23)_{ijk} = b(23)_{jk} g_i - B(2)_{ijk} - B(3)_{ijk} - B(0)_{ijk}$$
$$B(13)_{ijk} = b(13)_{ij} h_j - B(1)_{ijk} - B(3)_{ijk} - B(0)_{ijk}$$

Interaction of second order (three-factor effect)

$$B(123)_{ijk} = A_{ijk} - B(12)_{ijk} - B(23)_{ijk} - B(13)_{ijk} - B(1)_{ijk} - B(2)_{ijk} - B(3)_{ijk} - B(0)_{ijk} \quad (22)$$

If we suppose that

$$g_i = 1 / \sqrt{l}\ ,\ h_j = 1 / \sqrt{m}\ ,\ \text{and}\ e_k = 1 / \sqrt{n}\ ,$$

then the above quantities coincide with the mean, effects, and interactions of ANOVA model. Therefore, the naming above may be appropriate. Note that A_{ijk} is decomposed as follows:

$$A_{ijk} = B(0)_{ijk} + B(1)_{ijk} + B(2)_{ijk} + B(3)_{ijk} + B(12)_{ijk} + B(23)_{ijk} + B(13)_{ijk} + B(123)_{ijk} \quad (23)$$

This concept of interaction may be applied in a straightforward manner to multiarray data.

B. Nested configuration

The quantities $A^{(p+1)}_{ij\ldots k}$ computed in equation (19) of the algorithm defined in section III are considered to be the interaction of the highest order of $A^{(p)}_{ij\ldots k}$ relative to $\{g_{ip}\}$, $\{h_{jp}\},\ldots,\{e_{kp}\}$. Accordingly, any multiarray data can be decomposed into a nest. For the sake of simplicity, the nested decomposition of three-way data will be shown.

Let c_p and $\{g_{ip}\}$, $\{h_{jp}\}$,$\{e_{kp}\}$ denote the solutions of equations (18) at the p-th step of the the nested algorithm for the three-way data A_{ijk}. Next we obtain the following configuration:

$$\left.\begin{aligned} A_{ijk} &= B(0)_{ijk}+B(12)_{ijk}+B(23)_{ijk}+B(13)_{ijk} \\ &\quad +B(123)_{ijk} \\ B(123)_{ijk} &= C(0)_{ijk}+C(12)_{ijk}+C(23)_{ijk}+C(13)_{ijk} \\ &\quad +C(123)_{ijk} \\ C(123)_{ijk} &= \cdot\ \cdot\ \cdot\ \cdot\ \cdot \\ \cdot\ \cdot\ \cdot\ \cdot\ \cdot\ \cdot\ &\cdot\ \cdot\ \cdot\ \cdot\ \cdot\ \cdot \end{aligned}\right\} \tag{24}$$

where the quantities of B and those of C are computed from $\{A_{ijk}, g_{i1}, h_{j1}, e_{k1}\}$ and $\{B(123)_{ijk}, g_{i2}, h_{j2}, e_{k2}\}$ respectively. It is one of the characteristics of the decomposition that the main effect terms vanish as follows:

$$B(1)_{ijk} = B(2)_{ijk} = B(3)_{ijk} = 0 \tag{25}$$

This property holds generally for multiarray data.

V. SOME APPLICATIONS

A. Decomposition of the sum of squares

It is shown that the nested configuration (24) is an orthogonal decomposition. Therefore the sum of squares of A_{ijk} is partitioned into a nest. Let SS[ ] denote the sum of squares as in

$$SS[A] = \sum_{i,j,k} A_{ijk}^{2}.$$

We then obtain from (24)

$$\left.\begin{aligned} &SS[A]=SS[B(0)]+SS[B(12)]+SS[B(23)]+SS[B(13)] \\ &\qquad +SS[B(123)] \\ &SS[B(123)]=SS[C(0)]+SS[C(12)]+SS[C(23)]+SS[C(13)] \\ &\qquad\qquad +SS[C(123)] \\ &\quad SS[C(123)]= \;.\;.\;.\;.\;. \end{aligned}\right\} \quad (26)$$

The degrees of freedom may be partitioned as in the ANOVA model. However, the main effect terms vanish in the decomposition (26), so that the degrees of freedom for the main effect terms may be added to the first terms SS[B(0)], SS[C(0)], ... in equations (26). It is rather useful to know the relation of the core matrix with the degrees of freedom and the sum of squares. The following examples show the relation.

Example 2. Consider the same three-way data A_{ijk} of dimension 2 x 2 x 2 as in example 1. Using equation (17), the interaction terms are computed as below:

$$b(0) = \Sigma A_{ijk} g_{i1} h_{j1} e_{k1} = c_{111}$$
$$b(1)_i = \Sigma A_{ijk} h_{j1} e_{k1} = c_{111} g_{i1}$$
$$b(2)_j = \Sigma A_{ijk} g_{i1} e_{k1} = c_{111} h_{j1}$$
$$b(3)_k = \Sigma A_{ijk} g_{i1} h_{j1} = c_{111} e_{k1}$$
$$b(12)_{ij} = \Sigma A_{ijk} e_{k1}$$
$$b(23)_{jk} = \Sigma A_{ijk} g_{i1}$$
$$b(13)_{ik} = \Sigma A_{ijk} h_{j1}$$

Grand mean term

$$B(0)_{ijk} = b(0) g_{i1} h_{j1} e_{k1} = c_{111} g_{i1} h_{j1} e_{k1}$$

Main effect

$$B(1)_{ijk} = b(1)_i h_{j1} e_{k1} - B(0)_{ijk} = 0$$
$$B(2)_{ijk} = b(2)_j g_{i1} e_{k1} - B(0)_{ijk} = 0$$
$$B(3)_{ijk} = b(3)_k g_{i1} h_{j1} - B(0)_{ijk} = 0$$

Interaction of first order (two-factor effect)

$$B(12)_{ijk} = b(12)_{ij} e_{k1} - B(1)_{ijk} - B(2)_{ijk} - B(0)_{ijk}$$
$$= c_{221} g_{i2} h_{j2} e_{k1}$$
$$B(23)_{ijk} = c_{122} g_{i1} h_{j2} e_{k2}$$
$$B(13)_{ijk} = c_{212} g_{i2} h_{j1} e_{k2}$$

Interaction of second order (three-factor effect)

$$B(123)_{ijk} = A_{ijk} - B(12)_{ijk} - B(23)_{ijk} - B(13)_{ijk}$$
$$- B(1)_{ijk} - B(2)_{ijk} - B(3)_{ijk} - B(0)_{ijk}$$
$$= c_{222} g_{i2} h_{j2} e_{k2}$$

The sum of squares for each term is as follows:

$$SS[B(0)] = c_{111}^2$$
$$SS[B(12)] = c_{221}^2$$
$$SS[B(23)] = c_{122}^2$$
$$SS[B(13)] = c_{212}^2$$
$$SS[B(123)] = c_{222}^2 \,.$$

Example 3. Let the dimension of A be 2 x 2 x 3. The sums of squares are then easily calculated as follows:

$$SS[B(0)] = c_{111}^2$$
$$SS[B(12)] = c_{221}^2$$
$$SS[B(23)] = c_{122}^2 + c_{123}^2$$
$$SS[B(13)] = c_{212}^2 + c_{213}^2$$
$$SS[B(123)] = c_{222}^2 \,,$$

where $\{e_{k3}\}$ is assumed to be orthonormal to $\{e_{k1}\}$ and $\{e_{k2}\}$. Note that

$$c_{112} = c_{121} = c_{211} = c_{113} = c_{223} = 0.$$

Example 4. Let the dimension of A be 2 x 3 x 3. The sums of squares are calculated as follows:

$$SS[B(0)] = c_{111}^2$$
$$SS[B(12)] = c_{221}^2 + c_{231}^2$$
$$SS[B(23)] = c_{122}^2 + c_{123}^2 + c_{132}^2 + c_{133}^2$$
$$SS[B(13)] = c_{212}^2 + c_{213}^2$$
$$SS[B(123)] = c_{222}^2 + c_{233}^2$$

$$SS[C(0)] = c_{222}^2$$
$$SS[C(23)] = c_{233}^2$$

where $\{h_{j3}\}$ and $\{e_{k3}\}$ are assumed to be orthonormal to $\{h_{j1}\}$ and $\{h_{j2}\}$, $\{e_{k1}\}$ and $\{e_{k2}\}$ respectively. 7 c-terms among 18 vanish in this case.

B. Several notes on applications

(1) Let $\{X_{ijk}\}$ be a three-way contingency table and transform it to.

$$A_{ijk} = x_{ijk} / \sqrt{x_{i++}x_{+j+}x_{++k}} \cdot$$

In this case, the additive model proposed by Lancaster (5) is no other than the first step decomposition by our algorithm.

(2) Let B_{ijk} denote the interaction of the second order for the above A_{ijk}. The quantification method proposed by Iwatsubo (3) is no other than the singular value decomposition of B_{ijk}.

(3) Let the root square of each element of a multiple contingency table be $y_{ij\ldots k}$. If we decompose $y_{ij\ldots k}$ using our algorithm, we may select the approximate log-linear models.

(4) When the three-way data A_{ijk} is symmetric in the sense that

$$A_{ijk} = A_{ikj},$$

we may obtain a singular value decomposition by our algorithm without modification. The solved vectors h_{jp} and e_{kp} coincide.

(5) Table 1 shows the number of c-terms reduced in singular value decomposition for typical cases.

Table 1. number of c-terms reduced in singular value decomposition

l	m	n	lxmxn	# of c-terms reduced
2	2	2	8	3
2	2	3	12	5
2	3	3	18	7
3	3	3	27	9
r	r	r	r^3	$3r(r-1)/2$

REFERENCES

1. Bahadur, R.R. (1961). A representation of the joint distribution of responses to n dichotomous items. In "Studies in Item Analysis and Prediction" (H. Solomon ed.), pp.158-176. Stanford Univ. Press.
2. Hill, M.O. (1974). Correspondence analysis: a neglected multivariate method. Applied Statistics, 23, 340-354.
3. Iwatsubo, S. (1974). Two classification techniques of 3-way discrete data - quantification by means of correlation ratio and three-dimensional correlation coefficient - Koudoukeiryougaku, (Journal of Japanese Society for Behabiometrika) 2,54-65.
4. Kroonenberg, P.M. (1983). Three-mode Principal Component Analysis. DSWO Press, 398pp.
5. Lancaster, H.O. (1951). Complex contingency tables treated by the partition of χ^2. J. Royal Stat. Soc., Ser. B, 13, 242-249.
6. Tucker, L.R. (1966). Some mathematical notes on three-mode factor analysis. Psychometrika, 31, 279-311.

7. Yoshizawa, T. (1975). Models for quantification techniques in multiple contingency tables---the theoretical approach. Koudoukeiryougaku, 3, 1-11.

8. Yoshizawa, T. (1976). A generalized definition of interaction and singular value decomposition of multiway arrays. Koudoukeiryougaku, 4, 32-43.

9. Yoshizawa, T. (1977). Structure of Multiarray Data and Population Spaces. Ph. D. thesis, Tokyo University.

PARTIAL CORRESPONDENCE ANALYSIS AND ITS PROPERTIES

Haruo Yanai

The National Center for University Entrance Examination
2-19-23 Komaba, Meguro, Tokyo, JAPAN

I. INTRODUCTION

It is well known that correspondence analysis is one of the most popular data reduction technique for categorical data. Mathematically speaking, correspondence analysis can be defined in the context of canonical correlation analysis using two sets of categorical data, and the resulting measure of association can also be interpreted as the sum of the squared canonical correlations between the two sets (Takeuchi, Yanai & Mukherjee, 1982, Lebart, Morineau & Warwick, 1984). So far, Daudin(1980) proposed partial association measure as an extension of the association measure of two qualitative variables. With these developments in mind , the present author proposed the idea of partial correspondence analysis(Yanai,1986).

In this paper, we formulate partial correspondence analysis in terms of orthogonal projection operator(we call projector hereafter) and examine some mathematical properties of the method. Finally, we give a numerical method to demonstrate the validity of the method.

ISBN 0-12-215485-1

II. MATHEMATICAL PREPARATIONS

Let E^n be the n-dimensional Euclidian vector space and, for an n x p matrix X and an n x q matrix Y, let S(X) and S(Y) be the subspaces spanned by the column vectors of X and Y, respectively. Further, let S(X;Y) be the subspace spanned by the columns of (X;Y). Let P_X , P_Y and P_{XUY} be orthogonal projectors onto S(X),S(Y) and S(X;Y), respectively. Then

(1) $$P_X = X(X'X)^{-}X' \quad , \quad P_Y = Y(Y'Y)^{-}Y'$$

(2) $$P_{XUY} = (X;Y)\begin{bmatrix} X'X & X'Y \\ X'Y & Y'Y \end{bmatrix}^{-}\begin{bmatrix} X' \\ Y' \end{bmatrix}$$

Further, put $Q_X = I_n - P_X$ and $Q_Y = I_n - P_Y$. Then the following relationships hold.

Lemma 1:(Rao & Yanai, 1979)

(3) $$P_{XUY} = P_X + P_{Y/X} = P_Y + P_{X/Y}$$

where $P_{Y/X} = Q_XY(Y'Q_XY)^{-}Y'Q_X$ and $P_{X/Y} = Q_YX(X'Q_YX)^{-}X'Q_Y$

are orthogonal projectors onto $S(Q_XY)$ and $S(Q_YX)$, respectively.

Lemma 2: Suppose that each element of the vectors in X and Y is given as the deviation score from the mean of the respective variables. Then, vectors a and b maximizing

$$r_{Xa.Yb} = (Xa, Yb)/(\|Xa\| \; \|Yb\|)$$

are given by solving any of the following two equations.

(4) $(P_X P_Y)Xa = \lambda Xa$ and $(P_Y P_X)Yb = \lambda Yb$

(5) $(X'P_Y X)a = \lambda X'Xa$ and $(Y'P_X Y)b = \lambda Y'Yb$

We now consider the third set of variables $Z = (z_1,\ldots,z_r)$. Maximizing the correlation coefficient r_{fg} where $f = Xa + Zc$ and $g = Yb + Zc$ leads to the following Lemma.

Lemma 3: The following statements are equivalent.

(6) $(P_{XUZ} P_{YUZ})(Xa, Zc) = \lambda(Xa, Zc)$

(7) $$\begin{bmatrix} X'Y & X'Z \\ Z'Y & Z'Z \end{bmatrix} \begin{bmatrix} Y'Y & Y'Z \\ Z'Y & Z'Z \end{bmatrix}^{-} \begin{bmatrix} Y'X & Y'Z \\ Z'X & Z'Z \end{bmatrix} \begin{bmatrix} a \\ c \end{bmatrix} = \lambda \begin{bmatrix} X'X & X'Z \\ Z'X & Z'Z \end{bmatrix} \begin{bmatrix} a \\ c \end{bmatrix}$$

Regarding these equations, the following properties hold.

(i) $P_{X/Y} P_{Y/Z} = P_{XUY} P_{YUZ} - P_Z$

(ii) The maximum eigenvalue of (6) or (7) is the unit value of multiplicity $r = \text{rank}(Z)$.

(iii) $Zc = 0$ if $\lambda \neq 1$ and Zc is arbitrary if $\lambda = 1$.

(iv) The $(r + j)$th eigenvalue of (6) or (7) is equivalent to the j-th eigenvalue of any of the following equations.

(8) $(P_{X/Z} P_{Y/Z})(Q_Z Xa) = \lambda(Q_Z Xa)$

(9) $(X'Q_Z Y)(Y'Q_Z Y)^{-}(Y'Q_Z X)a = \lambda(X'Q_Z X)a$

Observe that the eigenvalues of the above equations (8) or (9) are the squared partial canonical correlations between

X and Y removing the effect of Z from both X and Y.

In view of these arguments, we may solve (7) instead of (9) in order to obtain partial canonical correlation coefficients between X and Y removing Z , and also the corresponding canonical variables.

III. PARTIAL CORRESPONDENCE ANALYSIS

We now consider the situation in which n subjects are measured on three items A,B and C, comprising p,q and r categories respectively. In such a case, response pattern matrices for the three items can be represented in terms of the three dummy matrices G_A, G_B and G_C, which satisfy

$$(10) \qquad G_A 1_p = G_B 1_q = G_C 1_r = 1_n$$

where $(1_n)' = (1,1,..,1)$ is the n-dimensional unit vector . Further, consider the situation when the two items A and B are given, together with an auxiliary item C which may mask the exact relationship between the items A and B. In such a case, partial canonical correlation analysis between G_A and G_B removing the effects of G_C may duely be applied. Replacing X,Y and Z in (6) by G_A ,G_B and G_C , respectively, we get

$$(11) \qquad (P_{AUC} P_{BUC})(G_A a, G_C c) = \lambda(G_A a, G_C C)$$

which is equivalent to

$$(12) \quad \begin{bmatrix} N_{AB} & N_{AC} \\ N_{CB} & D_C \end{bmatrix} \begin{bmatrix} D_B & N_{BC} \\ N_{CB} & D_C \end{bmatrix}^{-} \begin{bmatrix} N_{BA} & N_{BC} \\ N_{CA} & D_C \end{bmatrix} \begin{bmatrix} a \\ c \end{bmatrix} = \lambda \begin{bmatrix} D_A & N_{AC} \\ N_{CA} & D_C \end{bmatrix} \begin{bmatrix} a \\ c \end{bmatrix}$$

where $N_{AB} = G_A{}'G_B$, $N_{BC} = G_B{}'G_C$ and $N_{CA} = G_C{}'G_A$

and $D_A = G_A{}'G_A$, $D_B = G_B{}'G_B$ and $D_C = G_C{}'G_C$

and $N_{BA} = (N_{AB})'$, $N_{CA} = (N_{AC})'$ and $N_{CB} = (N_{BC})'$

In view of the correspondence between the equations (6) , (8) and (9), the (r + j)th largest eigenvalue of (11) gives the j-th largest eigenvalue of the following equation.

$$(13) \qquad (G_A{}'Q_C\, G_B(G_B{}'Q_C\, G_B)^{-}G_B{}'Q_C\, G_A)a = \lambda(G_A{}'Q_C\, G_A)a$$

Using the equation(3), the left sides of (13) turns out to be $(G_A{}'P_{BUC}\, G_A - G_A{}'P_C\, G_A)a$. Thus,

it follows that (13) is equivalent to

$$(14) \quad \left((N_{AB}, N_{AC})\begin{bmatrix} D_B & N_{BC} \\ N_{CB} & D_C \end{bmatrix}^{-} \begin{bmatrix} N_{BA} \\ N_{CA} \end{bmatrix} - H_{AC} \right)a = (D_A - H_{AC})a$$

$$\text{(where } H_{AC} = N_{AC}D_C^{-1}\, N_{CA}\text{)}$$

Apparently, it is easier to solve (12) in stead of (14). It should be noted, however, that the middle term of the left hand side(LHS) of the equation (12) is singular because of (10). Thus we may replace G_C by $G_{\bar{C}}$ which results by excluding any one column vector of G_C. Resultingly, the middle terms of LHS of (12) becomes nonsingular. Thus, we have

$$\begin{bmatrix} N_{AB} & N_{A\bar{C}} \\ N_{\bar{C}B} & D_{\bar{C}} \end{bmatrix}\begin{bmatrix} D_B & N_{B\bar{C}} \\ N_{\bar{C}B} & D_{\bar{C}} \end{bmatrix}^{-} \begin{bmatrix} N_{BA} & N_{B\bar{C}} \\ N_{\bar{C}A} & D_{\bar{C}} \end{bmatrix}\begin{bmatrix} a \\ \bar{c} \end{bmatrix} = \lambda \begin{bmatrix} D_A & N_{A\bar{C}} \\ N_{\bar{C}A} & D_{\bar{C}} \end{bmatrix}\begin{bmatrix} a \\ \bar{c} \end{bmatrix}$$

where $N_{B\bar{C}} = G_B{}'G_{\bar{C}}$, $N_{\bar{C}A} = G_{\bar{C}}{}'G_A$, $D_{\bar{C}} = G_{\bar{C}}{}'G_{\bar{C}}$, $N_{\bar{C}A} = (N_{A\bar{C}})'$ and $N_{\bar{C}B} = (N_{B\bar{C}})'$.

Our method can be extended into the following two ways. The first is the case when there exists a number of covariates $C_1,::,C_g$ which mask the correlation between A and B, in which case we may put $G_C = (G_{C1};G_{C2};::;G_{Cg})$. The second is the case when there are two covariates C and D which may relate to the item A and the item B , respectively. In such a case, following the idea of bipartial canonical correlation analysis introduced by Timm & Carlson(1976) , we can develop bipartial correspondence analysis as well.

IV. NUMERICAL EXAMPLE OF PARTIAL CORRESPONDENCE ANALYSIS

In Japan, since 1979 the Joint First-Stage Achievement Test(JFSAT) have been conducted in order to evaluate the basic abilities of the candidates who apply for national universities . We performed a small survey asking how university students evaluate the JFSAT. We showed in Table 1 the contingency Table tabulated in terms of five rating points(1 through 5) by means of self evaluation of the results of JFSAT(item A) , and the Second Stage Achievement Test(SSAT)(item B). The data comprises 273 freshmen and sophomores of a national university. Of 273 students, 182 belongs to natural science course(N.C.) , while the remaining 91 students belongs to humanities course (H.C.). With this classification of an auxiliary item C, the contingency tables 2 and 3 are established.

By means of (12), we obtained the following six eigenvalues(the figure in the parenthesis indicates the square root),i.e., $\lambda_1 = \lambda_2 = 1, \lambda_3 = 0.087(0.295), \lambda_4 = 0.021(0.145)$ $\lambda_5 = 0.007(0.084)$ and $\lambda_6 = 0.000(0.0130)$.

Table 1: Contingency table between items A and B (N_{AB})

JFSAT	SSAT fairly bad	bad	intermediate	well	fairly well	Sum
1:fairly bad	1	1	1	0	0	2
2:bad	4	5	4	3	0	16
3:intermediate	17	54	41	7	1	120
4:well	14	47	32	8	1	102
5:fairly well	2	12	9	7	3	33
Sum	38	118	87	25	5	273

Table 2:

Contingency Table (N_{AB})

JFSAT	N.C.	H.C.
1: fairly bad	2	0
2: bad	8	8
3: intermediate	70	150
4: well	74	28
5: fairly well	28	5
Sum	182	91

Table 2

Contingency Table (N_{BC})

SSAT	N.C.	H.C.
1: fairly bad	31	7
2: bad	85	33
3: intermediate	50	37
4: well	14	11
5: fairly well	2	3
Sum	182	91

Further, we showed the weights for the ratings of JFAST and SSAT corresponding to the six eigenvalues in Tables 4 and 5, respectively. The weights corresponding to the third eigenvalue reflect a dimension regarding evaluation of both the JFSAT and SSAT with respect to the five rating points.

Statistically, speaking , it is interesting to note that weights corresponding to the unit eigenvalue result in the constant values as in the case of the ordinary correspondence analysis. Regarding the properties, our method proposed in this paper may be regarded as an extension of ordinary correspondence analysis.

Table 4: Weights for the JFSAT

	1	2	3	4	5	6
1:fairly bad	0.066	0.082	-0.067	-0.335	0.603	-0.162
2:bad	0.066	0.082	0.003	-0.207	-0.113	0.020
3:intermediate	0.066	0.082	-0.020	0.022	0.020	0.044
4:well	0.066	0.082	0.003	0.020	-0.009	-0.086
5:fairly well	0.066	0.082	0.176	0.005	0.030	0.007

Table 5: Weights for the SSAT

	1	2	3	4	5	6
1:fairly bad	0.066	0.082	-0.070	-0.121	0.026	0.081
2:bad	0.066	0.082	-0.028	0.040	-0.050	0.040
3:intermediate	0.066	0.082	-0.018	0.011	0.055	-0.052
4:well	0.066	0.082	0.104	-0.097	-0.103	-0.060
5:fairly well	0.066	0.082	0.311	0.062	0.188	0.236

REFERENCES

1.Daudin,J.J.(1980) Partial association measure and an application to qualitative regression, Biometrika,67,3,581-590

2.Lebart, L., Morineau,A. & Warwick, K.M.(1984) Multivariate Descriptive Statistical Analysis, John Wiely, New York.

3.Rao,C.R. & Yanai,H.(1979) General Definition and Decomposition of projectors and Some Application to Statistical Problems, J. of Statistical Planning and Inference,3,1-17

4.Takeuchi,K, Yanai, H. and Mukherjee ,B.N.(1982) The Foundation of Multivariate Analysis, Wiely Eastern, New Delhi

5.Timm ,N.N. & Carlson, J.E.(1976) Part and Bipartial Canonical Correlation Analysis, Psychometrika, 41,159-176

6.Yanai,H.(1986) Some Generalizations of Correspondence Analysis in terms of Projectors, In " Data Analysis and Informatics IV" (E.Diday et al.,eds.) , North-Holland.

CORRELATION ANALYSIS OF N-WAY QUALITATIVE DATA AND ITS APPLICATIONS

Shuichi Iwatsubo

The National Center
for University Entrance Examination
19-23 Komaba 2-chome, Meguro-ku, Tokyo, JAPAN

I. INTRODUCTION

Consider an $N_1 \times \ldots \times N_k \times \ldots \times N_n$ table. The cells in the table can be written in the form

$$(i_1, \ldots, i_k, \ldots, i_n) \in I_1 \times \ldots \times I_k \times \ldots \times I_n \qquad (i_k = 1, \ldots, N_k \; ; \; k = 1, \ldots, n), \quad (1)$$

where i_k stands for the i_k-th element of the index set I_k.

When the probabilities falling into the cells $(i_1, \ldots, i_k, \ldots, i_n)$ $(i_k = 1, \ldots, N_k \; ; \; k = 1, \ldots, n)$ are given by

$$p_{1\ldots k\ldots n}(i_1, \ldots, i_k, \ldots, i_n) \quad (i_k = 1, \ldots, N_k \; ; \; k=1, \ldots, n) \quad (2)$$

let us call (2) <u>n-way qualitative data</u>. Furthermore, we introduce the marginals $p_k(i_k) = \sum_{(k)} p_{1\ldots k\ldots n}(i_1, \ldots, i_k, \ldots, i_n)$ and $p_{km}(i_k, i_m) = \sum_{(k,m)} p_{1\ldots k\ldots m\ldots n}(i_1, \ldots, i_k, \ldots, i_m, \ldots, i_n)$, where $\sum_{(k)}$ and $\sum_{(k,m)}$ represent the summations over all indices except for i_k and except for both i_k and i_m respectively.

If we can assign optimal scores $x_k(i_k)$ to the i_k-th element of I_k $(i_k = 1, \ldots, N_k \; ; \; k = 1, \ldots, n)$ by optimizing an object function, we have

ISBN 0-12-215485-1

$$P_r\,(\,X_1 = x_1(i_1),\ldots,\; X_k = x_k(i_k),\ldots,\; X_n = x_n(i_n)\,)$$
$$= p_{1\ldots k\ldots n}(i_1,\ldots,i_k,\ldots,i_n)$$
$$(i_k = 1,\ldots,N_k\;;\; k = 1,\ldots,n) \qquad (3)$$

and can investigate the latent structure of n-way qualitative data by observing Euclidean configurations of the elements by using the obtained scores. Therefore it solely depends upon the choice of object function whether or not the unknown structure can be successfully extracted by applying the optimal scoring method to a given n-way qualitative data. It is, however, very difficult problem to determine the most appropriate object function, since the structure of a given data is unknown in advance.

Here we confine the class of object functions to those which can serve to extract linear relationships among the n variables $(X_1,\ldots,X_n)$ and introduce an optimal scoring method based on this class of object functions.

II. OPTIMAL SCORING METHOD MAXIMIZING CANONICAL CORRELATION COEFFICIENT

The data analyses for n-way quantitative data were developed by Horst, Kettenring, Carrol and Chang and so on (1),(2),(3). On the other hand, Kyogoku proposed a new method for n-way qualitative data and applied it to the real 3-way qualitative data of the general elections in Japan (4). Iwatsubo also proposed the optimal scoring method for n-way qualitative data under the criterion of maximizing multiple correlation coefficient (5). One of the generalizations of correspondence analysis to n-way qualitative data was introduced by Saporta (6). The singular value decomposition of multi-way contingency table was scrutinized comprehensively by Yoshizawa (7).

In this article the optimal scoring method is introduced under the criterion of maximization of canonical correlation.

We partition the set of n variables ($X_1,\dots,X_n$) into two groups, the set of r variables ($X_1,\dots,X_r$) and the set of (n-r) variables ($X_{r+1},\dots,X_n$). The number of ways of the partition is given by $2^{n-1} - 1$ (e.g., 511 in the case of 10-way data).

Then we introduce the object function with the following form.

$$\eta = \sum_{i_1=1}^{N_1} \cdots \sum_{i_n=1}^{N_n} p_{1\dots n}(i_1,\dots,i_n) \, [\{ x_1(i_1) + \dots + x_r(i_r) \} - \{ x_{r+1}(i_{r+1}) + \dots + x_n(i_n) \}]^2$$

$$/ \sum_{i_1=1}^{N_1} \cdots \sum_{i_n=1}^{N_n} p_{1\dots n}(i_1,\dots,i_n)\{x_1(i_1)+\dots+ x_r(i_r)\}^2$$

$$= \{ \sum_{k=1}^{n} \underset{\sim}{x}_k{}^t P_{kk} \underset{\sim}{x}_k + 2 \sum_{\substack{k=1 \\ k \neq m}}^{r} \sum_{m=1}^{r} \underset{\sim}{x}_k{}^t P_{km} \underset{\sim}{x}_m + 2 \sum_{\substack{k=r+1 \\ k \neq m}}^{n} \sum_{m=r+1}^{n} \underset{\sim}{x}_k{}^t P_{km} \underset{\sim}{x}_m - 2 \sum_{k=1}^{r} \sum_{m=r+1}^{n} \underset{\sim}{x}_k{}^t P_{km} \underset{\sim}{x}_m \}$$

$$/ \{ \sum_{k=1}^{r} \underset{\sim}{x}_k{}^t P_{kk} \underset{\sim}{x}_k + 2 \sum_{\substack{k=1 \\ k \neq m}}^{r} \sum_{m=1}^{r} \underset{\sim}{x}_k{}^t P_{km} \underset{\sim}{x}_m \} , \qquad (4)$$

where P_{km} represents the N_k by N_m matrix whose (i_k,i_m) element is $p_{km}(i_k,i_m)$ for $k \neq m$, P_{kk} is the N_k by N_k diagonal matrix whose i_k-th diagonal element is $p_k(i_k)$ and $\underset{\sim}{x}_k$ represents the column vector of dimension N_k whose i_k-th component is $x_k(i_k)$.

Note that $P_{km}{}^t = P_{mk}$.

The optimal scores $x_k(i_k)$ ($i_k = 1,\dots,N_k$; $k = 1,\dots,n$) can be obtained by minimizing the object function (4), which reduces to the following equations

$$\partial\eta/\partial\underset{\sim}{x}_k = \underset{\sim}{0} \qquad (k=1,\dots,n). \qquad (5)$$

It follows from equations (5) that

$$(1 - \eta) P[r,r]\, x[r] = P[r,n-r]\, x[n-r] , \qquad (5)$$

$$P[n-r,n-r]\, x[n-r] = P[r,n-r]^t x[r] , \qquad (6)$$

where $1 - \eta$ is a parameter representing the squared canonical correlation coefficient and

$$P[r,r] = \begin{bmatrix} P_{11} \cdots P_{1r} \\ \vdots \quad \vdots \\ P_{r1} \cdots P_{rr} \end{bmatrix} , \quad P[r,n-r] = \begin{bmatrix} P_{1\ r+1} \cdots P_{1n} \\ \vdots \quad \vdots \\ P_{r\ r+1} \cdots P_{rn} \end{bmatrix} ,$$

$$P[n-r,n-r] = \begin{bmatrix} P_{r+1r+1} \cdots P_{r+1n} \\ \vdots \\ P_{n\ r+1} \cdots P_{n\ n} \end{bmatrix} , \quad x[r] = (\tilde{x}_1 \cdots \tilde{x}_r)^t$$

and $x[n-r] = (\tilde{x}_{r+1} \cdots \tilde{x}_n)^t$.

Considering the rank of P[r,r], P[r,n-r] and P[n-r,n-r], the formulae (6) and (7) lead to solving the characteristic equation (For further details, see (8)).

Note that $1 - \eta$ becomes a parameter representing the multiple correlation coefficient for $r = 1$.

III. OPTIMAL SCORING METHOD FOR THREE-WAY QUALITATIVE DATA

In this section, the optimal scoring method for three-way qualitative data is investigated in detail and some properties derived from the method are shown.

In the case of three-way data, the object function is given by the following form

$$\eta = \sum_{i_1=1}^{N_1} \sum_{i_2=1}^{N_2} \sum_{i_3=1}^{N_3} p_{123}(i_1,i_2,i_3)\{x_1(i_1)-x_2(i_2)- -x_3(i_3)\}^2 / \sum_{i_1=1}^{N_1} \sum_{i_2=1}^{N_2} \sum_{i_3=1}^{N_3} p_{123}(i_1,i_2,i_3)\, x_1(i_1)^2$$

$$= \{ \underset{\sim}{x}_1^t P_{11} \underset{\sim}{x}_1 + \underset{\sim}{x}_2^t P_{22} \underset{\sim}{x}_2 + \underset{\sim}{x}_3^t P_{33} \underset{\sim}{x}_3 - 2 \underset{\sim}{x}_1^t P_{12} \underset{\sim}{x}_2 - 2 \underset{\sim}{x}_1^t P_{13} \underset{\sim}{x}_3 + 2 \underset{\sim}{x}_2^t P_{23} \underset{\sim}{x}_3 \} / \underset{\sim}{x}_1^t P_{11} \underset{\sim}{x}_1 . \quad (7)$$

From $\partial\eta/\partial\underset{\sim}{x}_k = \underset{\sim}{0}$ ($k=1,\ldots,3$), we have the following equations.

$$P_{12} \underset{\sim}{x}_2 + P_{13} \underset{\sim}{x}_3 = (1 - \eta) P_{11} \underset{\sim}{x}_1 , \quad (8)$$

$$P_{22} \underset{\sim}{x}_2 + P_{23} \underset{\sim}{x}_3 = P_{21} \underset{\sim}{x}_1 , \quad (9)$$

$$P_{32} \underset{\sim}{x}_2 + P_{33} \underset{\sim}{x}_3 = P_{31} \underset{\sim}{x}_1 , \quad (10)$$

which lead to

$$(P_{12.3} P_{22.3}^{+} P_{21.3} + P_{13} P_{33}^{-1} P_{31}) \underset{\sim}{x}_1 = (1-\eta) P_{11} \underset{\sim}{x}_1 , \quad (11)$$

$$\underset{\sim}{x}_2 = P_2 P_{22.3}^{+} P_{21.3} \underset{\sim}{x}_1 , \quad (12)$$

$$\underset{\sim}{x}_3 = P_{33}^{-1} (P_{31} - P_{32} P_2 P_{22.3}^{+} P_{21.3}) \underset{\sim}{x}_1 , \quad (13)$$

where $P_{ij.k} = P_{ij} - P_{ik} P_{kk}^{-1} P_{kj}$, $P_{22.3}^{+}$ is the Moore-Penrose generalized inverse of $P_{22.3}$. The matrix P_2 is represented by

$$P_2 = I_2 - \underset{\sim}{1}_2^t \underset{\sim}{1}_2 P_{22} ,$$

where I_2 is the N_2 by N_2 identity matrix and $\underset{\sim}{1}_2$ is $N_2 \times 1$ unit vector.

The optimal scores for i_1 ($i_1=1,\ldots,N_1$) are given as the components of the characteristic vector with unit variance corresponding to the second largest root of the characteristic equation derived from the formula (11). The other optimal scores for i_2 ($i_2=1,\ldots,N_2$) and i_3 ($i_3=1,\ldots,N_3$) are obtained from the formulae (12) and (13).

The following properties are derived from the method.

Property 1. The components of $\underset{\sim}{x}_1$, $\underset{\sim}{x}_2$ and $\underset{\sim}{x}_3$ have zero means respectively iff $\eta = 0$, i.e. $\underset{\sim}{1}_1^t\, P_{11}\, \underset{\sim}{x}_1 = \underset{\sim}{1}_2^t\, P_{22}\, \underset{\sim}{x}_2 = \underset{\sim}{1}_3^t\, P_{33}\, \underset{\sim}{x}_3 = 0$.

From Property 1, the following property is easily shown.

Property 2. $1 - \eta = (s_{12} + s_{13}) \,/\, s_{11}$, (14)

$$s_{22} = s_{12} - s_{23} , \tag{15}$$

$$s_{33} = s_{13} - s_{23} , \tag{16}$$

where s_{ij} is the covariance of X_i and X_j and s_{ii} is the variance of X_i .

Let $r_{1.23}$ denote the multiple correlation coefficient between X_1 and (X_2,X_3) , which is represented by the following form

$$r_{1.23}^2 = (s_{22}s_{13}^2+s_{33}s_{12}^2-2s_{12}s_{13}s_{23})/s_{11}(s_{22}s_{33}-s_{23}^2) . \tag{17}$$

The right-hand side of the formula (17) is reduced to $(s_{12} + s_{13})/s_{11}$ by using the relations of the formulae (15) and (16), which lead to the following property.

Property 3. $1 - \eta$ is a parameter representing the squared multiple correlation coefficient between X_1 and (X_2 , X_3).

From the formula (11), we have

$$P_{11}^{-1} (P_{12.3}\, P_{22.3}^{+}\, P_{21.3} + P_{13}\, P_{33}^{-1}\, P_{31})\, \underset{\sim}{x}_1 = (1 - \eta)\, \underset{\sim}{x}_1 . \tag{18}$$

Let $A = P_{11}^{-1} (P_{12.3} P_{22.3}^{+} P_{21.3} + P_{13} P_{33}^{-1} P_{31})$. Then we have the following property.

Property 4. A is a stochastic matrix.

From Property 4, the characteristic equation derived from the formulae (11) or (18) has the largest characteristic value 1 which is trivial for the purpose of obtaining the optimal scores.

With regard to the object function, note that we have the following property.

Property 5. If we start from the object function

$$\eta = \sum_{i_1}^{N_1} \sum_{i_2}^{N_2} \sum_{i_3}^{N_3} p_{123}(i_1,i_2,i_3)\{x_1(i_1)-x_2(i_2)-x_3(i_3)\}^2$$

$$/ \sum_{i_1}^{N_1} \sum_{i_2}^{N_2} \sum_{i_3}^{N_3} p_{123}(i_1,i_2,i_3)\{x_2(i_2)+x_3(i_3)\}^2$$

$$= \{ \underset{\sim}{x}_1^t P_{11} \underset{\sim}{x}_1 + \underset{\sim}{x}_2^t P_{22} \underset{\sim}{x}_2 + \underset{\sim}{x}_3^t P_{33} \underset{\sim}{x}_3 - 2 \underset{\sim}{x}_1^t P_{12} \underset{\sim}{x}_2 - 2 \underset{\sim}{x}_1^t P_{13} \underset{\sim}{x}_3 + 2 \underset{\sim}{x}_2^t P_{23} \underset{\sim}{x}_3 \} / \{ \underset{\sim}{x}_2^t P_{22} \underset{\sim}{x}_2 + \underset{\sim}{x}_3^t P_{33} \underset{\sim}{x}_3 + 2 \underset{\sim}{x}_2^t P_{23} \underset{\sim}{x}_3 \}, \qquad (19)$$

the characteristic function coincides with that derived from the formula (11).

IV. EXAMPLES FOR THREE-WAY QUALITATIVE DATA

A. Data from Treatment of Anxiety-ridden Neurotic Patients

Eighty-nine anxiety-ridden neurotic patients (X_1) were interviewed and evaluated with regard to the nineteen symptomatic items in three grading points (X_2) during four weeks after pre-treatment conditions until post-treatment states (X_3) (9). We have the 89 patients by 57 categories by 5 interviews binary data $\{ n(i_1,i_2,i_3) / i_1=1,\ldots,89; i_2=1,\ldots,57; i_3=1,\ldots,5 \}$.

Considering that the variable X_2 is categorized, we have the following properties.

Property 6. $$r_{2.13}^2 = r_{12}^2 + r_{23}^2 , \qquad (20)$$

where r_{ij} represents a correlation coefficient between X_i and X_j.

Property 7. If the variable X_2 is categorized,

$$\max \{ r_{2.13} \} \geqq \max \{ r_{1.23} \} , \max \{ r_{3.12} \}. \qquad (21)$$

The optimal scoring method maximizing $r_{2.13}$ leads to the equations

$$(N_1 P_{21} P_{12} + N_3 P_{23} P_{32}) \underset{\sim}{x}_2 = \lambda P_2 \underset{\sim}{x}_2 , \qquad (22)$$

$$\underset{\sim}{x}_1 = N_1 P_{12} \underset{\sim}{x}_2 , \qquad (23)$$

$$\underset{\sim}{x}_3 = N_3 P_{32} \underset{\sim}{x}_2 , \qquad (24)$$

where λ is a parameter representing the squared multiple correlation coefficient between X_2 and (X_1 , X_3).

From the formulae (22) and (23), we get

$$x_{1k}(i_1) = (1/95)(1/r_{12}) \sum_{i_3}^{5} \sum_{i_2}^{57} n(i_1,i_2,i_3)\, x_{2k}(i_2). \qquad (25)$$

where $x_{1k}(i_1)$ and $x_{2k}(i_2)$ are the optimal scores corresponding to the k-th largest characteristic value.

Let $x_{1k}(i_1,i_3) = (1/95)(1/r_{12})\sum_{i_2=1}^{57} n(i_1,i_2,i_3)x_{2k}(i_2).$

Fig. 1a - 1d are the configurations of some patients with the coordinate ($x_{12}(i_1,i_3)$, $x_{13}(i_1,i_3)$) ($i_3 = 1,\ldots,5$), which suggest that several types of the improving process of neurotic anxiety are observed among the patients. (In Fig. 1a - 1d, " ++ ", " + " and " 0 " are the points which stand for the patients with the grades " ++ (seriously ill) ", " + (ill) " and " 0 (no symptom) over all diagnostic items respectively.)

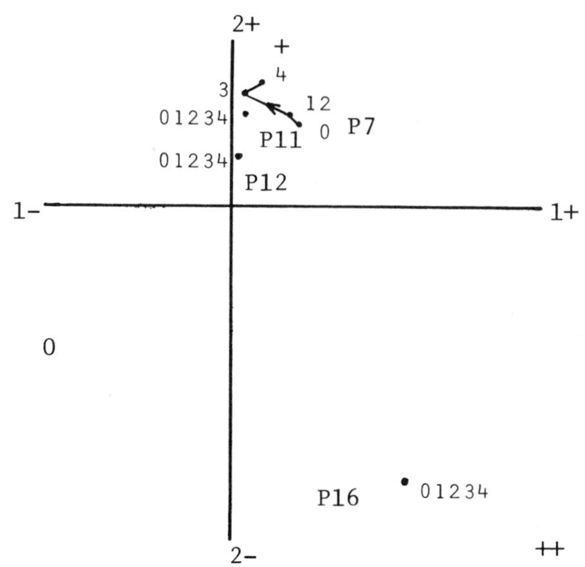

Fig. 1a. Configuration of patients
(no change)

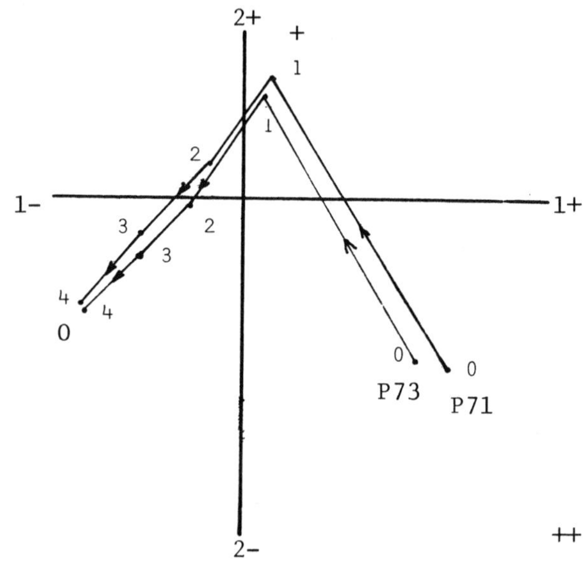

Fig. 1b. Configuration of patients
(well improved)

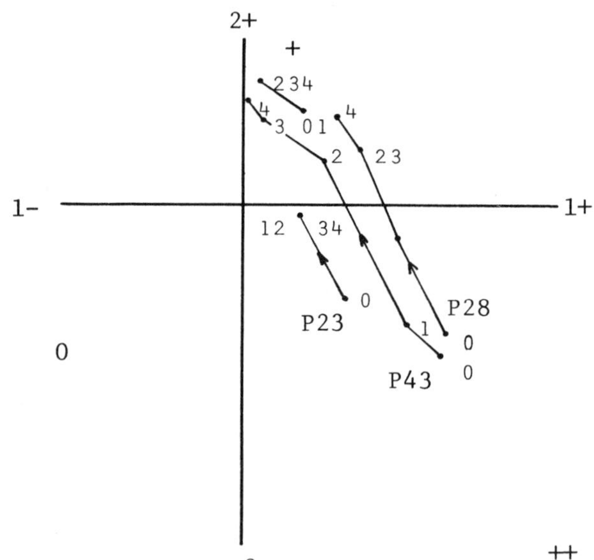

Fig. 1c. Configuration of patients
(improved)

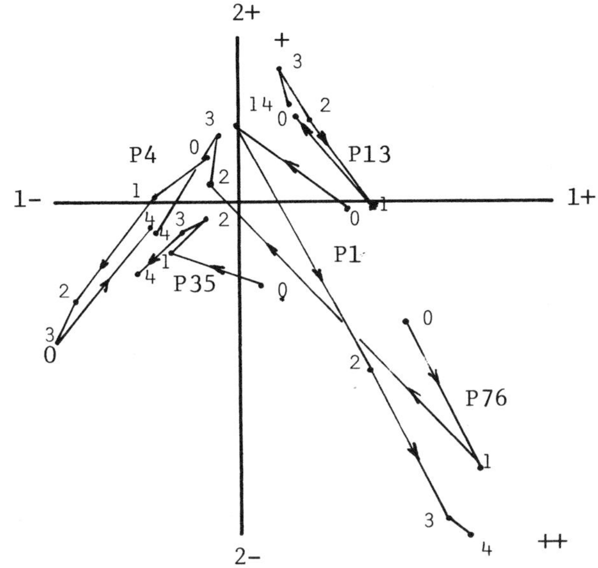

Fig. 1d. Configuration of patients (worse)

B. Data from the Distribution of Researchers in an Institute

A certain Research Institute in Japan used to have twelve research division (X_1). All the researchers could be classified into one of eight academic backgrounds (X_2). The distribution of researchers was examined during nineteen years from 1951 to 1969 (X_3). Then we have 12 divisions by 8 academic backgrounds by 19 years frequency table. The optimal scoring method maximizing $r_{12.3}$ was applied to the table. The two-dimensional Euclidean configurations of divisions, academic backgrounds and years are shown in Fig. 2a, 2b and 2c respectively. Fig. 2c suggests that the Institute began to concentrate its researches on the area of information sciences since 1961.

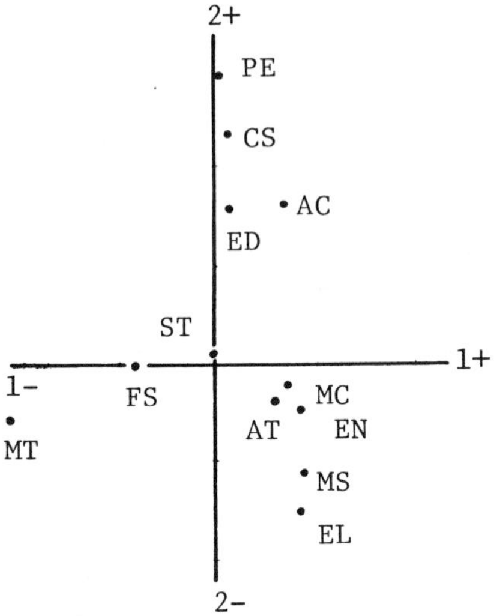

Fig. 2a. Configuration of research divisions (X_1)

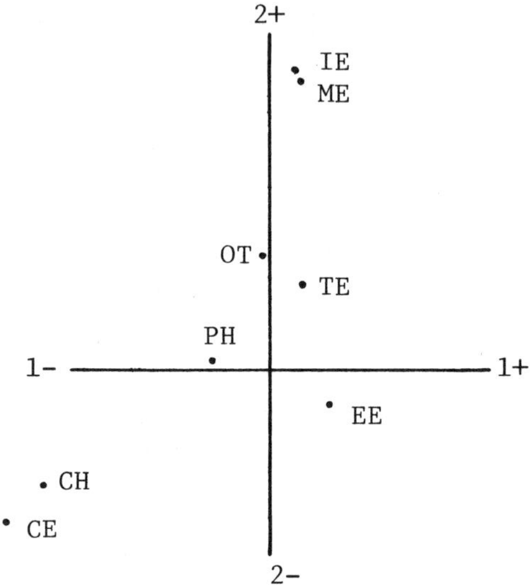

Fig. 2b. Configuration of academic backgrownds(X_2)

List of research divisions (X_1)

AC Automatic Control	AT Advanced Technology
CS Computer Science	ED Electric Device
EL Electronics	EN Energy
FS Fundamental Science	MC Machinery
MS Measurement	MT Materials
PE Production Engineering	ST Standards

List of academic backgrounds (X_2)

CE Chemical Eng.
CH Chemistry
EE Electrical Eng.
IE Instrumentation Eng.
ME Mecanical Eng.
OT Others (Mathematics, Psychology etc.)
PH Physics
TE Telecommunication Eng.

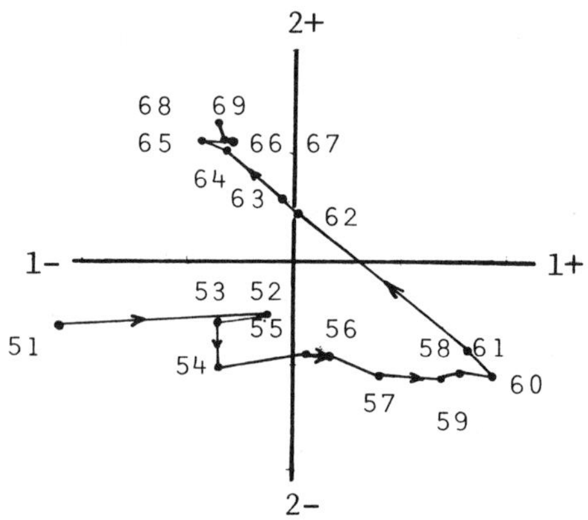

Fig. 2c. Configuration of years (X_3)

REFERENCES

1. Horst, P. (1961). Relations among m sets of measures, Psychometrika, 26, 129 - 149.
2. Kettenring, R. J. (1971). Canonical analysis of several sets of variables, Biometrika, 58, 433 - 451.
3. Carrol, J. D. and Chang, J-J. (1970). Analysis of individual differences in multidimensional scaling via an n-way generalization of " Eckert-Young " decomposition, Psychometrika, 35, 3, 283 - 319.
4. Kyogoku, J. (1967). On a method of quantifying N-way frequency tables (in Japanese), The Proceedings of the Institute of Statistical Mathematics, 15, 2, 140 -160.
5. Iwatsubo, S. (1971). An analysis of multidimensional qualitative data structure — quantification by multiple correlation coefficient — (in Japanese), Bulletin of the Electrotechnical Laboratory, 35, 7, 622 - 632.
6. Saporta, G. (1975). Liaisons entre plusieurs ensembles de variables et codage de données qualitatives, Thèse présentées à L'Université Pierre et Marie Curie (Paris VI) pour l'obtention du Doctorat Troisieme Cycle.
7. Yoshizawa, T. (1976). A generalized definition of interactions and singular value decompositions of multiway arrays (in Japanese), The Japanese Journal of Behaviormetrics, 4, 1, 32 - 43.
8. Iwatsubo, S. (1980). An optimal scoring method for n-way qualitative data and its applications, In " Data Analysis and Informatics ", (E.Diday et al. eds.) North-Holland.
9. Hayashi, S. and Iwatsubo, S. (1978). Study of the structure and the prediction of the prognosis of neurotic anxiety, 1978 Proceedings of the International Conference on Cybernetics and Society, 1, 105 - 110.

OUTLIERS AND INFLUENTIAL OBSERVATIONS IN QUANTIFICATION THEORY

Yutaka Tanaka
Tomoyuki Tarumi

Okayama University
2-1-1, Tsushima-naka, Okayama 700, JAPAN

I. INTRODUCTION

In the present paper we shall consider the methods for detection of outliers and influential observations in Hayashi's first and second methods of quantification (Quantification I, Quantification II). Quantification I, which is discussed in section 2, is mathematically an extension of regression analysis to the case where the design matrix is not of full rank. We can apply various measures for outliers and influential observations used in regression analysis in their original or modified forms. Quantification II, which is discussed in section 3, is a discriminant analysis based on qualitative data and it is mathematically equivalent to a certain type of correspondence analysis. Concerning Quantification II we proposed a method of sensitivity analysis in (6), and there we used the perturbation theory of eigenvalue problems. However, unfortunately it requires relatively high computing cost when the sample size and/or the number of categories are large. In this paper we consider some simplified versions of diagnostic measures which are easily calculated. The basic idea is to fix the scores for the outside variable and then use the measures for Quantification I as

ISBN 0-12-215485-1

the approximate measures in Quantification II. The properties of these measures are studied in section 4 and 5.

II. QUANTIFICATION I

A. Outline of Quantification I

Suppose we have n sets of observations $\{(y_i, x_i(0), x_i(11), \ldots, x_i(1c_1), \ldots, x_i(I1), \ldots, x_i(Ic_I)), i=1, \ldots, n\}$, where y is a quantitative explained variable (called an outside variable in quantification theory) and x's are dummy variables representing qualitative observations such that

$$x_i(jk) = \begin{cases} 1, & \text{if the individual } i \text{ responds to the category } k \text{ of the } j\text{-th factor item,} \\ 0, & \text{otherwise,} \end{cases} \qquad (1)$$
$$i=1, \ldots, n \; ; \; j=1, \ldots, I \; ; \; k=1, \ldots, c_j.$$

Each individual responds to one and only one category of each factor item. The variable x(0) is introduced to remove the constant term and always takes the value 1. In the above n is the number of individuals, I is the number of factor items, and c_j is the number of categories of the j-th factor item.

In Quantification I the matter of concern is to predict the quantitative outside variable y from the information of the factor items and to investigate the contribution of each item. For that purpose we assign numerical scores b_{jk}'s for the categories of the factor items operationally so that we can predict y_i as well as possible with the score of the individual i which is defined as the sum of the scores of the categories responded by the individual.

The statistical model can be expressed by

$$y_i = \mathbf{x}_i'\mathbf{b} + \varepsilon_i, \; i=1, \ldots, n. \qquad (2)$$

The vectors $\mathbf{x}_i' = (x_i(0), x_i(11), \ldots, x_i(Ic_I))$ and $\mathbf{b}' =$

$(b(0),b(11),\ldots,b(Ic_I))$ are both Kx1, where $K=c_1+\ldots+c_I+1$. The least squares estimate $\hat{\mathbf{b}}$ is obtained by solving the normal equation

$$X'X\hat{\mathbf{b}} = X'\mathbf{y} , \tag{3}$$

where $X=(\mathbf{x}_1,\ldots,\mathbf{x}_n)'$ is a nxK design matrix and $\mathbf{y}=(y_1,\ldots,y_n)'$ is a nx1 observation vector. In general the rank of X'X is r=K-I<K. Thus the solution is expressed by $\hat{\mathbf{b}} = (X'X)^-X'\mathbf{y}$, where $(X'X)^-$ is a generalized inverse of X'X. As is well known, since $(X'X)^-$ is not determined uniquely the parameter vector **b** is not estimable. However, the matrix $H = X(X'X)^-X'$ is invariant for the choice of $(X'X)^-$. Hence the predicted values $\hat{\mathbf{y}}=H\mathbf{y}$ and the residuals $\mathbf{e}=(I-H)\mathbf{y}$ are obtained uniquely.

The contribution of each item is measured by the range of the scores assigned to the categories $R_j = \max_k\hat{b}_{jk} - \min_k\hat{b}_{jk}$, or the partial correlation coefficient between $\{y_i\}$ and $\{\Sigma_k x_i(jk)b(jk)\}$ or, when the normality of errors is assumed, the F statistic for testing the hypothesis H_0: $b(j1)=\ldots=b(jc_j)=0$.

B. Measures for outliers and influential observations

The linear model (2) is equivalent to the ordinary regression model except that the design matrix X is not of full rank. Therefore, various measures for regression diagnostics can be used in their original forms or slightly modified forms. The symbol (i) indicates that the quantities with it are calculated without the i-th individual.

The following 1)- 4) can be applied in just the same forms as in regression analysis.

1) Residuals

$\mathbf{e}=(I-H)\mathbf{y}$ Ordinary residuals (4)

$r_i=e_i/(s(1-h_{ii})^{1/2})$ Standardized residuals (5)

$$t_i = e_i/(s(i)(1-h_{ii})^{1/2}) \quad \text{Studentized residuals} \tag{6}$$

where h_{ii} is the (i,i)-th element of the matrix H and

$$s^2 = \mathbf{e}'\mathbf{e}/(n-r) \tag{7}$$

$$s^2(i) = [(n-r)s^2 - e_i^2/(1-h_{ii})]/(n-r-1). \tag{8}$$

2) DFFITS

$$\begin{aligned} \text{DFFITS}_i &= (\hat{y}_i - \hat{y}_i(i))/\hat{\text{var}}(\hat{y}_i(i)) \\ &= (h_{ii}/(1-h_{ii}))^{1/2} e_i/(s(i)(1-h_{ii})^{1/2}) \end{aligned} \tag{9}$$

3) FVARATIO

$$\text{FVARATIO}_i = \hat{\text{var}}(\hat{y}_i(i))/\hat{\text{var}}(\hat{y}_i) = s^2(i)/(s^2(1-h_{ii})) \tag{10}$$

4) Cook's statistic D

$$D_i = h_{ii} e_i^2/(rs^2(1-h_{ii})^2) \tag{11}$$

We consider the influence curve which indicates the influence on coefficient vector **b**. Let

$$\begin{aligned} &\Sigma(F) = E(\mathbf{xx}'), \qquad \phi(F) = E(\mathbf{x}y), \\ &\mu(jk|F) = E(x(jk)), \\ &\underset{I\times K}{\Gamma'(F)} = \begin{bmatrix} 0 & \mu(11)\ldots\mu(1c_1) & 0 \ldots & & \\ 0 & \ldots 0 & \mu(21)\ldots\mu(2c_2) & 0 \ldots & \\ & \ldots & & & \\ 0 & \ldots & & & 0 \ \mu(I1)\ldots\mu(Ic_I) \end{bmatrix} \end{aligned} \tag{12}$$

Then the vector $\hat{\mathbf{b}}$ which minimize $E(y-\mathbf{x}'\mathbf{b})^2$ under the condition $\Gamma'\mathbf{b}=0$ is obtained by solving the following linear simultaneous equations.

$$\begin{bmatrix} \Sigma(F) & \Gamma(F) \\ \Gamma'(F) & 0 \end{bmatrix} \begin{bmatrix} \hat{\mathbf{b}}(F) \\ \lambda(F) \end{bmatrix} = \begin{bmatrix} \boldsymbol{\phi}(F) \\ 0 \end{bmatrix} \tag{13}$$

Transforming F to $(1-\varepsilon)F+\varepsilon\delta_{\mathbf{z}}$ $(\mathbf{z}=(\mathbf{x}'y)')$ and calculating

$$IC_{\mathbf{b},F}(\mathbf{x},y) = \lim_{\varepsilon\to+0} \{\hat{\mathbf{b}}((1-\varepsilon)F+\varepsilon\delta_{\mathbf{z}}) - \hat{\mathbf{b}}(F)\}/\varepsilon \ , \tag{14}$$

we obtain

$$IC_{\mathbf{b},F}(x,y) = \Psi(F)\mathbf{x}(y-\hat{y}) - (\Psi(F)R\Theta'(F)+\Theta(F)R'\Psi(F))\boldsymbol{\phi}(F), \tag{15}$$

where

$$\begin{bmatrix} \Psi(F) & \Theta(F) \\ \Theta'(F) & 0 \end{bmatrix} = \begin{bmatrix} \Sigma(F) & \Gamma(F) \\ \Gamma'(F) & 0 \end{bmatrix}^{-1} , \tag{2.16}$$

and

$$R' = \begin{bmatrix} 0 & x(11)\dots x(1c_1) & 0 \dots & & & \\ 0 & \dots\dots\dots & 0 & x(21)\dots x(2c_2) & 0 \dots & \\ & \dots & & & & \\ 0 \dots & & & & \dots & x(I1)\dots x(Ic_I) \end{bmatrix}. \quad (17)$$

As sample versions we can define the three types EIC_i, $EIC_{(i)}$ and SIC_i as discussed in (1). Substituting $\Sigma(\hat{F})=(1/n)X'X$ for $\Sigma(F)$, $\boldsymbol{\phi}(\hat{F})=(1/n)X'\mathbf{y}$ for $\boldsymbol{\phi}(F)$ and

$$\Gamma(\hat{F}) = \frac{1}{n}\begin{bmatrix} 0 & f(11)\dots f(1c_1) & & & \\ 0 & \dots & 0 & f(21)\dots f(2c_2) & \\ & \dots & & & \\ 0 & \dots & & \dots\ 0 & f(I1)\dots f(Ic_I) \end{bmatrix} \quad (18)$$

for $\Gamma(F)$ (f(jk) denotes the frequency of the category k of the j-th factor item), and calculating $\Psi(\hat{F})$ and $\Theta(\hat{F})$ from (18) , we can express the empirical influence curve EIC_i as

$$EIC_i=\Psi(\hat{F})\mathbf{x}_i(y_i-\hat{y}_i)-(\Psi(\hat{F})R_i\Theta'(\hat{F})+\Theta(\hat{F})R_i'\Psi(\hat{F}))\boldsymbol{\phi}(\hat{F}). \quad (19)$$

The empirical influence curve with the i-th individual deleted, which is expressed by $EIC_{(i)}$, is obtained as follows by putting $F=\hat{F}_{(i)}$ and $(\mathbf{x}',y)=(\mathbf{x}_i',y_i)$.

$$EIC_{(i)}=\Psi(\hat{F}_{(i)})\mathbf{x}_i(y_i-\mathbf{x}_i'\Psi(\hat{F}_{(i)})\boldsymbol{\phi}(\hat{F}_{(i)}))$$
$$-(\Psi(\hat{F}_{(i)})R_i\Theta'(\hat{F}_{(i)})+\Theta(\hat{F}_{(i)})R_i'\Psi(\hat{F}_{(i)}))\boldsymbol{\phi}(\hat{F}_{(i)}) \quad (20)$$

where, if we neglect $O(n^{-2})$,

$$\boldsymbol{\phi}(\hat{F}_{(i)}) =\boldsymbol{\phi}(\hat{F})-(\mathbf{x}_i y_i-\boldsymbol{\phi}(\hat{F}))/n,$$
$$\Psi(\hat{F}_{(i)})=\Psi(\hat{F})+[\Psi(\hat{F})\mathbf{x}_i\mathbf{x}_i'\Psi(\hat{F})+\Psi(\hat{F})R_i\Theta'(\hat{F})+\Theta(\hat{F})R_i'\Psi(\hat{F})-\Psi(\hat{F})]/n,$$
$$\Theta'(\hat{F}_{(i)})=\Theta'(\hat{F})+[\Theta'(\hat{F})\mathbf{x}_i\mathbf{x}_i'\Psi(\hat{F})+\Theta'(\hat{F})R_i\Theta'(\hat{F})-\Theta'(\hat{F})]/n. \quad (21)$$

Finally the sample influence curve SIC_i is obtained by putting $F=\hat{F}$ and $\varepsilon=-1/(n-1)$.

$$SIC_i=(n-1)[\Psi(\hat{F})\boldsymbol{\phi}(\hat{F})-\Psi(\hat{F}_{(i)})\boldsymbol{\phi}(\hat{F}_{(i)})]. \quad (22)$$

Several statistics including Cook's D and DFFIT have been proposed to summarize the coefficient vector. Here we propose a Cook's D_i-like measure which indicates the coefficient changes of a specified item.

$$D_i(j) = (\hat{\mathbf{b}}_{j(i)}-\hat{\mathbf{b}}_j)'(X_j'X_j)(\hat{\mathbf{b}}_{j(i)}-\hat{\mathbf{b}}_j)/((K-I)s^2), \quad (23)$$

where j denotes the item number. If we wish to evaluate the influence on a specified category, we can use the following DFBETAS-like measure.

$$\text{DFBETAS}(jk;i) = (\hat{b}_{jk}-\hat{b}_{jk(i)})/(s_{(i)}\ (\Psi(\hat{F}))_{jk,jk}\ ), \quad (24)$$

where jk indicates the category k of the j-th factor item. The above summarized statistics are closely related to SIC_i. We may define the similar statistics based on EIC_i or $EIC_{(i)}$.

III. QUANTIFICATION II

A. Outline of Quantification II

Suppose we have response patterns $\{(z_i(1), \ldots, z_i(g); x_i(11), \ldots, x_i(1c_1), \ldots, x_i(I1), \ldots, x_i(Ic_I)), i=1,\ldots,n\}$ of n individuals to I+1 items, where each individual responds to one and only one category of the outside variable and also of the factor items. To represent such qualitative observations we introduce the dummy variable $z_i(l)$ which takes the values 1 or 0 depending on whether the individual i responds to the category l or not, in addition to $x_i(jk)$'s defined in (1) .

In Quantification II the matter of concern is to analyze the relationship between the qualitative outside variable and the factor items and discriminate the categories of the outside variable from the information of the factor items. For that purpose numerical scores $\mathbf{b}'=(b(11),\ldots,b(1c_1),\ldots, b(I1),\ldots,b(Ic_I))$ for the categories of I factor items are introduced so that the categories of the outside variable can be discriminated as well as possible by using these scores. If we also introduce the scores $\mathbf{t}'=(t(1),\ldots,t(g))$ for the outside variable, the problem is formulated as the canonical correlation analysis between $\{z_i(l)\}$ and $\{x_i(jk)\}$.

The optimal scores **b** and **t** can be obtained as the solution of the optimization problem to maximize $\mathbf{t}'Z'X\mathbf{b}$ subject to the constraints $\mathbf{1}'Z\mathbf{t}=\mathbf{1}'XE_j\mathbf{b}=0$ where the matrices Z and X are nxg and nxc ($c=\Sigma c_j$) matrices such as $Z=(z_i(l))$ and $X=(x_i(jk))$, respectively. The matrix E_j is defined as $E_j = \mathrm{diag}(0,\ldots,0,\ 1,\ldots,1,\ 0,\ldots,0)$ where 1's are located from the $c_1+\ldots+c_{j-1}+1$-th to the $c_1+\ldots+c_j$-th elements.

Some calculations by using Lagrange multipliers lead to the following two eigenvalue problems with common eigenvalues.

$$Z'X(X'X)^{-}X'Z\mathbf{t} - \eta^2 Z'Z\mathbf{t} = 0 \qquad (25)$$

$$X'Z(Z'Z)^{-}Z'X\mathbf{b} - \eta^2 X'X\mathbf{b} = 0. \qquad (26)$$

The optimal scores **b** and **t** are obtained as the eigenvector $\mathbf{b}_1$ and $\mathbf{t}_1$ associated with the largest eigenvalue η_1^2 except for $\eta^2=1$. If the one dimensional scores $\mathbf{b}_1$ and $\mathbf{t}_1$ are not enough for the discrimination we may use the eigenvectors associated with the eigenvalues $\eta_2^2>\eta_3^2>\ldots$. In actual data analysis we need not solve both of (25) and (26), because there exist the transition formulas as follows.

$$\mathbf{b} = (1/\eta)(X'X)^{-}X'Z\mathbf{t}, \quad \mathbf{t} = (1/\eta)(Z'Z)^{-1}Z'X\mathbf{b} \qquad (27)$$

B. Measures for outliers and influential observations

In order to detect influential observations and to evaluate the stability of the result of analysis we proposed a method of sensitivity analysis in Quantification II (6). There we applied the perturbation theory of eigenvalue problems with multiple eigenvalues as well as simple eigenvalues. The basic idea was to change slightly the weight w_i for the i-th individual from $w_i=1$ to $w_i=1-\varepsilon$ and to calculate the influence on the result by using the perturbation theory.

From the formulation of Quantification II it is obvious that Quantification II is equivalent to Quantification I, if the scores **t** for the outside variable are known . Thus with the fixed scores **t** we can use the measures for outliers and influential observations in Quantification I also in Quantification II. Since the first equation in (26) corresponds to the least square solution of Quantification I with the outside variable Z**t**, the above approach is closely related to the one step estimation technique in the alternating least squares procedure.

IV. NUMERICAL INVESTIGATION

As an illustration we analyzed the data of a collective medical examination of 52 persons who got sick with the four kinds of diseases within a year after the examination (3).

First we applied the ordinary Quantification II taking the kind of diseases as the outside variable. The correlation ratios (eigenvalues) which represent the ratios of the explained variations on the axes to the total variations are $\eta_1^2=0.319 > \eta_2^2=0.181 > \eta_3^2=0.072 > \eta_4^2=0$ in order of magnitude.

Then to detect outliers and influential observations we calculated several measures. Fig. 1 and Fig. 2 show the index plots of the measures of type A (measures based on the perturbation theory of eigenvalue problems) and type B (measures for Quantification I), respectively. Looking at these two figures it can be said that, although the movements of the measures of Type A and Type B are broadly similar, there are some difference between them. Actually from Fig.1 (Type A) we can find that the individuals No. 25 and 51 are

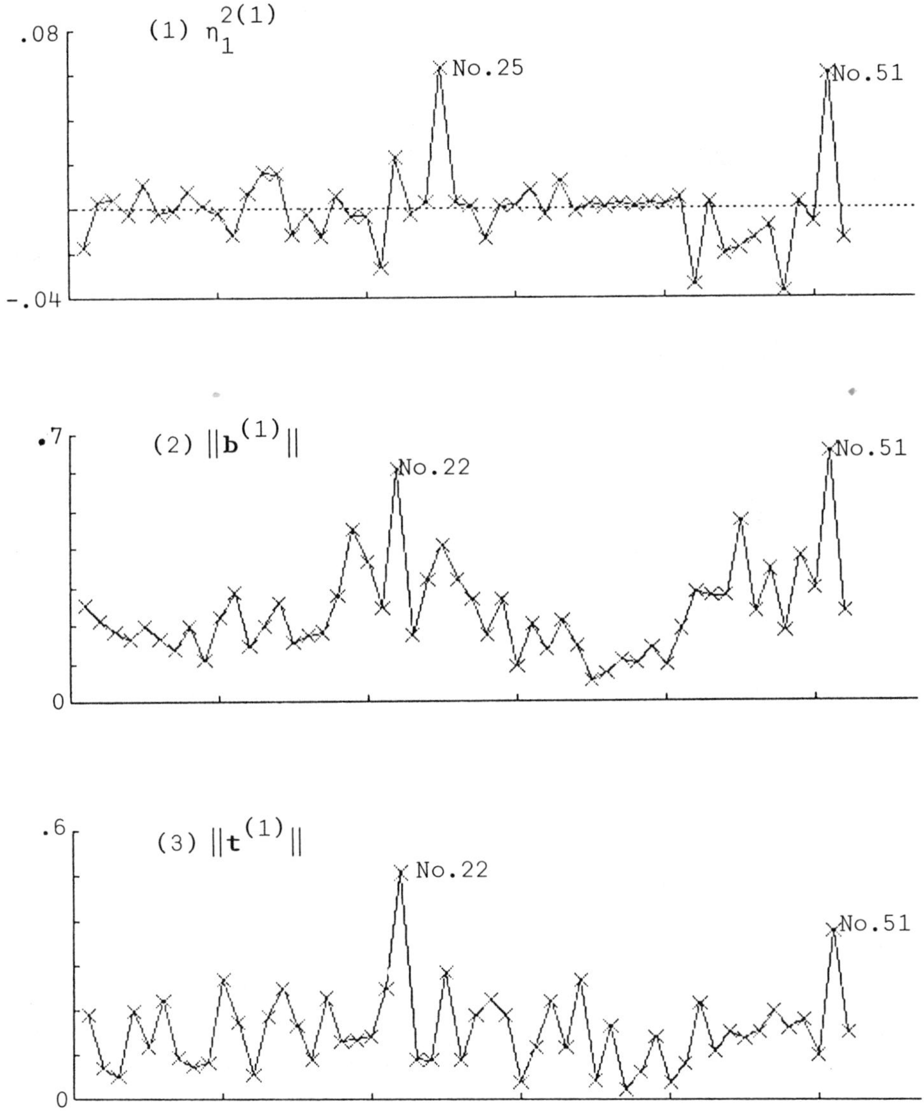

Fig.1. Type A measures for outliers and influential observations

(Note. The superscript (1) indicates the first differential coefficient.)

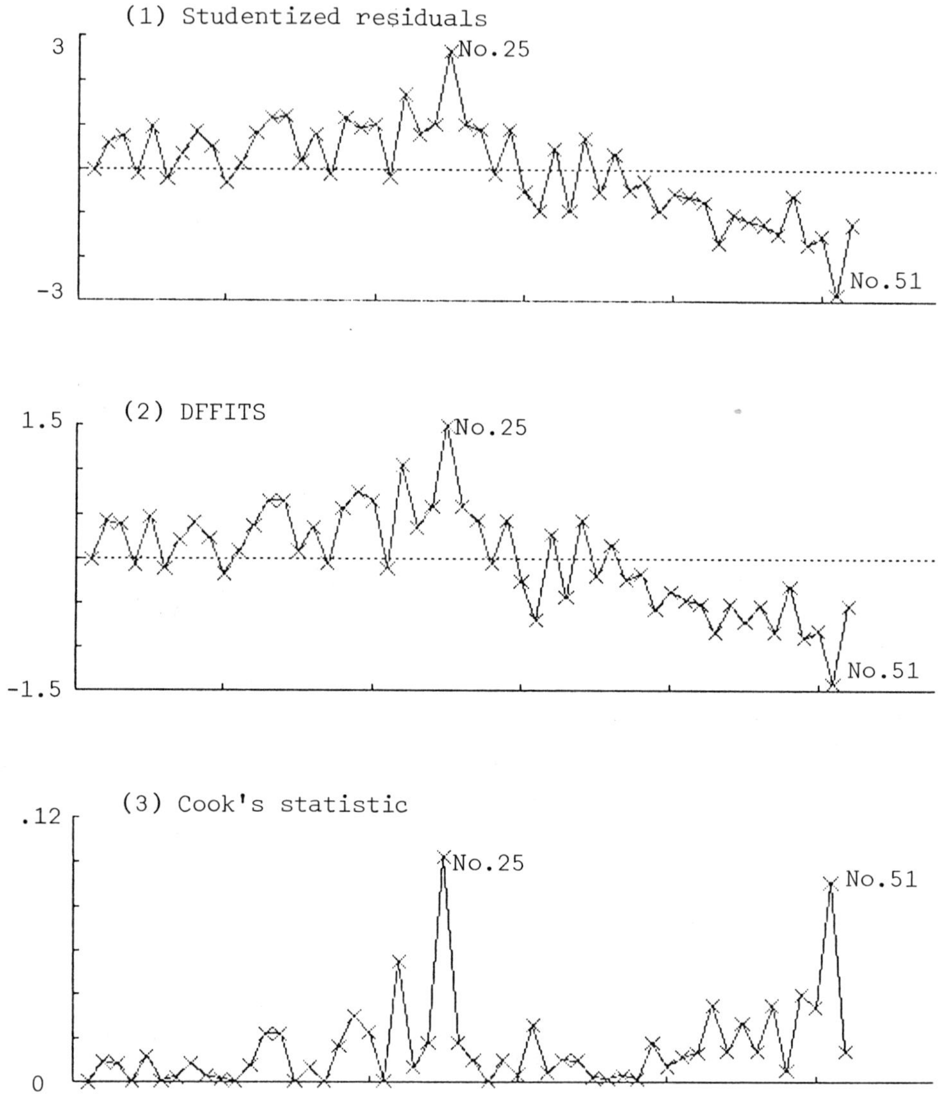

Fig.2. Type B measures for outliers and influential observations

influential to the eigenvalue η_1^2, while the individuals No. 22 and 51 are influential to the scores $\mathbf{b}_1$ and $\mathbf{t}_1$. However, looking at the three measures in Fig.2 (Type B) it is a little difficult to find individual No. 22 among the three which are found with Type A measures. If we apply Quantification II to the data with the 25-th and 51-th individuals deleted, we obtain the eigenvalues $\eta_1^2=0.541>\eta_2^2=0.190>\eta_3^2=0.061>\eta_4^2=0$, and furthermore, if we also delete No. 22, the eigenvalues become $\eta_1^2=0.580>\eta_2^2= 0.281>\eta_3^2=0.061>\eta_4^2=0$.

V. DISCUSSION

We introduced the simplified measures by fixing the scores for the outside variable. This technique can be considered as a one step estimation technique of the alternating least squares method based on the following equations.

$$\mathbf{b} = (1/\eta)(X'X)^{-}X'Z\mathbf{t} \tag{28}$$

$$\mathbf{t} = (1/\eta)(Z'Z)^{-1}Z'X\mathbf{b} \tag{29}$$

Combining these two equations we obtain

$$\mathbf{b} = (1/\eta^2)(X'X)^{-}X'Z(Z'Z)^{-1}Z'X\mathbf{b}. \tag{30}$$

This is the renewal formula of the power method to solve the eigenvalue problem. In the sense that a half of the renewal formula (30) is performed, our technique may be called a half step estimation technique of the power method, and therefore, the accuracy will be less than in the case of the one step estimation technique of the power method. However, as the formula (28), the half of (30), just corresponds to the least square solution in Quantification I and various measures have already been introduced in Quantification I, this technique has some merits besides the accuracy.

We shall discuss the properties of this technique more

precisely. From (28)

$$\mathbf{b}+\varepsilon\mathbf{b}^{(1)}=\{((X'X)^{-}X'Z)^{(1)}\mathbf{t}+(X'X)^{-}X'Z\mathbf{t}^{(1)}\}/(\eta+\varepsilon\eta^{(1)}),$$

$$\mathbf{b}^{(1)}\propto((X'X)^{-}X'Z)^{(1)}\mathbf{t}+(X'X)^{-}X'Z\mathbf{t}^{(1)}\ , \tag{31}$$

where the superscript (1) denotes the first order differential coefficient at $\varepsilon=0$. In our simplified measures based on the fixed scores for the outside variable the second term of the right hand side is neglected. Substituting $\mathbf{t}=(Z'Z)^{-1/2}\mathbf{v}$ into the eigenvalue problem (27), we obtain

$$A\mathbf{v}-\eta^2\mathbf{v}=0,\quad \mathbf{v}'\mathbf{v}=1\ , \tag{32}$$

where

$$A=(Z'Z)^{-1/2}Z'X(X'X)^{-}X'Z(Z'Z)^{-1/2}. \tag{33}$$

If it can be assumed that the eigenvalues of interest are distinct, then by the similar technique as used in (4), we can derive the following.

$$\mathbf{v}_j^{(1)}=-(A+\mathbf{v}_j\mathbf{v}_j'-\eta_j^2I)^{-1}(A^{(1)}-\eta_j^{2(1)}I)\mathbf{v}_j \tag{34}$$

Application of the spectral decomposition of A and some calculations finally lead to

$$\begin{aligned}\mathbf{t}_j^{(1)}=&((Z'Z)^{-1/2})^{(1)}(Z'Z)^{1/2}\mathbf{t}_j\\&-(Z'Z)^{-1/2}\{\Sigma_{k\neq j}(\eta_k^2-\eta_j^2)^{-1}\mathbf{v}_k\mathbf{v}_k'\\&+\mathbf{v}_j\mathbf{v}_j'\}(A^{(1)}-\mathbf{v}_j'A^{(1)}\mathbf{v}_jI)(Z'Z)^{1/2}\mathbf{t}_j.\end{aligned} \tag{35}$$

Thus we can evaluate the neglected part by substituting (35) into (31). Obviously the first term of the right hand side of (35) is generally small, because Z'Z is a diagonal matrix whose diagonal elements are equal to the frequencies of the categories of the outside variable and these frequencies are usually large. The second term may be large or small depending on the values $|\eta_k^2-\eta_j^2|$. It may become very large when there are some eigenvalues η_k^2's which are close to η_j^2. This fact suggests that our simplified measures will work well when the eigenvalues of interest are isolated from each other and from the remaining eigenvalues. Therefore, it may be safe to pick up a little larger number of the candidates for outlying or influential observations with Type A measures and investigate their influences accurately with Type B measures afterwards.

REFERENCES

1. Cook, R.D. and Weisberg, S.(1982). Residuals and Influence in Regression, Chapman and Hall.
2. Hayashi, C.(1952). On the prdiction of phenomena from qualitative data and the quantification of qualitative data from the mathematico-statistical point of view, Ann. Inst. Statist. Math. 3, 69.
3. Komazawa, T.(1982). Quantification Theory and Data Processing, Asakura Publishing Co. (In Japanese)
4. Radhakrishnan, R. and Kshirsagar, A.M.(1981). Influence functions for certain parameters in multivatiate analysis, Commun. Statist.-Theor. Meth., A10, 515.
5. Tanaka, Y.(1984). Sensitivity analysis in Hayashi's third method of quantification, Behaviormetrika, No.16, 31.
6. Tanaka, Y. and Tarumi, T.(1986). Sensitivity analysis in Hayashi's second method of quantification, J. Japan Statist. Soc. 16, 37.

CONVERSATIONAL DATA ANALYSIS SYSTEM
- Selection of Items and Pooling of Categories -

Toshiro Haga

Faculty of Engineering, Science University of Tokyo
1-3, Kagurazaka, Shinjuku-ku, Tokyo 162, JAPAN

I. INTRODUCTION

A statistical analysis program, "Conversational data Analysis (CDA)," has been designed to maximize interactive and conversational functions in every stage of analysis.

CDA has been designed for nonstatisticians, and consists of simple methods, none of which is sohpisticated. Users can, however, analyze data with free combination of these simple methods. CDA analyzes quantitative variables and categorical variables, and contains preliminary analysis of multivariate data and three methods of multivariate analysis: regression analysis, discriminant analysis, and principal component analysis.

The first method of the quantification theory by Professor C. Hayashi is a type of regression analysis, after conversion of the categorical variable to a set of dummy variables. A categorical variable is called an "item" in the quantification theory. The second method of the quantification theory corresponds to discriminant analysis. Many computer programs have been developed for the first and the second methods of the quantification theory. Only categorical variables are used as regressor variables in the aforementioned two methods. However, in CDA, categorical variables can be used together quantitative variables.

In an ordinary regression analysis, one regressor

ISBN 0-12-215485-1

variable has one parameter (regression coefficient), but one item has more than one parameter. This means that selection of items is more important than that of quantitative variables, and also that the number of categories is important. Most of previous programs have no function for item selection. After conversion of the categorical variable to a set of dummy variables, ordinary regression analysis can be applied. However, ordinary variable selection is not suitable for dummy variables. A set of dummy variables derived from an item should enter or leave the regression equation at one time. CDA can select items as quantitative variables.

CDA was originally developed in FORTRAN for mainframe computers. Afterwards it was converted to BASIC for microcomputers, and was enhanced with graphical capabilities of result representation.

II. PRELIMINARY ANALYSIS OF MULTIVARIATE DATA

A simple example of the regression of Y on one quantitative variable X and one item C with 5 categories is used to explain the functions of CDA. The number of observations is 30.

Since it is a system for nonprofessional users, CDA must be simple to use. Users select a function from the menu shown on the CRT display, and input a mnemonic code for the function. Some functions require parameter(s). Simple explanations of functions and required parameters are provided by entering "?".

Fundamental statistics such as average, standard deviation, skewness, kurtosis and correlation matrix are given by entering "F".

```
) List, Fund.-stat, Graph, Monit, Reg, Disc, PCA or End
F
Variable    mean minimum maximum std.dev.   C.V.    skew    kurt
  1   X   9.614   8.030  11.580   0.923   0.096   0.031  -0.216
  3   Y   9.334   5.930  12.680   1.693   0.181   0.057  -0.408
Correlation Matrix
          1   X   3   Y
  1   X  1.0000
  3   Y  0.6149  1.0000
```

To see the features of the variables, graphical representations are more convenient. One of the typical outputs of CDA is given by specifying "G". Distributions of X, C and Y, scatter diagrams, and stratified histograms are shown in compact forms.

```
) List, Fund.-stat, Graph, Monit, Reg, Disc, PCA or End
G
```

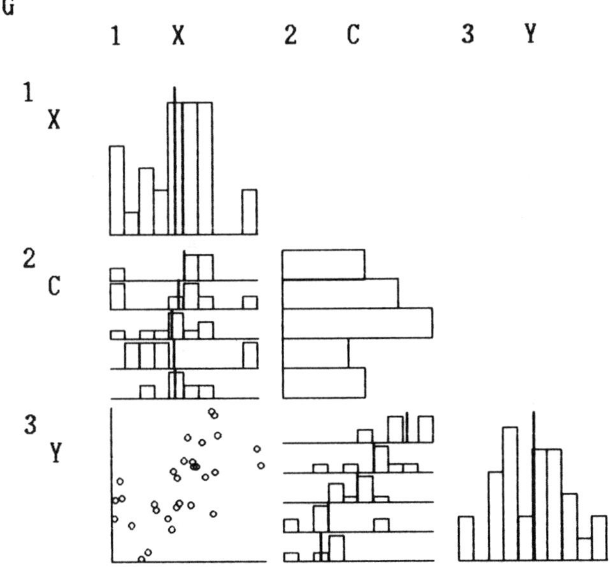

Details of distribution of variables and relation between two or three variables can be obtained by the monitoring subsystem. It contains:

1) one quantitative variable: histogram, normal Q-Q plot, time series plot,
2) two quantitative variables: scatter diagram,
3) one categorical variable and one quantitative variable: stratified histogram, with means and standard deviations of each category,
4) two categorical variables: cross table, Chi-square test of contingency table and a graph,
5) two quantitative variables and one categorical variable: stratified scatter diagram.

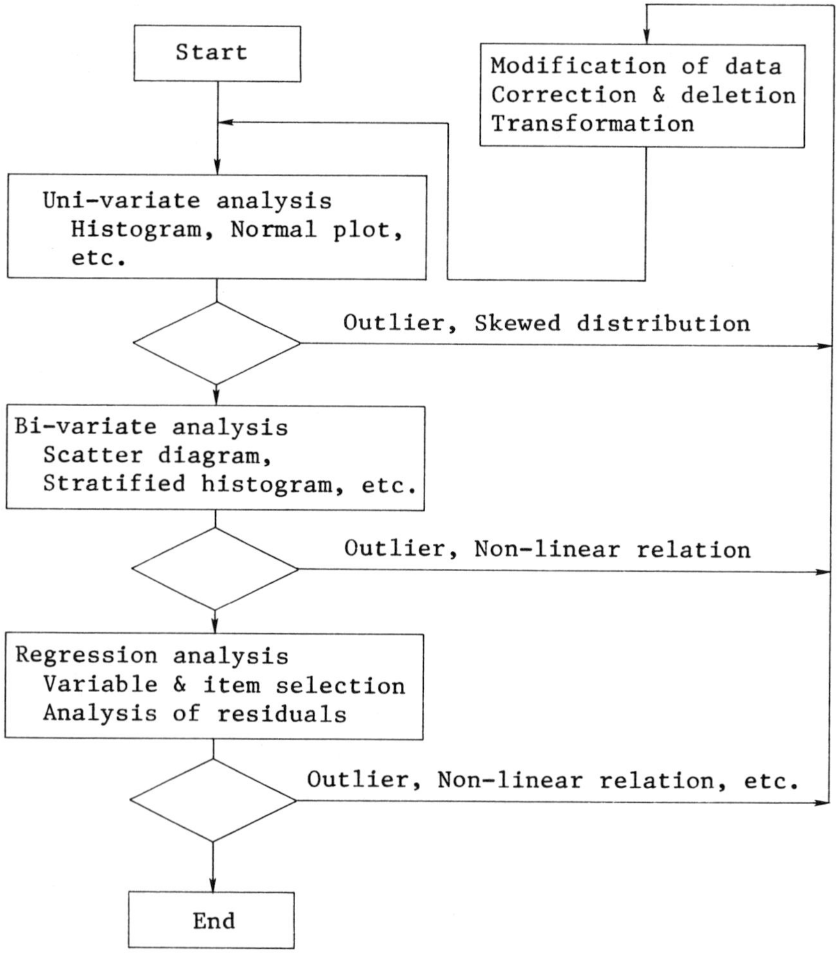

If there are some outliers, the distribution is skewed or the relation is nonlinear, preliminary analysis should be repeated after deletion of abnormal observations or transformation of variables. In this trial-and-error process, the conversational analysis program is very powerful. (cf. Fig. 1)

III. CONVERSATIONAL SELECTION OF VARIABLES AND ITEMS

In regression analysis, a categorical variable is converted to dummy variables. The number of dummy variables is obtained by subtracting 1 from the number of categories. For item C, four dummy variables are generated.

By specifying "R" for regression analysis, sum of squares (S.S.) of y, degree of freedom (d.f.) and standard deviation are given under St. df(t) and sig(t).

```
) List, Fund.-stat, Graph, Monit, Reg, Disc, PCA or End
R
              St                              df(t)  sig(t)
          83.118                                 29   1.693

Variable       Se     D(S)        F         b
  0 CON  2696.6382613.520 911.858    9.334
  1  X     51.694 -31.424  17.021
  2  C     33.992 -49.126   9.033
```

In regression analysis, selection of variables from candidates for the regressor is very important. Almost all of the programs for regression analysis have a function of variable selection, such as stepwise selection by Fin and Fout. In such programs, the user specifies values of Fin and Fout; after that, variables are selected by the computer. In CDA, the user selects a variable by himself concerning information about candidates given by CDA and knowledge of the object, such as science, engineering or economics. CDA gives the informations for selection of variables and items as shown above. Se is the sum of squares of the error after including a variable or dummy variables generated from an item. D(S) is the difference of Se before and after including a variable or an item. F-ratio of X and C are calculated in the following manner.

```
X:    F =   31.424    / (51.694/28) = 17.021,
C:    F = (49.126/4) / (33.992/25) =  9.033.
```

The user selects a variable or an item to be included in regression equation, considering Se, F-ratio and specific knowledge on the object. The variable X is selected based on its F value.

```
R) Sel., B-&-Cat.Sc., Res., Gra., Time, Nplot, Pred., List or End
S 1
 enter          Se      R^2     R*^2    R**^2 df(e)  sig(e)
  1  X      51.694    0.378    0.356    0.335    28   1.359
Variable        Se     D(S)        F        b
  0 CON     52.299    0.605    0.328   -1.512
  1  X      83.118   31.424   17.021    1.128
  2  C       8.601  -43.093   30.060
```

Constant and regression coefficients are shown in the column of b, giving the regression equation,

$$Y = -1.512 + 1.128X,$$

and its S.S. of error, three coefficients of determination (ordinarily, adjusted by d.f. and doubly adjusted by d.f.), d.f. of error and standard deviation of error are given.

Adding item C to the above equation, S.S. of error will be 8.601 decreasing by 43.093, which gives F-ratio as 30.060.

By adding item C, four dummy variables are included in regression equation at a time.

```
R) Sel., B-&-Cat.Sc., Res., Gra., Time, Nplot, Pred., List or End
S 2
 enter          Se      R^2     R*^2    R**^2 df(e)  sig(e)
  2   C      8.601    0.897    0.875   0.855    24    0.599
Variable        Se     D(S)        F       b
  0 CON      9.124    0.523    1.458   1.472
  1   X     33.992   25.391   70.847   1.021
  2   C     51.694   43.093   30.060  -1.358  -1.947  -3.279  -3.674
```

Four regression coefficients of dummy variables, shown in column b and row C, correspond to the category scores of the 2nd, 3rd, ... categories. The category score of the 1st category is defined as 0. The regression equation is,

$$
\begin{aligned}
Y = 1.472 + 1.021x &+ 0.000\ (C = 1) \\
&- 1.358\ (C = 2) \\
&- 1.947\ (C = 3) \\
&- 3.279\ (C = 4) \\
&- 3.674\ (C = 5)
\end{aligned}
$$

To test significance of the differences between category scores, command "B" is used.

```
R) Sel., B-&-Cat.Sc., Res., Gra., Time, Nplot, Pred., List or End
B
Variable          b     sig(b)          t     std(b)     tol.
  0 CON       1.472      1.219      1.208
  1   X       1.021      0.121      8.417      0.556  0.9871
Var. Cat.      diff. of score \ t-value
  2   C         1        2        3        4        5
          1     \     -3.871   -5.797   -8.140   -9.679
          2  -1.358      \     -1.947   -5.114   -6.602
          3  -1.947   -0.588      \     -3.702   -5.169
          4  -3.279   -1.920   -1.332      \     -0.983
          5  -3.674   -2.315   -1.727   -0.395      \
```

The numbers located in the lower side and the upper side of the diagonal show the difference of category scores and their Student's t values respectively. Difference of category scores between the 4th and the 5th categories is -0.395, and its t values, -0.983, is less than t(24;0.05), that is, the difference between 4th and 5th categories is not significant.

IV. REGRESSION DIAGNOSIS AND ANALYSIS OF RESIDUAL

In the regression models, it is necessary to check the following three assumptions on error:

1) randomness and independency
2) homogeneous variance
3) normal distribution.

For that purpose, CDA provides various kinds of graphs of residuals, such as:

1) scatter diagram of estimated y
2) scatter diagram of qualitative regressor
3) stratified histogram of categorical regressor
4) time series plot
5) normal Q-Q plot.

```
R) Sel., B-&-Cat.Sc., Res., Gra., Time, Nplot, Pred., List or End
R
  I  y(obs)  y(est)  resid.  t-val.       h  pred.err
  2  10.930  11.820  -0.890  -1.831   0.276  -1.229
 23   8.480   9.323  -0.843  -1.535   0.111  -0.949
 27   6.260   7.226  -0.966  -2.003   0.270  -1.323
 29   8.700   7.904   0.796   1.507   0.180   0.971
  PRESS =    13.823
```

V. POOLING OF CATEGORIES

New categorically variable "C-P" is derived by pooling the 4th and 5th categories of item C, and is added to the data file. A powerful data handling program, "CDADAT.BAS," prepared to make a new data file and to modify a existing data file, is used in this transformation.

The results of regression analysis on new data file are as follows:

```
) List, Fund.-stat, Graph, Monit, Reg, Disc, PCA or End
R
                St                              df(t)  sig(t)
            83.118                                 29   1.693

Variable        Se     D(S)         F         b
  0 CON  2696.6382613.520 911.858    9.334
  1  X     51.694 -31.424  17.021
  2  C     33.992 -49.126   9.033
  3 C-P    34.270 -48.849  12.354
R) Sel., B-&-Cat.Sc., Res., Gra., Time, Nplot, Pred., List or End
S 1 3
 enter          Se      R^2      R*^2     R**^2 df(e)  sig(e)
  1  X     51.694   0.378   0.356    0.335    28    1.359
 enter          Se      R^2      R*^2     R**^2 df(e)  sig(e)
  3 C-P     8.948   0.892   0.875    0.859    25    0.598
Variable        Se     D(S)         F         b
  0 CON     9.481   0.533   1.489    1.486
  1  X     34.270  25.322  70.750    1.019
  2  C     *** TOL CHECK ***
  3 C-P    51.694  42.746  39.811   -1.359  -1.947  -3.498
R) Sel., B-&-Cat.Sc., Res., Gra., Time, Nplot, Pred., List or End
B
Variable         b     sig(b)          t     std(b)     tol.
  0 CON      1.486      1.218      1.220
  1  X       1.019      0.121      8.411      0.556  0.9872
Var. Cat.      diff. of score \ t-value
  3 C-P          1         2         3        4
           1     \      -3.874   -5.803  -10.446
           2  -1.359     \       -1.949   -7.092
           3  -1.947  -0.589      \       -5.498
           4  -3.498  -2.140   -1.551     \
R) Sel., B-&-Cat.Sc., Res., Gra., Time, Nplot, Pred., List or End
R
   I  y(obs)  y(est)  resid.  t-val.        h pred.err
   2  10.930  11.817  -0.887  -1.822    0.276  -1.226
  19   5.930   6.844  -0.914  -1.713    0.143  -1.067
  23   8.480   9.323  -0.843  -1.535    0.111  -0.949
  24   7.490   6.589   0.901   1.708    0.163   1.077
  29   8.700   7.906   0.794   1.502    0.180   0.969
  PRESS =      13.087
```

Comparison of the two regression equations is shown below.

	R^2	R*^2	R**^2	df(e)	sig(e)	PRESS
before pooling	0.897	0.875	0.855	24	0.599	13.823
after pooling	0.892	0.875	0.859	25	0.598	13.087

By pooling categories, the coefficient of determination after doubly adjusted by degrees of freedom (R**^2) increases, although the standard deviation of error (sig(e)) and PRESS decrease.

Adequacy of the pooling of categories can be checked with histogram of residuals stratified with the item C.

```
R) Sel., B-&-Cat.Sc., Res., Gra., Time, Nplot, Pred., List or End
G 2
 horiz. =  2    C  vert. = residual
    1.496 :   :    :     :    :
    1.197 :   :    :     :    :
    0.897 :   :    :*    :*   :
    0.598 :*  :*   :*    :    :*
    0.299 :*  :**  :**   :**- :*
    0.000 :*- :**- :---- :    :
   -0.299 :*  :    :***  :    :*-
   -0.598 :*  :*   :*    :*   :*
   -0.897 :   :*   :*    :    :*
   -1.197 :   :    :     :    :
   -1.496 :   :    :     :    :
                1       2       3       4       5
average  -0.000  -0.000   0.000   0.219  -0.175
sigma     0.479   0.559   0.591   0.695   0.626
```

VI. CONVERSATIONAL PROCESSING AND BATCH PROCESSING

In conversational analysis, results are shown on the CRT screen, and commands received from the keyboard are recorded in the command file. The command file can be modified by the common editor. In batch processing, CDA receives commands from the file instead of the keyboard, and sends results to the screen, the printer or the output file depending on the user's request. By combination of conversational and batch processing, we can save printing time in conversational analysis. The file of results can be used as an input to the word processor for report writing. (cf. Fig. 2)

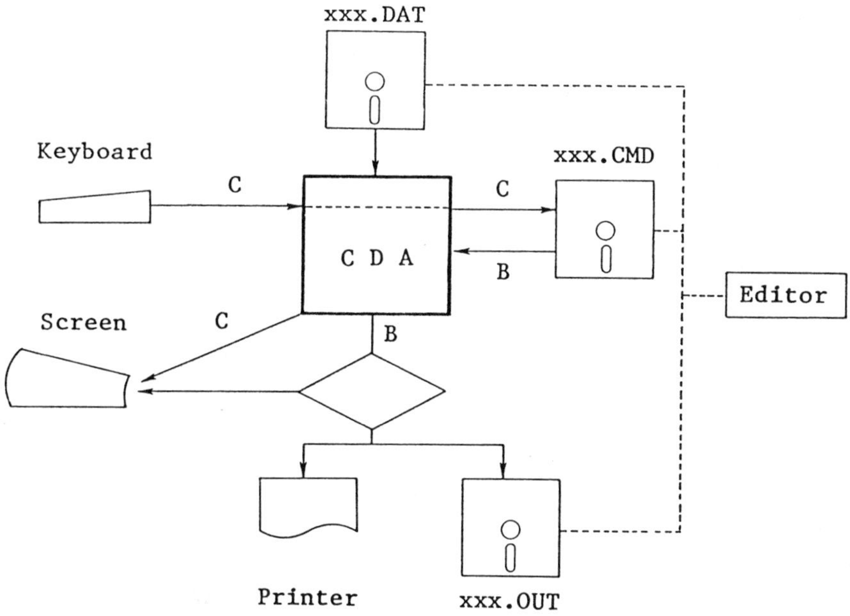

ANALYSIS AND COMPARISON OF DIFFERENT TABLES

Brigitte Escofier

INSA, 20 avenue des Buttes de Coësmes 35043 RENNES CEDEX
IRISA, Capus de Beaulieu 35042 RENNES CEDEX
FRANCE

I. INTRODUCTION

More and more frequently the advice of statisticians is taken for problems concerning the simultaneous study of different data tables. One wishes to analyse those tables together and compare them. The comparison of different tables may appear in quite various ways, depending on one's aims, and on the type of tables.

For tables whose rows represent the same set of individuals and the columns different sets of (numerical or qualitative) variables, we propose Multiple Factor Analysis. We first point out the problems and the aims of this study. M.F.A. provides coherent answers to all of these problems. The solutions obtained by M.F.A. are compared to classical results. (Principal Component Analysis, Multiple Correspondence Analysis, Generalised Canonical Analysis).

For frequency tables, we propose another method : Correspondence Analysis using a division into subtables.

In both methods, we analyse the table resulting from placing side by side the different tables. The algorithm is based on splitting into subtables. For very large data tables, we can

ISBN 0-12-215485-1

use an approximation, which allows the study of tables whose processing by classical programs is impossible or too costly.

II. MULTIPLE FACTOR ANALYSIS

A. Notations

1. The data tables

We first study the case of numerical variables. Let I denote the set of individuals (rows of the table) and K denote the set of variables (the columns). The set K is split into J groups denoted by K_j. The data table denoted by X is thus the union of J subtables each denoted by X_j.

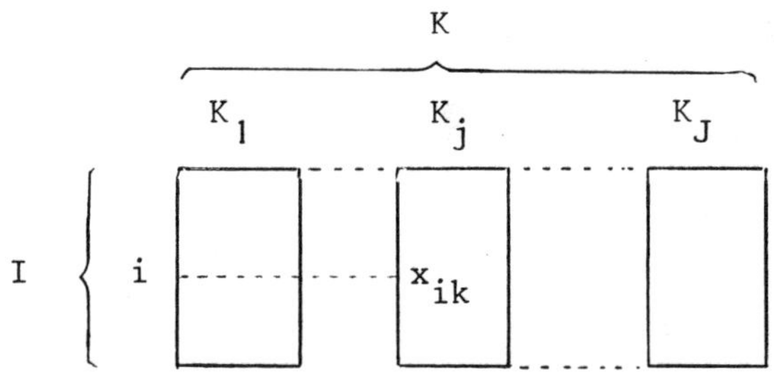

Figure 1 : The table X and the J subtables X_j

The notations, I, J, K or K_j represent either the set or the number of its elements. Let x_{ik} be the coefficient of the ith-row-k^{th}-column and x_k be the variable (i.e. the column) k.

The variables are centered and reduced (if wanted).
A weight denoted by p_i is assigned to each individual i ($\sum_i p_i = 1$) ; a weight m_k is assigned to each variable k (mk = 1 generally).

2. The cloud of variables N(K)

This cloud lies in a I-dimensional space R^I. The scalar product of two reduced variables x_k and x_ℓ is the correlation coefficient between these variables :

$$< x_k, x_\ell > = \sum_i x_{ik} \, x_i \, P_i$$

The N(K) cloud is divided into J subclouds $N(K_j)$, each one corresponding to the jth group.

3. The cloud of individuals

The cloud associated with X, lies in a K-dimensional space. In this space, R^K, the metric is based on the weight of the variable. The distance between two individuals i and i' is given by :

$$d^2(i,i) = \sum_{j \in J} \sum_{k \in K_j} (x_{ij} - x_{i'j})^2 \, m_k \; .$$

For each subtable X_j , there is an associated cloud of individuals $N_j(I)$ which lies in R^{K_j}. The R^K space is the direct sum of J orthogonal subspaces which are isomorphic each to R^{K_j} and the projection of N(I) into R^{K_j} is exactly $N_j(I)$.

This remark is of importance : the J clouds $N_j(I)$ lie in the same space can thus be compared with one another.

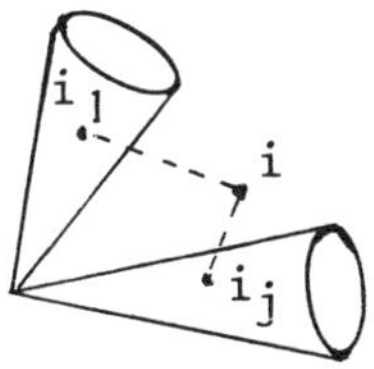

Figure 2 : $R^K = \oplus R^{K_j}$. The projection $N_j(I)$ of N(I) into R^{K_j}

B. Principal Component Analysis of the Weighted Groups

1. The problem

One can apply a classical P.C.A. Nevertheless in this case X is made up of groups of variables and one then can wish to balance the influence of the different groups : a given group (as a given variable) must not prevail. In the case of groups of variables the problem is more complex (than in the case of variables) : the influence of a group depends on both the number of variables and their correlation. If group j is made up of many highly correlated variables, the inertia of $N(K_j)$ will be large in the direction of these variables and thus might cause the first principal component of N(K).

2. The solution

To balance the groups of variables, we divide the initial weight of each group j-variable by the first eigenvalue of group j P.C.A. With these new weights the inertia of the first principal component of X_j equals 1 ; the inertia of the projections of $N(K_j)$ (and $N_j(I)$) on any other axis is less than 1. This method leads to a balanced influence of each group on the 1st principal component. If one group has different directions of large inertia, it can influence different components.

If a group is made up of two orthogonal variables u and v, the weight assigned to each variable will be 1. If another group is made of u and u, their weight will be 1/2. This group will thus be equivalent to a one-variable group, with weight equal to 1.

In what follows, this redefinition of weights is assumed. This is an essential ingredient of Multiple Factor Analysis. This method provides an overall solution to the problems of the analyse of groups of variables.

C. Global comparison of groups

In this paragraph we study the structure of the set of groups by defining a distance over it. This is really of interest if one studies the case of a large number of groups and specially to bring out a possible evolution process in the case of time-dependent groups.

To compare the groups, we compare the clouds $N_j(I)$: two groups j and j' are close if the clouds $N_j(I)$ and $N_{j'}(I)$ look alike, that is to say if the distance between corresponding points of each are similar.

Note that the redefinition of weight has "normalized" the clouds $N_j(I)$, (it has modified the metric over the spaces R^{K_j}). Homothetic clouds are thus made identical. In the example of paragraph 1.2.2. (two groups $K_1 = (u,v)$, $K_2 = (u,u), \langle u,v\rangle = 0)$), the clouds $N_1(I)$ and $N_2(I)$ are superposable in the common dimension u (since u is a maximum inertia dimension in both clouds).

It is well-known that the knowledge of the distances between all pairs of points is enough to build the whole cloud. Each group of variables can thus be characterized by the distances between all pairs of points of $N_j(I)$. For the sake of simplicity the IxI matrix of scalar products $\langle O_i, O_{i'}\rangle$ is used rather than the distances. It is easily checked the matrix defined by group j reads :

$$W_j = X_j\, M_j\, X_j' = \sum_{k \in K_j} m_k\, x_k\, x_k'$$

where X_j' and x_k' denote the transposed of X_j and x_k, and M_j the diagonal matrix of the weights of the variables of group j.

1. The R^{I^2} space

The matrix W_j, lying in a I^2 dimensional space R^{I^2} represents the group j. The distance between two matrices W_j and $W_{j'}$ is defined by :

$$d^2(W_j, W_{j'}) = \sum_{i,i'} p_i\, p_{i'} \left(W_j(i,i') - W_{j'}(i,i')\right)^2$$

(the influence of (i,i') is assigned the weight $p_i\, p_{i'}$).

A group may consist of only one variable u. The R^{I^2}-vector associated with u is then the matrix uu'. The scalar product in the R^{I^2}-space of uu' and vv' is the square of the scalar product of u and v in the R^I-space :

$$\langle uu',vv' \rangle = \sum_{i,i'} p_i p_{i'}\, u(i)\, u(i')\, v(i)\, v(i')$$
$$= (\langle u,v \rangle)^2$$

If u is a unit vector, so is uu'.

The scalar product of W_j and a unit vector uu' is the inertia of the projection of $N(K_j)$ on u (in the R^I-space) :

$$\langle W_j,uu' \rangle = \sum_{i,i'} p_i p_{i'}\, u(i)\, u(i') \sum_{k \in K_j} m_k x_k(i)\, x_k(i')$$
$$= \sum_{k \in K_j} m_k (\langle u,x_k \rangle)^2$$

In M.F.A. we shall use $\langle W_j,uu' \rangle$ as a multiple weighted correlation coefficient between w_j and u.
(It is less than 1 or equals 1, cf. §. B-2)

When K_j consists of orthogonal variables, the inertia of $N(K_j)$ is the same for all directions of the subspace spanned by $N(K_j)$, and $\langle W_j,uu' \rangle$ equals the square of the multiple correlation coefficient between W_j and u. Let P(u) be the projection of u on the subspace spanned by $N(K_j)$ and θ be the angle between u and P(u) ; the multiple correlation coefficient equals $\cos^2\theta$. The inertia of $N(K_j)$ on u = Inertia of $N(K_j)$ on $P(u) \times \cos^2\theta$.

2. Projection of the clouds of groups

We project the cloud of W_j's on axis of R^{I^2}, orthogonals to one another. We could project on the maximum inertia axis (as usually in P.C.A.), but we could not then relate the axis to the initial data (variables or individuals) ; of what use is the knowledge of proximity of two groups, if this proximity cannot be interpreted ? On the other hand the relationship between two groups of variables is complex and multidimensional (if each group consists of different variables). Only one coefficient is much too poor : in the example of §. B-2). ($K_1 = (u,v)$, $K_2 = (u,u)$, $\langle u,v \rangle = 0$), the relationship between K_1 and K_2 is composed of 2 terms. The one showing that they have a common dimension, the other that there is a specific direction of K_1.

We project the W_j on the uu' axes (i.e. defined by one-variable groups) : the interpretation of these axes and the projections of W_j are easy : each axis is related to a one-dimensional correlation. We maximize the sum of coordinates of groups (unlike the usual criterion, the sum of their squares) : the group coordinates are positive, this is a reliable criterion ; the computation is much simpler ; last but not least the graphic representation of groups fits with those of individuals and of variables.

The sum of the coordinates of W_j on uu' equals the inertia of N(K). The maximizing vector of this projected inertia criterion is F_1, the 1st principal component of X : the 1st axis, in R^{I^2}, is F_1F_1'. Orthogonality of two vectors u and v in R^I is equivalent to orthogonality of uu' and vv' in R^{I^2} : the sequence of maximizing orthogonal axes is defined by the sequence of principal components of X. The graphical representation of the groups thus fits, for each axis, with the graphical representation of individuals and of variables.

In the example of §. B-2 ($K_1 = (u,v)$, $K_2 = (u,u)$, $\langle u,v \rangle = 0$) there are two principal components from M.F.A. the one u with inertia 2, the other v with inertia 1. On the graphical representation of groups, both W_1 and W_2 have a unit coordinate on uu' : u is a maximum inertia direction for both groups. On vv', W_1 has a unit coordinate while W_2 has a zero coordinate : v is a specific direction of W_1.

3. Relationship between the three graphical representations : individuals, variables, groups

The coordinate of W_j on $F_s'F_s$ (where F_s is the sth principal component of X) is the sum of the inertias of the projection of $N(K_j)$ on F_s : the group's contribution to the inertia of the component. It points out the groups which determine a given component.

On the contrary, the coordinate of W_j on an axis is the w.m.c.c. (weighted multiple correlation coefficient, cf. §. II.C-2) between W_j and variable F_s : the graphs of individuals and of variables characterize F_s. This coordinate depends on the angle between F_s and a direction of $N(K_j)$ and on the inertia of $N(K_j)$ in that direction. In the following section we introduce the notion of "common factor" with which we can distinguish between inertia and angle.

D. Common Factors

With reference to INDSCAL, we call "common factor" the directions of clouds $N_j(I)$ for which the projection of the J $N_j(I)$ are (almost) homothetical.

In practice it is often much more interesting to look for such factors than to compare the groups globally ; it should be noticed that M.F.A. offers a relationship between those two aspects.

To determine these common factors (if any) we first compute "mean factors" ; then we compute for each mean factor F_s and each j a projection F_s^j of $N_j(I)$ as correlated as possible with F_s. If the correlation coefficient between F_s and the F_s^j is close to 1, then F_s is a common factor to the K_j.

1. Mean Factors

In M.F.A. the mean factors F_s are orthogonal variables which maximize the sum of the J w.m.c.c. between F_s and each group. It appears clearly that the mean factors are the principal components of X. The principle is Carroll's Generalized Canonical Analysis except that Carroll maximizes the sum of the squares of multiple correlation coefficients. In G.C.A. all directions of $N(K_j)$ share the same importance and thus we can get directions of least inertia. This is not what happens with M.F.A. where inertia takes part in the calculation of the factors.

2. Projections of $N_j(I)$ associated with the mean factors

In the R^K-space, a component F_s is the projection of N(I) on the a_s-axis. The clouds $N_j(I)$ lie in $\mathbb{R}^k$ (cf. §. A-3) ; F_s^j is the projection of $N_j(I)$ on a_s.

Let $\tilde{X}_j$ the I*K-table consisting of X_j and filled up with zeros ; X_j contains the coordinates of the points of the $N_j(I)$ cloud in $\mathbb{R}^K$. We compute F_s^j by applying the transition formula to the rows of X_j :

$$F_s^j = \tilde{X}_j \; M \; a_s$$

Remark : The equality : $a_s = 1/\sqrt{\lambda s}\; X' \; D \; F_s$ yields :

$$F_s^j = (1/\sqrt{\lambda s}) \; W_j \; D \; F_s$$

In the R^I-space, vector F_s^j is the image of F_s through operator W_jD. When the variables of group j are orthogonal, it can be shown that W_j D is the orthogonal projection on the subspace spanned by $N(K_j)$, as in the case of G.C.A. When this hypothesis does not hold the inertia of the projection of F_s is less than the inertia of $W_jD\; F_s$. The choice for W_jD instead of the projection will also be justified by the representations of the clouds $N_j(I)$ as we will see in the next section.

E. Superposed graphs of the clouds $N_j(I)$

Another aspect of the comparison between groups of variables is the pointwise comparison of the J clouds of individuals : we study the evolution of each individual in the differents clouds ; to point out, for instance, individuals with similar evolution, or on the contrary, outliers.

For this, we superpose plane projections of the $N_j(I)$ (this is possible since the clouds lie all in the same space R^K). For the graphs to be fit for use, the projections of the J clouds must look alike : the J projections of one individual must be close and those of distinct individuals must hold apart.

Note that each point i of cloud N(I) (in fact its homothetic in the ratio 1/J) is the isobarycenter of the J points which represent the same individual in the clouds $N_j(I)$. The axes of inertia of N(I) thus yield a good solution : they maximize the inertia "between" individuals. Unlike the case of Discriminant Analysis (where the inertia "within" individuals is also minimized) the axes have an important inertia, and moreover, the plane projections of the $N_j(I)$ are equipped with the usual metric.

F. Qualitative variables

Up to now we have only considered the case of numerical variables, but our method also applies to qualitative variables. A qualitative variable defines a partition of the set of individuals. When several qualitative variables are implied, the table consisting of the characteristic variables of classes of partitions is called a complete disjunctive table (C.D.T.). Multiple correspondence analysis, the usual method in analysis of qualitative variables, is a correspondence analysis of the C.D.T. It can be shown that such an analysis is equivalent to classical P.C.A. of this table with the following modifications which assigns to each characteristic the appropriate weight : weight $(I-I_k)/I$ is assigned to the characteristic x_k where $I_k = \sum_i x_{ik}$. The cloud of individuals N(I) is then exactly the same as in M.C.A. and the clouds of columns have the same quadratic form of inertia. We are thus brought back to the preceding case (we had introduced the notion of weighted variables, not so much interest for numerical variables). C.D.T. and numerical tables can thus be dealt with in the same analysis and be compared. In the case where qualitative variables are implied, the interpretation of results is made easier by projecting barycenters of each class in the mean cloud N(I) and in the J clouds $N_j(I)$.

G. M.F.A. Implementation and approximation

M.F.A. consists in two steps. The one is a separate analysis of each subtable : P.C.A. for the numerical table and P.C.A. of weighted characteristic (or an M.C.A.) for qualitative tables. The second step is an analysis of table X in which the variables of a subtable are weighted by the inverse of the first eigenvalue resulting from the analysis of this subtable.

Approximation

Note that an alternative to the P.C.A. of variables is a P.C.A. of principal components of subtables. The fact is that the principal components of X_j are the coordinates of $N_j(I)$ over a new orthonormal basis of X_j ; thus the principal components of the J tables X_j are the coordinates of N(I) over a new basis of R^K. When the dimension of table X is very large we only retain the first components of each group . We can show that the results thus obtained are quite close the exact ones when the inertia of neglected components is small with regard to the eigenvalues and to the difference between successive eigenvalues of the analysis of X.

At the end of this step we get the graphs of N(I) and N(K). We then compute :

- the projections of groups of variables summing up the inertia of variables of each group.

- the superposed projections, F_s^j of the J clouds $N_j(I)$.
- the correlations between F_s and F_s^j. This makes it clear whether F_s can be considered as a common factor to the groups.

III. ANALYSIS OF FREQUENCY TABLES

We now study and compare J frequency tables for which the set of rows I is identical.

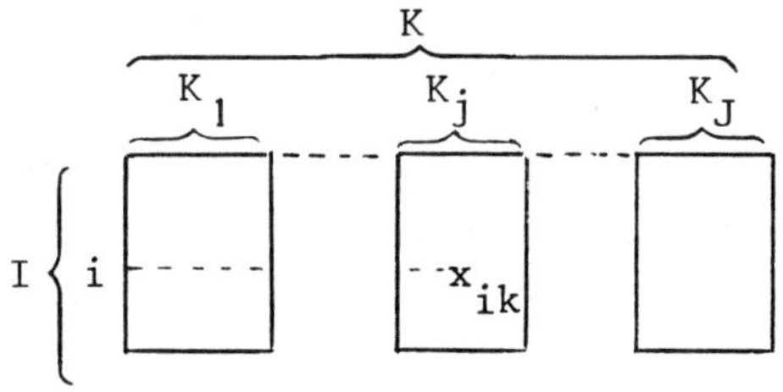

$$n = \sum_{i \in J} \sum_{k \in K} x_{ik}$$

$$n_j = \sum_{i \in I} \sum_{k \in K_j} x_{ik}$$

Figure 3 : The frequency tables X_j and X

The discrepancy between such tables depends on their margins on I, and on the dispersion of the profiles of columns of each table. In the cloud N(K) of profiles of columns of X, the margin of each subtable X_j is at the barycenter of the subcloud $N(K_j)$.

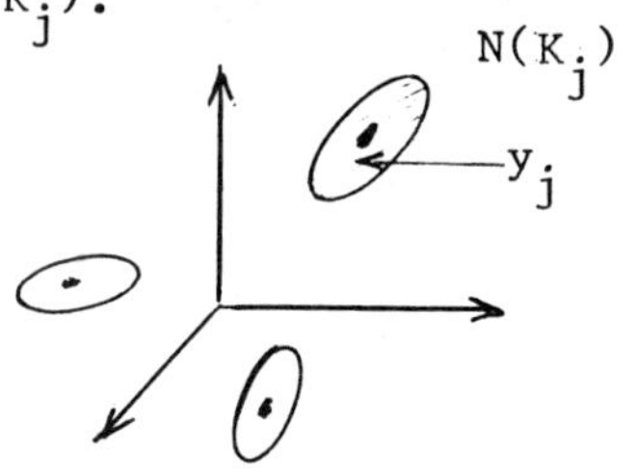

$$y_{ij} = \sum_{k \in K_j} x_{ij}/n_j$$

y_j = margin of K_j

Figure 4 : The subclouds $N(K_j)$ of N(K) in R^I space

The inertia of cloud N(K) is the sum of both between and within inertias (which are respectively the inertia of the cloud of barycenters and the inertia of each cloud around their own barycenters). In Correspondence Analysis of table X, the influence of a subtable depends on two things : the position of its barycenter and the inertia of $N(K_j)$. Both of them depend on the data structure without problems due to redundant variables as in numerical tables : in C.A. the inertia of $N(K_j)$ is up to a scaling factor, the χ^2 of table K_j.

A. Indices for table comparison

Correspondence Analysis using a division into subtables is a C.A. of table X with additional indices concerning the X_j. The main indices are :

. Projection of barycenters of clouds $N(K_j)$

The J-column table consisting of the margins of subtables is built and X is incremented by J supplementary columns.

. Contributions of the J+1 tables

For each factor we compute the contribution of barycenters and of each subcloud to the inertia of the factor. When the former is the larger, the factor distinguishes the margins of subtables; on the contrary when the inertia between is small and the inertia of each subcloud is large, the factor is a common dispersion direction of the K_j.

. Quality of representation of the J+1 tables

It can be measured by the ratio of projected inertia to total inertia as in the case of individuals. Note that the inertia of each subtable is computed around its barycenter and that the barycenter of the cloud of margins N(J) lies at the origin.

B. Approximation

The cloud of row profiles of X denoted by N(I) lies in a K-dimensional space R^K ; the clouds of profiles of X_j, denoted by $N_j(I)$, lie in R^{K_j}, K_j-dimensional spaces. R^K is the direct sum of orthogonal subspaces, isomorphic to R^{K_j} ; the metric over such a subspace is up to a scaling factor $\sqrt{n_j/n}$, the metric over R^{K_j}. The orthogonal projections of N(I) on R^{K_j} differ from $N_j(I)$ by the ratio between the margins of X_j and of X. The set of all inertial axes of the J clouds $N_j(I)$ make up a orthogonal basis of R^K. The coordinates of N(I) over this basis are :

$$Z_s^j(i) = \sqrt{n_j/n}(y_{ij}/m_i)\, F_s^j(i)$$

where m_i is the margin of X and F_s^j the sth factor of C.A. of X_j. (The trivial factor $F_1^j(i) = 1$ is included). P.C.A. of this set of function Z_s^j is equivalent to C.A. of X since the clouds N(I) associated with each case are identical. When the dimension of X is large, we can (as in M.F.A.) neglect the Z_s^j functions defined by the last factors of each group. The results look close to the exact one when the inertia of neglected factors is small.

CONCLUSION

The two methods presented are implemented and have been tested in many examples. Computing algorithms are quite simple. The approximation method, retaining the first factors of each subtable, allows the analysis of tables of any dimension. M.F.A. has revealed its efficiency for comparison of groups of variables, either numerical or qualitative ; it allows us to deal with simultaneous analysis of groups of numerical variables and groups of qualitative variables.

The great number of indices computed in correspondence analysis using subtables makes easier the interpretation of results of C.A. of tables structured into subtables. Nevertheless, all the aspects of comparison of frequency tables are not dealt with. For example, C.A. of the table of the margins will point out their discrepancies better. On the contrary to compare the within dispersion of the $N_j(I)$ we have proposed another method, in which the between dispersion vanishes.

REFERENCE

ESCOFIER B. et PAGES J. (1984). L'analyse factorielle multiple. Une méthode de comparaison de groupes de variables. DATA ANALYSIS AND INFORMATIC (North Holland).

ESCOFIER B. et PAGES J. (1985). Le traitement des variables qualitatives et de tableaux mixtes par analyse factorielle multiple. DATA ANALYSIS AND INFORMATIC.

ESCOFIER B. (1983). Correspondence Analysis using a division into subtables. In cluster Analysis and Data Analysis, North Holland. M. Jambu and M.O. Lebaux.

USE OF TABULATED DATA IN DATA ANALYSIS

Shoichi Ueda

Ryukoku University
Fushimi-ku, Kyoto, JAPAN

I. INTRODUCTION

In any problem, the accuracy and the availability of data determine how deeply the data analysis can say about the problem. It is true that the target of research is to explore outcomes as much as possible, but, exsistence of a limit must be denoted. Especially in the problem of social science, the data are too dirty to apply any standardized statistical procedures. Being hoped to do so, a very careful preprocedure must be performed to clean up contamination, for instance, by dividing data into a number of subsets within which the data are enough homogeneous to lead a convincing interpretation. Note that the real difficulties in data analysis lie in such a preprocedure.

The cross-tabulation is one of useful preprocedure and data setup. In this note, let us consider its logic as a step of data analysis putting emphasis on the feature of "designing analysis" to reach a persuasive output.

ISBN 0-12-215485-1

II. CROSS-TABULATION AS A PROCEDURE OF DATA ANALYSIS

In this note, PCA (principal component analysis) and cross-tabulation are considered both as a procedure of data analysis.

Suppose that a data matrix X_{ki} are obtained for units U_i. Then, the PCA derive scores X_{*i} as a linear combination of raw data, where the coefficients W_{k*} are eigen vectors of covariance matrix $V(X'X)$. Namely,

$$X_{*i} = \Sigma W_{k*} X_{ki} \tag{1}$$

The total amount of information of raw data $I(X) = \mathrm{tr}\, V(X'X) = \Sigma V(X_k)$ is redistributed by this procedure as $I(X) = \Sigma V(X_*)$, where new components ordered according to the magnitude of contribution. Therefore, we may choose several major components to simplify the data expression.

If individuals are classified according to a facter C, then information are summarized as

$$\bar{X}_{kj} = \Sigma C_{ij} X_{ki} / \Sigma C_{ij} \tag{2}$$

where

$$C_{ij} = \begin{cases} 1 & \text{if } U_i \text{ belong to } C_j \\ 0 & \text{otherwise} \end{cases}$$

and C_j are subgroups defined by C.

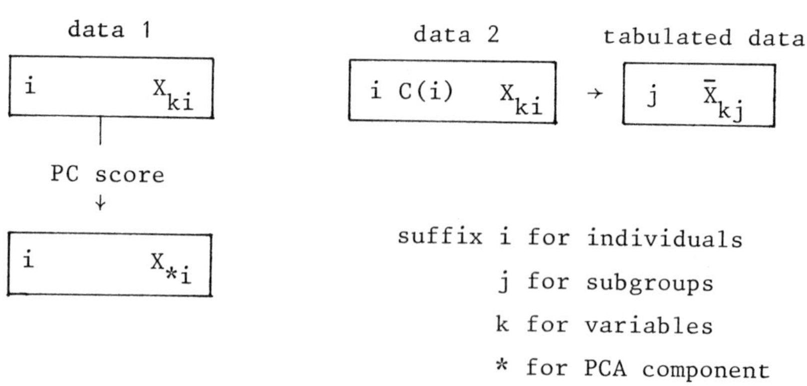

FIG. 1a Data Compression by PCA

FIG. 1b Data Compression by by Cross-Tabulation

Since $V(\bar{X}_k) < V(X_k)$, the amount of information decrease through this procedure, namely $I(\bar{X}_k) < I(X_k)$. Note that this decrease does not mean loss of adequacy as discussed later.

The data matrix are condensed by this procedure along with one direction (row wise compression) while PCA along with another direction (column wise compression). Each of them works to condense data with different logical meaning, hence, we may combine these two kinds of data-condense. Let us consider the route of column-wise compression followed by row-wise compression (FIG. 2a).

The first step is cross-tabulation and same as FIG. 1b. In the second step, principal component scores are derived similarly to FIG. 1a, but, the base data are matrix of mean values $\bar{X}_{kj}$. Therefore,

$$\bar{X}_{*j} = \Sigma \bar{W}_{k*} \bar{X}_{kj} \tag{3}$$

with weight vectors $\bar{W}_{k*}$ determined as eigen vectors of covariance matrix $V(\bar{X}'\bar{X})$.

The PCA scores derived from the route of FIG. 2a are

$$\bar{X}_{*j} = \Sigma \bar{W}_{k*} \Sigma C_{ij} X_{ki} \tag{4}$$

whereas that derived from the route of FIG. 2b are

$$\hat{X}_{*j} = \Sigma C_{ij} \Sigma W_{k*} X_{ki} \tag{5}$$

where $\bar{W}_{k*}$ and W_{k*} are eigen values of covariance matrices $V(\bar{X}'\bar{X})$ and $V(X'X)$ respectively. Both of them are linear

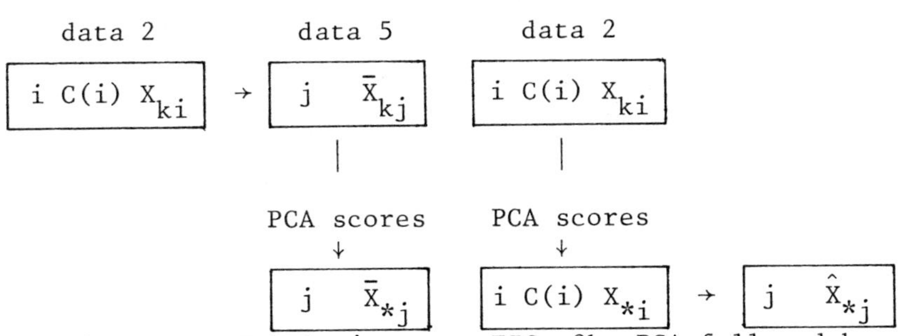

FIG. 2a Cross-Tabulation followed by PCA

FIG. 2b PCA followed by Cross-tabulation

combinations of raw data X_{ki} but the coefficients are generally different and hence $\bar{X}_{*j}$ and $\hat{X}_{*j}$ are different.

The mainpoint here is to discuss this difference and show how cross-tabulation works as a preprocedure of data analysis.

The procedure of cross-tabulation contains several steps to be distinguished logically.

Firstly, the observation units are classified and an identification code is assigned to each unit U_i. Utilizing this additional information, it become possible to find a relationship between facter C and variables X_k. Corresponding to this classification, the raw data X_{ki} are decomposed into two variables $\bar{X}_{kj}$ and E_{ki}. The former represent between class (or average level) variation and the later within class variation. This is the data 3 in FIG. 3, where the number of variables are twiced.

It is reasonable to drop within class information if we can limit the scope of analysis to the level of between classes. The data 4 and 5 in FIG. 3 are corresponding to such a scope.

Note that in data 4, a set of values represent each observation units U_i even though same values are dupplicated for every members of each class C_j. If our attention lies on between class variation, an alternative way is to use data 5 where each class C_j are represented by single set of values. Although their size are different class by class, we had

data 2	Decompose	data 3	Limit Scope	data 4	Change Unit	data 5
i C(i) X_{ki}	→	i C(i) $\bar{X}_{kj}$ E_{ki}	→	i C(i) $\bar{X}_{kj}$	→	j $\bar{X}_{kj}$

FIG. 3 Logic of Cross-Tabulation

changed the level of scope, hence it become natural to treat each class with same weight regardless size.

Due to this difference, the PCA scores are different for data 4 and 5. If we treat data 5 with weight proportional to square root of class size, then two scores coincide. But, such a treatment is not desirable since it tends to hide interpretable difference that come from change of scope.

The nature of principal component score of data 3 is not clear since the base data contains information with different level. If a kind of hierarchical level is clearly defined on observation unit and applied common to all variables, then between class variables $\bar{X}$ and within class variables E are orthogonal, hence, the covariance matrix $V((\bar{X}|E)'(\bar{X}|E))$ has off-diagonal submatrix [0], namely, forms like

$$\begin{vmatrix} V(\bar{X}'\bar{X}) & 0 \\ 0 & V(E'E) \end{vmatrix}$$

Under this assumption, PCA procedure is able to separate informations with different levels as different components. Therefore, a subset of principal component scores of data 3 coincide with that of data 4.

However, this assumption is not generally satisfied. For instance, different classifications are applied to each variable and off-diagonal part of covariance matrix contain non-zeros. It could be treated as a kind of error term in some cases, but not always so. For example, we may adopt variables with different level intentionally to describe a cross-level relationship, hence the off-diagonal terms are essential part to be analyzed.

We must, therefore, analyze a mixed-level data set in spite of difficultes in analysis and interpretation.

The principal component score of data 2 are same as that of data 1. This means that information C(i) are included with intention but not utilized. Hence, certain steps are necessary to utilize them.

III. DESIGNING ANALYSIS

Now we come back to data 1. The data matrix X_{ki} is a standard form in the sense of mathematical model. If we can assume that the observation units are a set of random sample or replication in the word of experimental design, then the procedure could be supported by a probability concept.

But, the term "model" must be used in relation to real phenomena to be analyzed. In this sence, the model of data 1 is too simple to describe real phenomena. The principal component scores derived from this data provide a summarized description of individual variation but nothing more than that. Presumably, researchers could not be satisfied with this proto-type and would include certain instruments, for example,

a: Allocate facters on observation units so that individual variations are decomposed into several components

b: Build in variables with which variation of observed values or their components are explained

c: Set up data for different times or areas in order to confirm generality of findings or to find real change

We may call these setup "design of analysis".

The principle of statistical method in social science are more or less data-oriented. It differs from hypothesis-oriented analysis where everythings are designed and performed to test a specified hypothesis. But, we need to construct a leading principle to diminish an ambiguity of exploratory procedures. Discussion of this note is a trial in this means.

Section 4

Applications: Extraction and Interpretation of Information in Multidimensional Data

MULTIVARIATE DESCRIPTIVE TECHNIQUES APPLIED TO THE PROCESSING OF LONGITUDINAL SAMPLE SURVEY DATA

Ludovic Lebart

Centre de Recherche pour l'Etude et l'Observation des conditions de vie (CREDOC)
140 Rue du Chevaleret, 75013, Paris, FRANCE

This contribution aims at presenting an assessment of the use and interpretation of Multivariate Descriptive Techniques(MTD) applied to longitudinal socio-economic sample surveys : a full scale experiment is discussed. It consists of the processing of the data set provided by the National Survey about the Aspiration and the Living Conditions of the French, from 1978 to 1984.

I. THE INITIAL PROBLEMS

From the beginning, the nature of the questions implied a descriptive and exploratory approach :

- What kind of organization or affinities can be detected among the opinions in relation to socio-economic problems ?
- How are the <u>opinions</u> conditioned by the <u>situations</u> ?
- How do these opinions develop or change throughout the considered period of time ? Is the structure of opinion stable ?

ISBN 0-12-215485-1

II. PRESENTATION OF THE DATA SET

The data set consist of seven independent samples of 2000 in size, each one being a representative of the French population over the age of 18, for each year from 1978 to 1984, (8). The questionnaires contained about 200 questions, mainly relating to opinions (about economic situations, family, environment, science, society, justice. The responses lead to the codification of categorical (or nominal) variables.

III. PRESENTATION OF THE METHODS

The two complementary techniques widely used to describe large set of categorical data are : Multiple Correspondence Analysis (MCA), whose names and origins are also multiple (4), (5), (1), and (9), (10) for references ; and Hybrid Clustering (HC) (12), (7), a technique comprising three steps :

a) Preliminary clustering, using aggregation around variable centers (k-means method, or dynamic cluster method, (2)).
b) Hierarchical clustering of the previously obtained clusters, (using Ward's criterium), and truncation of the dendrogram, which determines the number of final clusters and a provisional set of centers for the final partition.
c) Reallocation (using a k-means type method) of the individuals in order to improve the quality of the partition.

IV. THE FUNDAMENTAL NOTION OF ACTIVE VARIABLE (AV)

A typology (be it obtained by MCA or by HC) make sense only if the set of "active variables" is clearly specified.

A typology can be defined only from a specific point of view, characterized by this set. The AV set must satisfy a criterium of homogeneity whose rationale is intuitive : the computations of distances must make sense, so that the resulting groupings also make sense. It would be awkward, for instance, to mix objective and subjective questions : the results would prove ambiguous. Whereas the homogeneous set of AV allows for the definition of a specific point of view, the set of "supplementary variables" (SV) will allow for the a posteriori characterization or identification of the structural features produced. This set needs not be homogeneous, since its elements intervene one by one.

In the synthesis presented in this paper, there will be 14 nominal AV (about 60 categories of responses). These questions are aimed at giving an overall description of attitudes and opinions. All the remaining variables have the status of SV.

List of the Active Variables (Summary) :

1. Two questions relating to the perception of the evolution in personal and global living conditions.
2. Three questions about the image of the family, about the meaning of the marriage, the activity of women.
3. Three questions about environment and science.
4. Three questions about health (satisfaction towards personal state of health, role of the health care system).
5. Three more general questions about society : opinions on justice, social change, collective facilities and services.

The results obtained will depend on the choice of these questions : the patterns issuing from the statistical processing will be partially created and partially discovered : partially created because, among the questions, in their initial form, there are logical connections, redundancies and synonymies which induce a "structure of questions" prior to any "structure of responses" ; partially discovered because the results (i.e. : the planar maps, the clusters) go far beyond the a priori knowledge of the questionnaire. However, one cannot hope to discover an intrinsic structure independent from the content of the AV set. It would be equivalent to saying that the entire range of potential questions have been exhausted by such a set, supposed to be a representative "sample of questions" : this is for the time being, no more than a fascinating field of research where very little progress has been made (see (6), (3), (11)).

Based on the 14 responses to the active questions, the respondents are positioned in a spatially continuous "swarm of points". There are no clear-cut groupings in this continuum, but it is possible to divide it into 8 major zones by using the HC algorithm. The centroids of these zones are plotted as SV categories in the planar map of Figure 1, where most of the AV categories are represented.

V. THE "SWARM" OF AV AND THE "GRID" OF SV CATEGORIES

Eight stable zone are described below with the help of fictitious "modal individuals" (M.I.) possessing their most salient socio-demographic features.

Zone 1 (Right part of Figure 1) Modernists (about 18 %) :
People with modernists ideas on the family. They are in favour of environmental protection, technological progress ; active social life, mobile, and not very worried.
M.I. : A young parisian who does not have children, with a high level of education, an average standard of living.

Zone 2 (Top right) Dissatisfied/Isolated (about 11 %) :
Pessimistic, dissatisfied, critical, socially isolated people.
M.I. : A worker or jobless person, with an increasing number of serious problems of different types (housing, family, health).

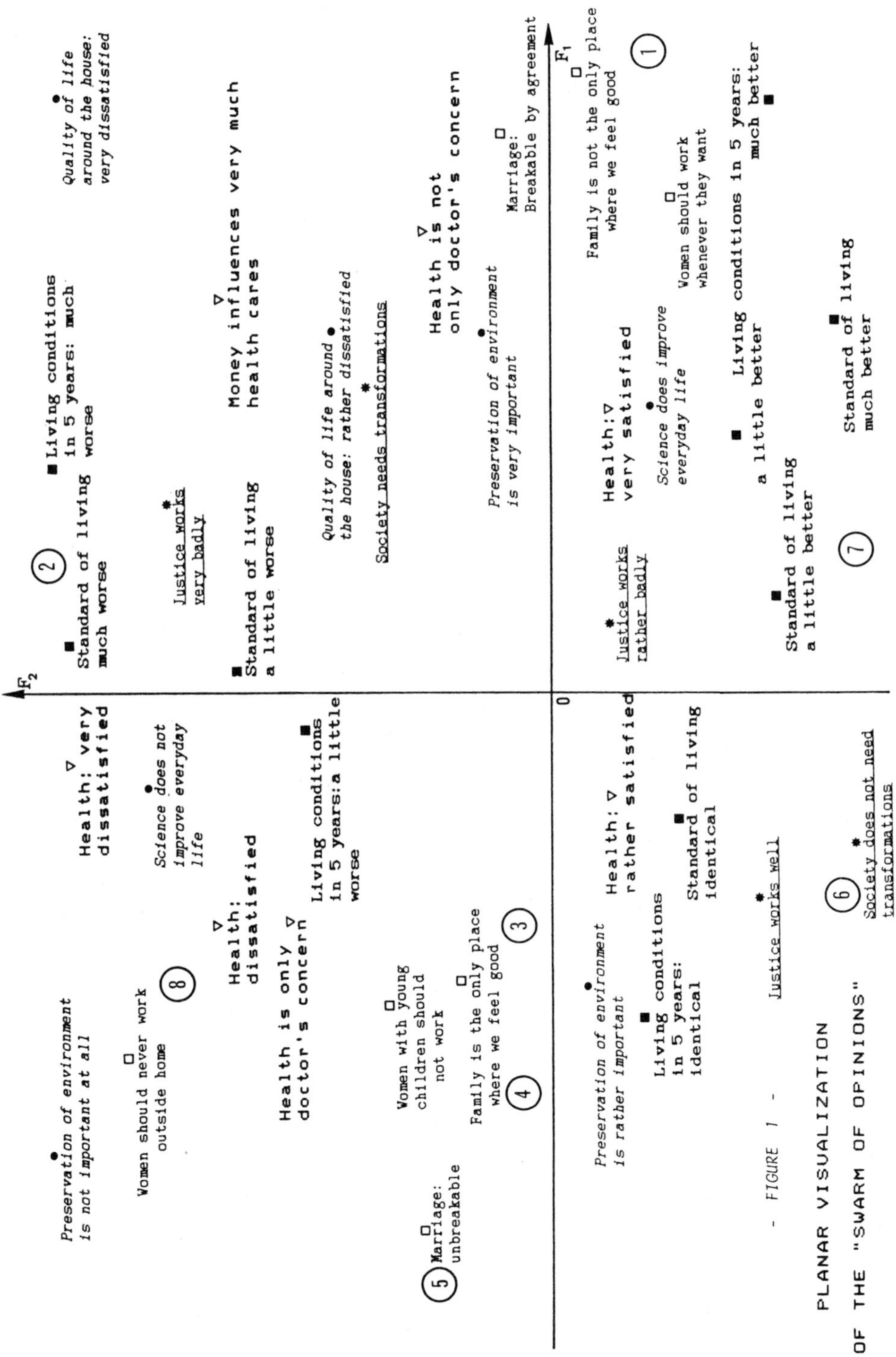

- FIGURE 1 - PLANAR VISUALIZATION OF THE "SWARM OF OPINIONS"

Figure 2

HIERARCHICAL CLUSTERING OF THE 8 ZONES

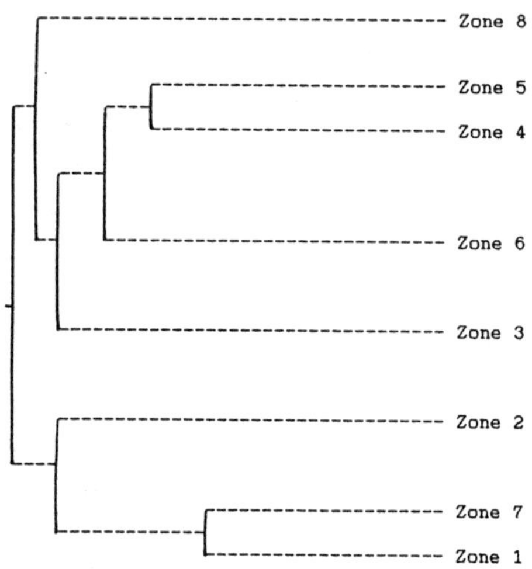

Zone 3 (Center left) Moderate pessimists (about 16 %). Unstable intermediate zone (center of Figure 1) : satisfaction vis-à-vis life-style, health ; divided opinions about family. M.I. : A person from a rural area, 50 years old, married, with children ; average level of education and standard of living.

Zone 4 (Left of Figure 1) Traditionalists(about 15 %). Very traditionalist opinions on the family, but quite critical in relation to justice and society. M.I. : a retired person who has had children, a low level of education, in a rural area.

Zone 5 (Left) "Don't know" (about 8 %). Lack of interest in the survey and poor understanding. M.I. : An old woman (often a widow) with a low level of education in a rural area.

Zone 6 (Bottom-left) Conservatives (about 13 %). No change needed in society, justice works well, satisfaction towards living conditions and science. M.I. : A person about 50, high income and amenities, medium level of education, house-owner,

Zone 7 (Bottom-left) Moderate optimists (about 12 %).

Hesitations, not very marked opinions, fairly general feeling of satisfaction. M.I. : A person about thirty years old, from "province", active, with children, average qualification.

Zone 8 (Top-left) "No answer" (about 5 %).

No-responses relate mainly to the issue of family.

M.I. : A single man of any age, no diploma, very low income.

Note : One could think that zones 5 and 8 are close to each other and that they could be agglomerated in a single region of "No answer, don't know". As is shown in Figure 1 and as the dendrogram (Figure 2) confirms, zone 5 is much closer to zone 4. This dendrogram is used here to remedy the deformations of distances due to the planar approximation of Figure 1.

The figure 3 highlights the regular display of the crosstabulated SV "age x qualification x standard of living". The left-hand "grid", in solid lines, is for individuals without diploma, the right-hand grid, (dotted lines) being for high school diploma. Identified by the letter "L", the vertices of the two grids correspond to Low standard of living categories. The letter "M" indicates Medium and the letter "H" High standard of living. Moreover, the 4 vertical elements of each grid correspond to the 4 age-groups, whose identifiers on Figure 3 are the lowest ages (i.e. : 45 designates the class "45-64 years old"). We can read the following : At a given age and at a given level of education, dissatisfaction or pessimism is voiced when the standard of living level decreases. A a given age and constant standard of living, "modernism" increases with diploma level. At a given education level and at constant standard of living, "traditionalism" increases with age.

If we superimpose Figure 1 and 3, we can see that most opinion zones (whose centroids are the circled numbers) are reached by the grid. Thus the "opinion profiles" of the vertices cover the main configurations of opinions. One can found "situations" likely to "explain" all patterns of opinions.

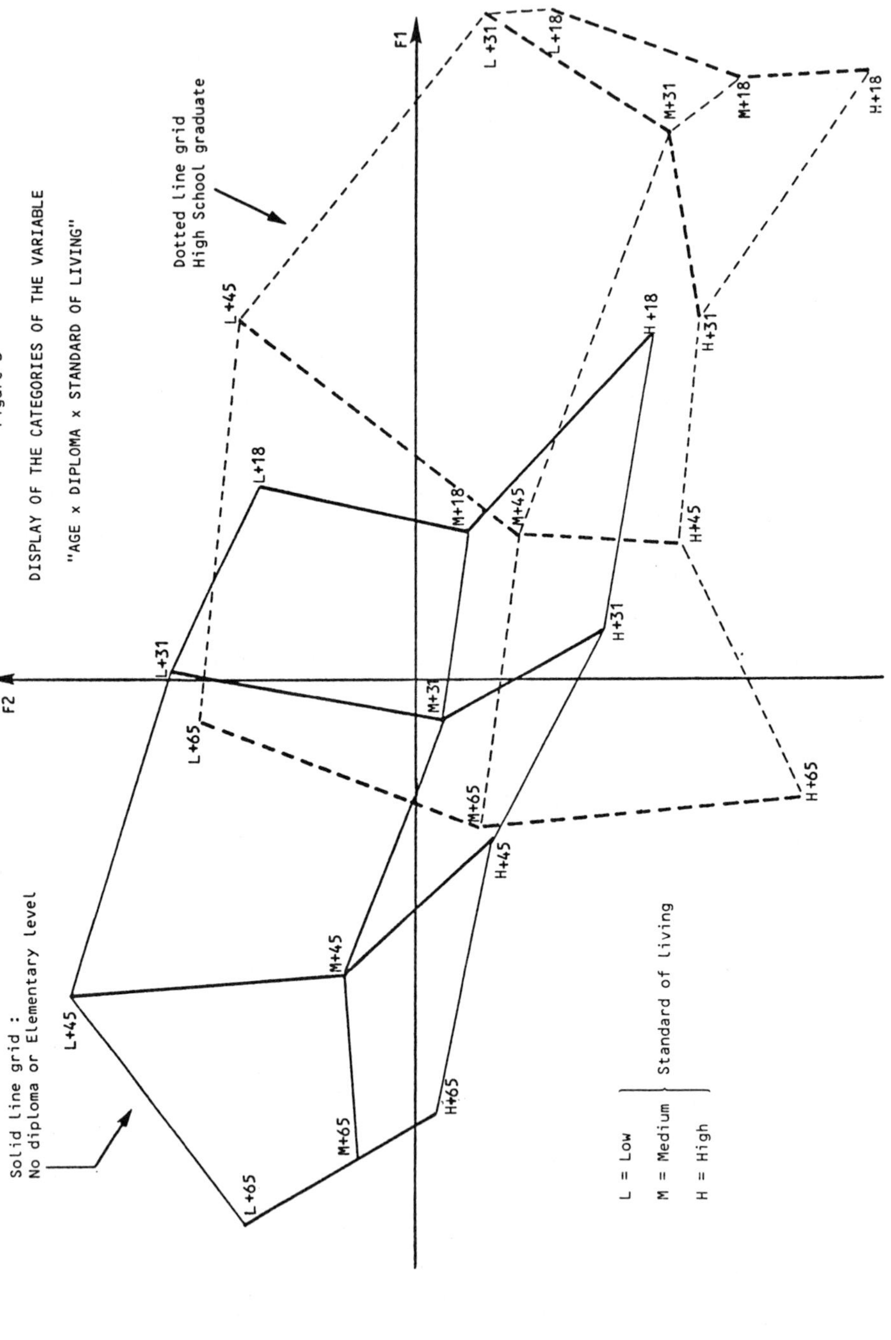

Figure 3

DISPLAY OF THE CATEGORIES OF THE VARIABLE

"AGE x DIPLOMA x STANDARD OF LIVING"

VI. CHANGES IN THE PATTERNINGS OF OPINIONS

The pattern of opinions shown in Figure 1 and 3 applies to the period 1978-1984. To detect structural shifts occurring with the passing of time, we may plot on these figures the points representing the years, and points representing SVs obtained by cross-tabulating years and questions : the trajectory of the whole country between 1978 and 1984 will be represented by the thick line in the center of Figure 4.

a) "Modernists" opinions spread until 1981 (year-points moving to the right in Figure 4), with a remarkable increase in 1981 (year of a change of government in the country).
b) After 1981, and particularly between 1983 and 1984, dissatisfaction grows (year-points moving upward).

To obtain a graphical evaluation of the intensity and the general direction of the change of opinions in the country, we can take the centroïd of the firts two years, the centroïd of the last two years, and join them with a straight line (oriented upwards to the right of Figure 4). This "average vector", describes the change of the whole country.

We can now plot, for each response separately, the first two and the last two years, and compare the obtained vector to the "average vector". In so doing, we will draw on the planar map as many vectors as there are response-points.

See for instance on the left, the case of people with no diploma : there is a drastic increase in dissatisfaction (the segment being longer and more vertical than the average one). The same observation can be made for "housewives" and women aged 40 to 60, on the left. On the contrary, satisfaction increased among parisians (on the right in Figure 4) and particularly those under 40 (more on the right). In the same way, satisfaction increases among people with diplomas (right hand side). It should be noted that holders of high school diplomas

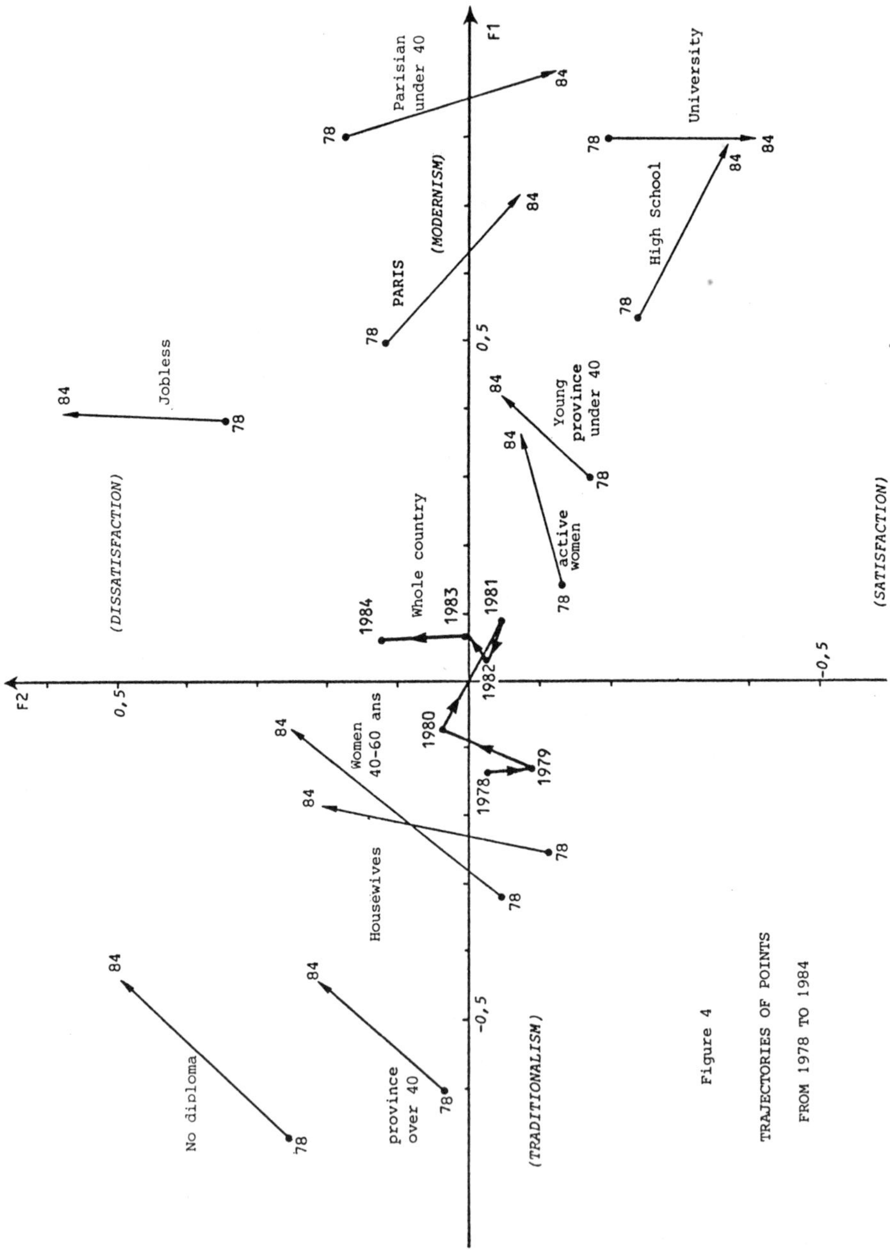

Figure 4

TRAJECTORIES OF POINTS

FROM 1978 TO 1984

only increase their "modernism level" (shift to the right) to reach in 1984 the 1978 level of people with university diplomas. The latter do not appear to have changed, just as if a "saturation" point had been reached.

VII. REMARKS : Existence and autonomy of structures

M.D.T. does not produce results in the usual meaning of this word in statistics : it produces a new kind of data ; the nature of initial information has changed : it is more readily available, but more vulnerable as well. More critical reflexion is made possible, but "spontaneous sociology" can take advantage of these suggestive outputs, while the use of the computer provides in some cases a technical alibi.
M.D.T. is generally intended to discover something, and not to prove anything. However, there are some favorable situations in which graphical displays and typologies are worth being published : as is the case here, when the stability (and hence the relevance) of the structures highlighted is tested in independent samples. What we have so far called a structure is more of a stable, but passive pattern : "passive" indicates that the structure under study can be merely apparent ; that it can be a compromise between heterogeneous structural features. Let us say to conclude that the observer is rather like a martian watching a rugby match from his flying saucer ... a number of regular features are striking, others more subtle, but the real reasons behind those observed developments are to be found elsewhere. To obtain a deep insight is a fundamental but preliminary step. One of the actual challenge of MDT is to fill the "methodological gap" between the powerful descriptions enabling the user to observe, to criticize, to test the consistency of data, and the construction of models.

REFERENCES

1. Benzécri J.P. (1973). L'Analyse des Données. Dunod, Paris.
2. Diday E. (1971). La Méthode des Nuées Dynamiques. Rev. de Stat. appl. 19, 2, 19-34.
3. Escoufier Y. (1970). Echantillonage dans une Population de Variables Aléatoires Réelles. Publi. Inst. Stat. Univ. Paris, 19, 4, 1-47.
4. Guttman L. (1941). The Quantification of a Class of Attributes. In "The prediction of personal adjustment" (P. Horst ed.) SSCR, New-York.
5. Hayashi C. (1956). Theory and Examples of Quantification. (II). Proc. of The I.S.M. 4, (2) 19-30, Tokyo.
6. Hotelling H. (1933). Analysis of a Complex of Statistical Variables into Principal Components. J. Etu. Psy. 24, 498-520.
7. Lebart L., Morineau A., Warwick K. (1984). Multivariate Descriptive Statistical Analysis. J. Wiley, New-York.
8. Lebart L. (1986). Qui pense quoi ? Structure et évolution des opinions en France de 1978 à 1984. Consommation N°4.
9. Nishisato S. (1986). Quantification of Categorical Data : A bibliography 1975-1986. Microstats, Toronto.
10. Tenenhaus M., Young F.W. (1985). An Analysis and Synthesis of Multiple Correspondence Analysis, Optimal Scaling, Dual Scaling, and other methods for quantifying categorical Multivariate Data. Psychometrika, 50, 1, 97-104.
11. Wachter K.W. (1978). The strong Limit of Matrix Random Spectra for Sample Matrices of Independent elements. Ann. Prob. 6, 1, 1-18.
12. Wong M.A. (1982). A Hybrid Clustering Method for Identifying High Density Clusters. J.A.S.A. 77, 380, 841-847.

MULTIDIMENSIONAL ANALYSIS OF OPINION SURVEY DATA

Meiko Sugiyama

NHK Broadcasting Culture Research Institute
2-1-1, Atago, Minato-ku, Tokyo 105, JAPAN

I. FORWORD

Opinion survey data are different in kind from the so called measurement data; they are mainly qualitative data (qualitative attributes) rather than quantitative data (quantitative variables). In the opinion survey, questions which can be answered quantitatively, such as "How many hours do you watch TV?" or "How much is your salary?", are few; rather, in most cases, they take a form in which the answer is selected from among some choices as shown below.

Question:

Are you now doing any of the things which you think have something to do with religion or faith? If so, please give some of the examples from the following list.

Answer:

(A) Make it a rule to practice religious conduct, such as attending religious services, religious worship, and missionary works, or on occassion, offer prayers or chant

ISBN 0-12-215485-1

sutras.

(B) Visit a grave once or twice a year.

(C) On occassion, read religious books, such as the Bible or the Buddhist Scriptures.

(D) Have visited shrines and temples to pray for business prosperity, security of life, success in an entrance examination and so forth, within the past couple of years.

(E) Keep a talisman, such as an amulet or a charm, or a mascot near me.

(F) Have drawn a fortune or consulted a diviner, or had my fortune told, within the past couple of years.

Questions of this kind are classified into two types: one is the single answer type, where it is asked to give one answer only from some choices in a list, and the other is the multiple answer type, where it is asked to give as many answers as appropriate. In either case, symbols A, B, ... are mere notations and do not mean any order, any quantity, etc., so these answers give qualitative data.

In the analysis of opinion survey data, it is the main objective to pursue mutual relations between these qualitative data. Simple tabulation for each question, cross tabulation between an attribute and a question choice, cross tabulation between question choices, triple-cross tabulation between three question choices, etc. are carried out. To examine the relation between four or more questions, the triple cross tabulation is no more sufficient. But, in the case of quadruple or quintuple cross tabulation, even if it is calculated, it will be difficult to read the results. Take, for example, the case of the question on religious behaviors shown above. If it is attempted to examine the relation between choices A through F, it is necessary to calculate the multiple cross tabulation in various combinations, such as A and B, A and C, ..., E and F, A and B and C, ..., and even if all these combinations of multiple cross tabulation were calculated, the results would be so complicated that it might be difficult to overlook the behavior of the whole. In such cases, the technique of multi-dimensional analysis is very useful.

Then, this multi-dimensional analysis, although it may be said in a word, has various models, which are different in the conditions of application and require different methods of utilization of the results, etc., as will be described in detail in the next chapter.

II. METHODS OF ANALYSIS

Methods of analysis for the multi-dimensional analysis are divided roughly into two cases; one where there are outside criteria and the other where there are no outside criteria.

Analysis techniques for the case where there are outside criteria are shown in Table 1.

These analysis techniques are further divided into two kinds according as the problem is "quantity estimation" or "quality classification".

First, in the case of the problem of "quantity estimations", known data, which can be expressed as quantitative

Table 1 Analysis techniques for the case where there are outside criteria(1)

Problem	Outside criteria	Explanatory item	Analysis technique
Quantity estimation	Quantitative variable	Quantitative variable	Multiple linear regression analysis
		Qualitative attribute	Quantification Theory No.I
Quality classification	Qualitative attribute	Quantitative variable	Discriminant analysis
		Qualitative attribute	Quantification Theory No.II

variables, such as TV audience time length, audience rating, and income, are given as outside criteria, and quantitative variables, such as age and production cost, or qualitative attributes, such as sex and program item, are given as explanatory items, and from the explanatory items, the outside criteria are estimated or predicted. When the explanatory items are quantitative variables, the technique of multiple linear regression analysis is used, and when the explanatory items are qualitative attributes, Quantification Theory No.I is used.

<Multiple Linear Regression Analysis>

Suppose that a quantitative variable Y(i) is given as an outside criterion, and as the explanatory item, a quantitative variable $X_j(i)$ is obtained. Then, put

$$Y'(i) = \sum_{j=1}^{m} a_j X_j(i) + c$$

and obtain a_j and c at which the following becomes minimum.

$$E = \sum_{i=1}^{n} (Y(i) - Y'(i))^2$$

where

i = subject No. (i = 1, 2, ..., n-person)

j = item No. (j = 1, 2, ..., m-item)

<Quantification Theory No.I>

Suppose a quantitative variable Y(i) is given as an outside criterion, and as the explanatory item, an answer $\delta_{jk}(i)$ to the qualitative attribute is obtained. Then, put

$$Y'(i) = \sum_{j=1}^{m} \sum_{k=1}^{k_\ell} \delta_{jk}(i) X_{jk} + c$$

and obtain category score X_{jk} and c at which the following becomes minimum.

$$E = \sum_{i=1}^{n} (Y(i) - Y'(i))^2$$

where

i = subject No. (i = 1, 2, ..., n-person)

j = item No. ($j = 1, 2, \ldots\ldots, m$-item)

k = choice No. ($k - 1, 2, \ldots\ldots, k_\ell$-category)

$\delta_{jk}(i)$ = answer of the ith subject to k-category of j-item

$$\begin{cases} \delta_{jk}(i) = 1 \ldots \text{ When answered} \\ \delta_{jk}(i) = 0 \ldots \text{ When not answered} \end{cases}$$

Then, in the case of the porblem of "quality classification" in Table 1, known qualitative attributes, which are given as classifications into two to several classes each, such as "conservative - reformist" and "old aged - middle aged - young aged", are given as outside criteria, and quantitative variables or qualitative attributes are given as explanatory items, and from the explanatory items, the outside criteria are classified. Again, the analysis technique is different according as the explanatory items are quantitative variables or qualitative attributes. In the former, the discriminant analysis is used, and in the latter, Quantification Theory No.II is used.

<Discriminant Function>

Suppose a qualitative attribute $T(i)$, shown by t classes, is given as an outside criterion, and as the expoanatory item, a quantitative variable $X_j(i)$ is obtained. Then, put

$$\alpha(i) = \sum_{j=1}^{m} a_j X_j (1)$$

and obtain a_j at which the following correlation ratio of α to T becomes maximum.

$$\eta^2 = \sigma_b{}^2/\sigma^2$$

where

i = subject No. (i = 1,....., n-person)

j = item No. (j = 1,....., m-item)

σ^2 = total variance of $\alpha(i)$

σ^2 = between class variance of t classes of $\alpha(i)$

<Quantification Theory No.Ⅱ>

Suppose a qualitative attribute T(i), shown by t classes, is given as an outside criterion, and as the expoanatory item, an answer $\delta_{jk}(i)$ to the qualitative attribute is obtained. Then, put

$$\alpha(i) = \sum_{j=1}^{m} \sum_{k=1}^{k_\ell} \delta_{jk}(i)X_{jk}$$

and obtain X_{jk} at which the following correlation ratio of α to T becomes maximum.

$$\eta^2 = \sigma_b{}^2/\sigma^2$$

where the notation is the same as before.

As described above, opinion survey data are mainly qualitative attributes since the answer is selected from some choices. Therefore, rather than the multiple linear regression analysis and the discriminant function, Quantification Theory No.I and Quantification Theory No.Ⅱ are used in many cases.

Analysis techniques for the case where there are no outside criteria are shown in Table 2.

Table 2 Analysis techniques for the case where there are no outside criteria[(2)]

Properties of relation R_{ij} between items	Analysis technique
Vague similarity (this is a quantity expression, but can not be said metrical)	Quantification Theory No.IV (e_{ij} type)
Paired comparison	Space arrangement based on Hayashi's paired comparison
Rank order (the order of all of n things is determined)	Nonmetric method...Coombs' method
Rank order (the rank order of the relation between two things)	Shepard's method Kruskal's method Guttman's SSA Young, de Leeuw, Takane's method
Grouping with rank order	Hayashi's MDA-OR
Mere grouping without rank order	Hayashi's MDA-UO
Similarity and dissimilarity (metrical case)	Togereson's method K-L type quantification
Frequency	Latent structure analysis Quantification Theory No.III (Patterning) (including MSA and POSA)
Correlation coefficient	Component analysis method, factor analysis method

These analysis techniques are further divided into

various kinds in accordance with the properties of relation R_{ij} between i-item and j-item. For example, if R_{ij} is vague similarity (this is a quantity expression, but can not be said metrical), e_{ij} type quantification of Quantification Theory No.IV is used. In addition, R_{ij} can be paired comparison, rank order, grouping with rank order, were grouping without rank order, similarity and dissimilarity, frequency, and corretation coefficient, and in accordance with the properties of R_{ij} in these cases, various analysis techniques have been developed.

For the analysis of opinion survey data, Quantification Therory No.III (usually called "pattern classification") has been used for a long time and usefull for the structure analysis of consciousness. Moreover, in increased cases, the individual score obtained from the calculation of Quantification Theory No.III is used to draw the Guttman's POSA (Partial Order Scalogram Analysis) chart and find out the flow of answer through some questions.

<Quantification Theory No.III>

Suppose an answer $\delta_{jk}(i)$ to a qualitative attribute is obtained, then
put

$$Y(i) = \sum_{j=1}^{m} \sum_{k=1}^{k_\ell} \delta_{jk}(i)X_{jk} \Big/ \sum_{j=1}^{m} \sum_{k=1}^{k_\ell} \delta_{jk}(i)$$

and obtain X_{jk} and $Y(i)$ at which the following correlation coefficient becomes maximum.

$$\rho = \frac{C_{XY}}{\sigma_X \sigma_Y}$$

where the notation is the same as before.

Moreover, the Kruskal's method (MDSCAL) is also used frequently. In accordance with the definition of the stress, this method is divided into two methods, new and old. The following description refers to new MDSCAL.

<Kruskal's method ..... new MDSCAL>

Suppose there are n objects (1, 2,...i,...j,..., n) and the measured value δ_{ij} between objects i and j is obtained. Then, obtain the space arrangement of n points $x_1, x_2, \ldots, x_n$ in the t-dimensional space, so as to show the best stress to the measured value δ_{ji}. For this purpose, let the stress be

$$S = \sqrt{\sum_{i<j} (d_{ij} - \hat{d}_{ij})^2 \Big/ \sum_{i<j} (d_{ij} - \bar{d})^2}$$

$$\bar{d} = \frac{1}{\sum_{i<j}} \sum_{i<j} d_{ij}$$

and obtain the space arrangement at which the stress S becomes minimum. The measured value δ_{ij} is the value of dissimilarity, non-consistency, crosstalk, etc. and d_{ij} is the distance between points i and j in the t-dimensional space. And $\hat{d}_{ij}$ is a value which is approximately in the same order as the measured value δ_{ij}.

III. Application Examples

Some of application examples of the multidimensional analysis of opinion survey data will be introduced and the features of the techniques used will be described.

A. Quantification Theory No.I
... TV Audience Prediction(3), (4)

The individual audience of TV programs is predicted from such items as ① family audience rating, ② broadcasting time, ③ broadcast events, ④ brodcasting day of week, etc. This corresponds to the problem of "quantity estimation" in the case where there are outside criteria, and all items are assumed to be qualitative attributes. The family audience rating of item ① is the quantity data in %, but as shown in Fig.1, it is handled as the qualitative attributes deliberately by making categories of 2% each, such as 0%, 1 to 2%, 3 to 4%,..... At a look, this seems to have discarded the valuable information, but when it is unknown whether the item has a linear relation to the outside criterion, it is an effective means.

In Quantification Theory No.I, the individual audience rating of an i-program is denoted by Y_i, and to the category $(1, 2...j_k...)$ in each of some items $(1, 2...j...)$ which are considered to contribute to that individual audience rating, a category score (X_{jk}) corresponding to the degree of the contribution is given.

By calculation in accordance with Quantification Theory No.I separately in Nov., 1974, Jun., 1975, Dec., 1975 and Jun., 1976, the correlation coefficient ρ between the measured individual audience rating and the reproduced individual audience rating, which is as high as 0.96, 0.96, 0.95 and 0.95, respectively, was obtained.

Then, from the trend at the past four times, it was attempted to predict the individual audience rating at the next time. That is, the average of scores from Nov., 1974 to Jun., 1976 was assumed as the predicted score of Nov., 1976.

		Prediction of Nov., 1976 (average from No., 1974 to Jun., 1976)
		Score
Family audience rating (Kanto)	0%	0.0
	1to 2%	0.2
	3to 4%	0.5
	5to 6%	1.0
	7to 8%	2.3
	9to10%	3.1
	11to12%	3.7
	13to14%	4.5
	15to16%	5.9
	17to18%	6.7
	19to20%	8.1
	21to25%	9.8
	26to30%	11.6
	31to35%	13.3
	36to40%	15.7
	Above 41%	20.0
Broadcasting time	6:00	5.4
	7:00	11.6
	8:00	-0.5
	9:00	1.1
	10:00	0.6
	11:00	0.5
	12:00	7.2
	13:00	0.8
	14:00	0.4
	15:00	0.4
	16:00	0.4
	17:00	0.6
	18:00	3.5
	19:00	9.4
	20:00	7.0
	21:00	5.6
	22:00	2.3
	23:00	0.4

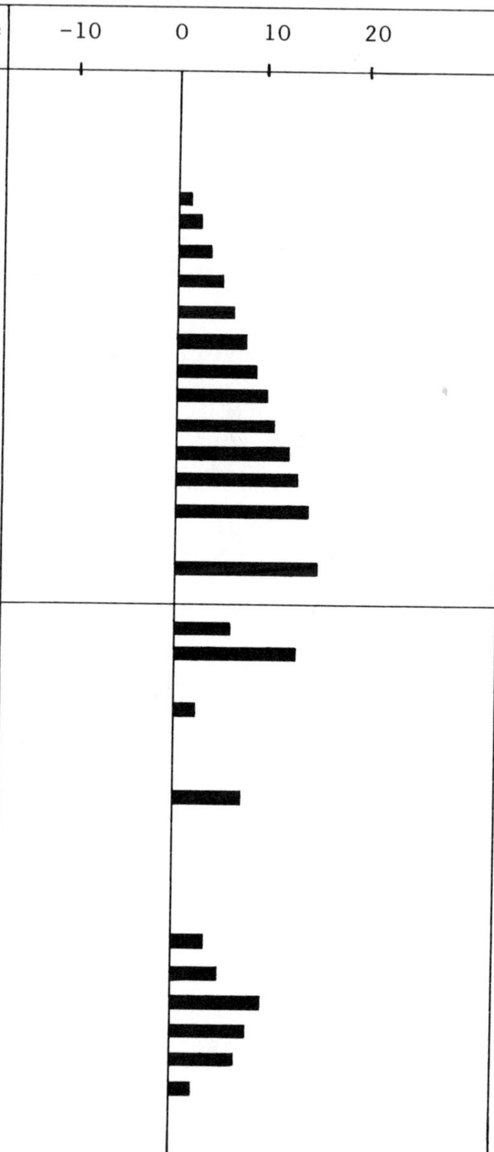

Fig.1 Category quantity of each item for individual audience rating -- nationwide NHK general TV --

The correlation diagram for the individual audience rating, which was predicted from this score, and the individual audience rating of Nov., 1976, which was obtained by the actual survey, is as shown in Fig.2. The correlation coefficient between measurement and prediction is $\rho = 0.97$, indicating that the prediction accuracy is fairly good. In Fig.2, program names are written for those programs for which the prediction was badly deviated.

In the individual audience prediction for TV programs, there will be the following three problems.

① Method of formulation
② Setting of program characteristics
③ Classification of program characteristics

As the method of formulation of ①, the multiple linear regression analysis is conceivable and noce, it was actually tried. The correlation coefficient between measured and reproduced values by the multiple linear regression analysis was $\rho = 0.87$, while the same by Quantification Theory No.I was $\rho = 0.96$, which indicates that the latter gave better results.

It can also be judged from the correlation coefficient, which was obtained, as to what should be set as the program characteristics of ② and whether it is necessary to add any items. Further, to investigate on what program characteristics should be added, it is advisable to get a hint from the characteristics of those programs which are badly deviated from the correlation diagram in Fig.2.

The method of classification of program characteristics of ③ is of great importance to improve the correlation coefficient, but there is no effective means other than repeating trial and error to decrease internal variance and

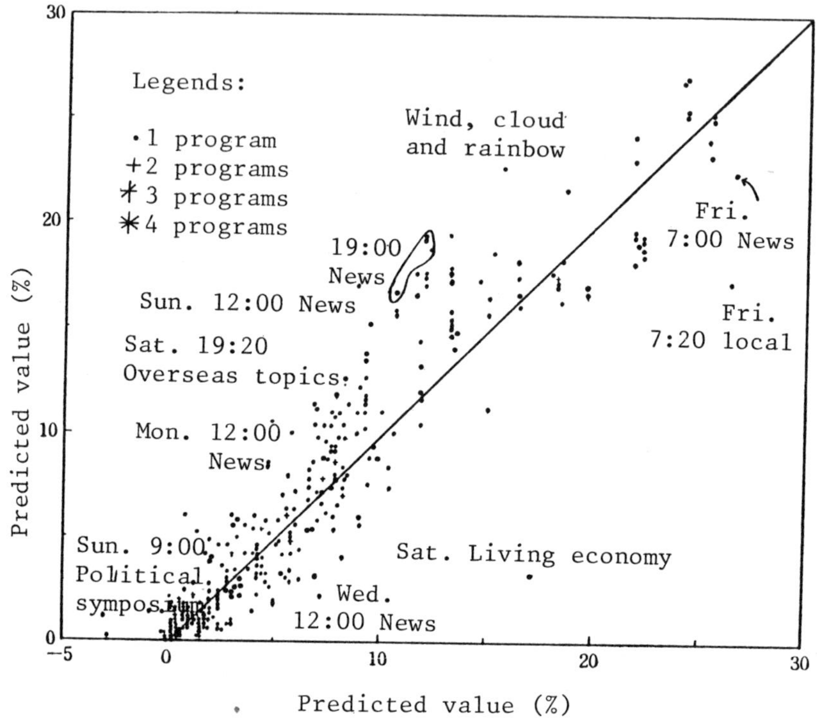

Fig.2 Individual audience prediction (Kanto NHK general TV, Nov., 1976)

increase external variance.

B. Quantification Theory No.II

... TV Audience Prediction on The Individuals Basis(5)

The purpose of this survey is to predict TV program audience of each subject of survey from the various circumstances of the individual. To this end, the relation of the TV program audience with various audience items was clarified. The analysis technique used is Quantification Theory No.II .

Audience items taken up are ① whether the time is convenient for TV audience, ② whether favorite event, ③ whether having liking for the broadcast station, ⑤ whether having watched the program at the preceding time, and ⑥ whether having watched the program at the followint time, and each of these items was divided into categories (2 or 3 classes).

And, each category was given a quantity so that those with TV audience and those without TV audience can be best separated, that is, the correlation ratio η becomes maximum.

The partial correlation coefficient γ between TV audience and each audience item was in a range between 0.00 and 0.35, and among 6 itmes, $\gamma = 0.34$ for the preceding program of ⑤ and $\gamma = 0.35$ for the following program were high.

In both cases of the preceding program of ⑤ and the following program of ⑥, each category score was the highest for "having watched own station", next for "having watched other station", and the lowest for "not having watched".

The results of analysis on the NHK general TV programs from broadcast beginning up to end indicate the correlation ratio of $\eta = 0.62$. This value can never be said high, and it is found difficult to predict the individual audience from these 6 items.

C. Quantification Theory No. III
... Pattern Analysis of Women's News Contact(6)

Since "Analysis of Audience Strata by Application of Quantification Theory" in 1960, Quantification Theory No. III has been very useful for the opinion survey data analysis, and has been familiarized under the name of "Pattern Analysis". At this point, an example of pattern analysis by Quantification Theory No. III, based on the answer (multiple choice response) to "What news (or news show) do you watch frequently" in women's survey, will be taken up.

Table 3 TV audience prediction on the individuals basis (NHK general TV, from broadcast beginning up to end, 10451 subjects, Mar., 1967, Metropolitan areas)

Item Category	Category score	Partial correlation coefficient
① Coincidence to convenient time		γ = 0.21
Yes	0.41	
No	-0.27	
② Coincidence to favorite event		γ = 0.12
Yes	0.19	
No	-0.17	
③ Liking for station		γ = 0.07
Liking	0.10	
Other	-0.09	
Disliking	-0.30	
④ Different channel audience		γ = 0.00
Yes	0.00	
No	0.01	
⑤ Preceding program audience		γ = 0.34
Own station	1.07	
Other station	0.05	
No audience	-0.40	
⑥ Following program audience		γ = 0.35
Own station	1.09	
Other station	-0.37	
No audience	-0.33	
Correlation ratio = 0.62		

In Fig.3, 27 news programs are arranged on a plane based on the similarity of the audience condition. That is, the position of each news program is obtained in accordance with Quantification Theory No.III in such a manner that those news programs, for which the audience condition of the same subject throughout the survey is similar, are arranged nearby and those news programs, for which the audience condition is not similar, are arranged far away.

From the results, it has been found that, as encircled in Fig.3, there are the following five types of way in which women watch TV news.

<Types of TV Audience>

Information type

1. NHK news type
2. Commercial TV show type

Entertainment type

3. Mid-night type
4. Entertainment follow type
5. Special type ... very little contact

However, "NHK news type" and "Commercial TV show type" are located nearby, so if they are combined together, may be named "Information type". Other types may be called "Entertainment type", because the women rather watch spot news incidentally before or after they watch mainly entertainment programs, or they watch news programs early morning or midnight for the reason of their living time. Roughly speaking, in Fig.3 of the program arrangement, the horizontal axis (X) discriminates NHK or commercial TV, and the vertical axis (Y) discriminates information or entertainment.

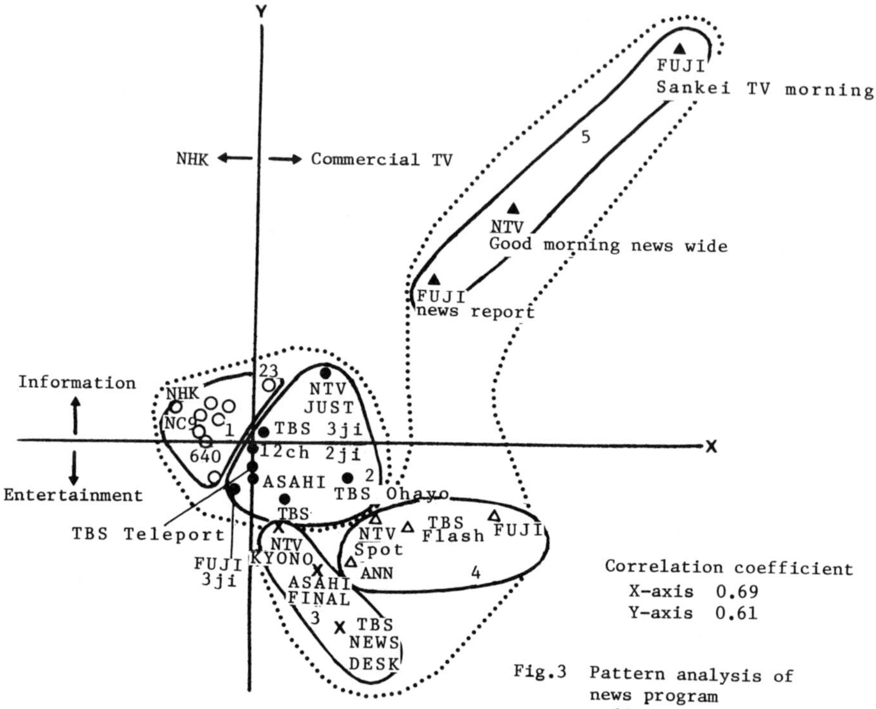

Fig.3 Pattern analysis of news program audience

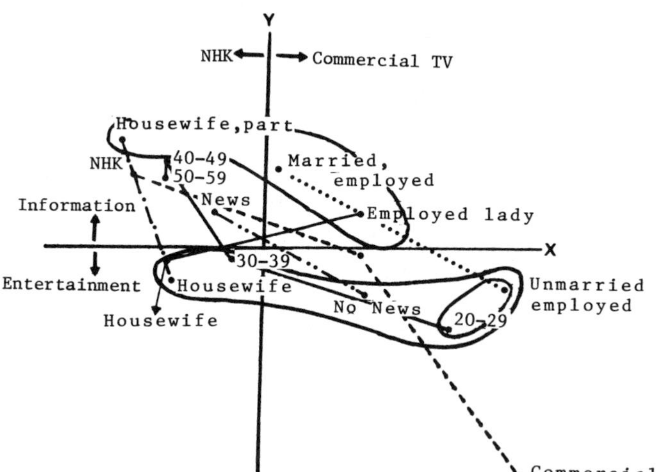

Fig.4 Pattern analysis of news program audience -- by woman's attributes

Similarly to Fig.3 which was drawn based on the similarity of the audience condition in the case of news programs, a figure for subjects (358 persons) can be drawn based on the similarity of the audience condition, by arranging those subjects, who are similar, nearby and those subjects, who are not similar, far away. In this way, Fig.4 was drawn by calculating the average values by attributes of each subject (age stralum, employed or housewife, married or unmarried, occupation, information type or non-information type, NHK type or commerical TV type). From the relative positions in Fig.3 and Fig.4, the following can be known.

① Subjects who have any occupation, except for the unmarried, desire to "contact news information" in any way.

② Homewives are inclined to NHK.

③ NHK type subjects like programs of political affairs such as straight news, news studio 102, etc., while commercial type subjects watch news before or after entertainment programs.

The example shown above refers to the case of multiple choices, but Quantification Theory No.III can handle single-choice questions, too. Therefore, it is a widely applicable technique to study the relation between questions in the opinion survey.

On the other hand, it is to be noted that any categories with less answers tend to be located far away because they are not similar to other categories. If this is known beforehand, it is better to exclude these categories or combine them into other catgories. Further, "Student" by occupation and "At school" by educational history, which are quite the same category to the subject even if the questions are different, cannot be handled simultaneously, and it is necessary to take up either one only.

D. Quantification Theory No. III
... A Woman's Place in Society--Comparative Attitudes in Japan, West Germany and The United States(7)

Quantification Theory No. III was applied to analyze the survey data collectively on three countries Japan, West Germany and the United Sates.

The problem at this point is to see how the people's consciousness about women's way of living are different in these countries.

Questions and choices, which were prepared about the women's way of living, are 6 questions and 26 choices shown in Table 4. These questions were common to all subjects of three countries. The results of Japan were obtained by calculating answers in % for 13 strata in total of 12 strata of subjects by sexe and age and one for the whole of Japan. There were 39 strata in total of three countries.

Therefore, the survey data were obtained as a 26 x 39 matrix. To analyze these data, Quantification Theory No. III was applied. From the results of this analysis, Fig.5 for the arrangement of 26 responses, and Fig.6 for the arrangement of 39 strata, were obtained.

According to Fig.5, three basic patterns emerged from this analysis: "reliance on husband" (left upper), "woman's autonomy" (left lower), and "intermediate" (right). The vertical axis in this figure means "reliance on husband" in the upper and "woman's autonomy" in the lower. The horizontal axis in the same figure discriminates "positive attitude" at left and "vague attitute" at right.

Table 4 Summary of attitudes toward women

A. Sexual relations between unmarried young adults	A1 On condition of marriage	A2 On condition of engagement	A3 On condition of love	A4 On condition of closeness	A5 Unconditional
B. Marriage for women	B1 All should marry	B2 It is better to marry if possible	B3 Not necessary to marry	B4 It is better not to marry	
C. Jobs for women	C1 No jobs	C2 Jobs until marriage	C3 Jobs until children	C4 Job re-entry	C5 Continuous jobs
D. Possibility of divorce	D1 Not possible	D2 Virtually not possible	D3 Somewhat possible	D4 Possible	
E. Sharing of housework by husband	E1 No sharing	E2 Prefer not sharing	E3 Prefer sharing	E4 Sharing required	
F. Husband's right to final decision	F1 Husband always has right to decide	F2 Husband usually should have right to decide	F3 Husband should not necessarily have right to decide	F4 Husband does not have right to decide	

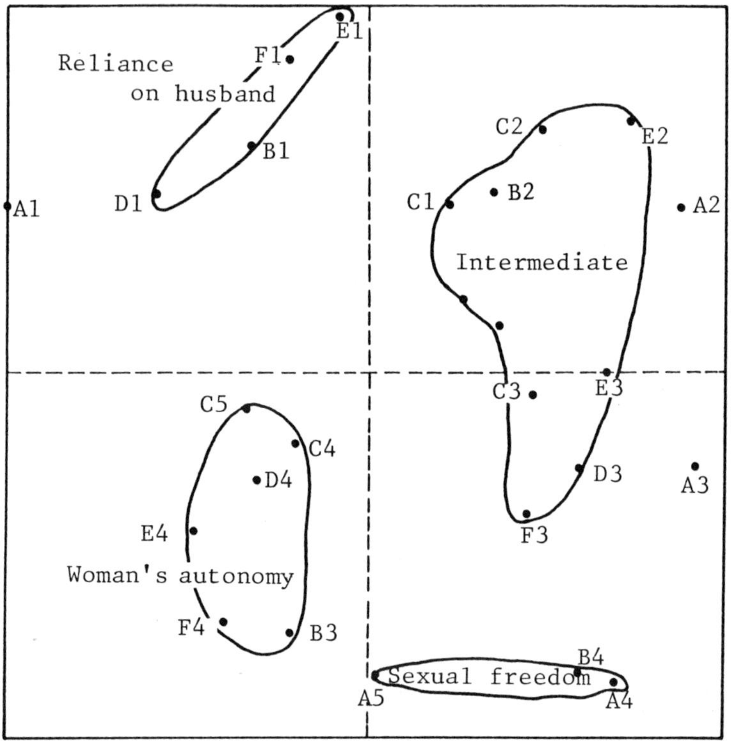

A1-5 Sexual relations between unmarried young adults
B1-4 Marriage for women
C1-5 Jobs for women
D1-4 Possibility of divorce
E1-4 Sharing of housework by husband
F1-4 Husband's right to final decision

Source: NHK International Survey (Japan, U.S.A. 1980, West Germany 1981)

Fig.5 Overall patterns

And, a flow of consciousness from "positive reliance on husband" clockwise through "negative reliance on husband" and "negative worman's autonomy" to "positive woman's autonomy" is suggested.

Then, let us examine what characteristics the people who construct "reliance on husband", "woman's autonomy" and "in-

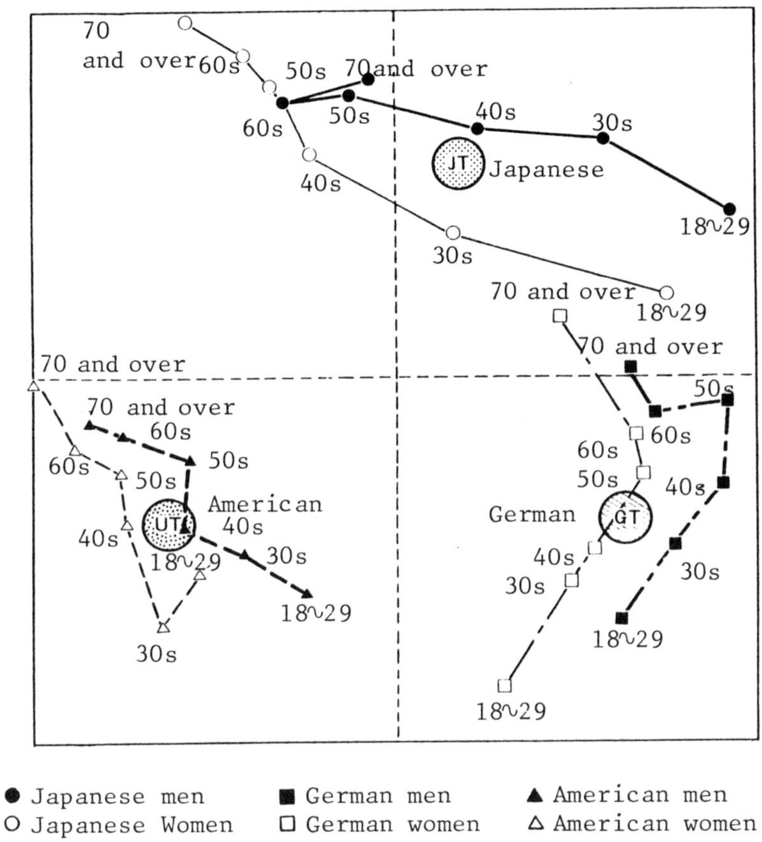

Source: NHK International Survey (Japan, U.S.A. 1980, West Germany 1981)

Fig.6 Response patterns by country, sex, and age

termediate" have.

In Fig.6 showing the arrangement of 39 strata, first the difference by country is prominent. The Japanese occupy the region of "reliance on husband", the Germany the region of "intermediate", and the Americans the region of "woman's autonomy", respectively.

Within each country, there is an age difference, but little sex difference. Especially, Japan shows a larger age difference than other two countries. It is interesting to note that, inspite of the "woman's way of living being discussed, the age difference is larger than the sex difference.

The technique used here can be utilized when it is desired to analyze several surveys, different not only in curvey area, but also in survey time and survey object, collectively.

E. POSA Based ON Quantification Theory No. III ... Analysis of Religious and Faith Behaviors(2)

Guttman's POSA (Partial Order Scalogram Analysis) is said as an effective means of analyzing complex phenomena. However, the method of its solution cannot be known concretely, and so, the POSA analysis will be executed by making use of Quantification Theory No. III.

The POSA aims at distributing the majority of response patterns into a space of the smallest possible dimensions, by eliminating markedly different and infrequent response patterns, with a slight sacrifice of the producibility.

The data to be analyzed are those obtained by opinion survey regarding "Religion and Faith" of the Japanese mentioned before.

For three categories, D (Praying for business prosperity and success in an entrance examination), E (A good-luck amulet and talisman) and F (a written oracle and divination), which mean utilitarian faith among the religious and faith behaviors, calculations in accordance with Quantification Theory No. III have been made. The subject scores obtained, that is, eight kinds (= 2^3) of response patterns are arranged in Fig.7. (Yes: 1, No: 2)

Getting a clue from Fig.7, the POSA diagram is drawn in Fig.8, by excluding (1, 2, 1) 109 persons of relatively small

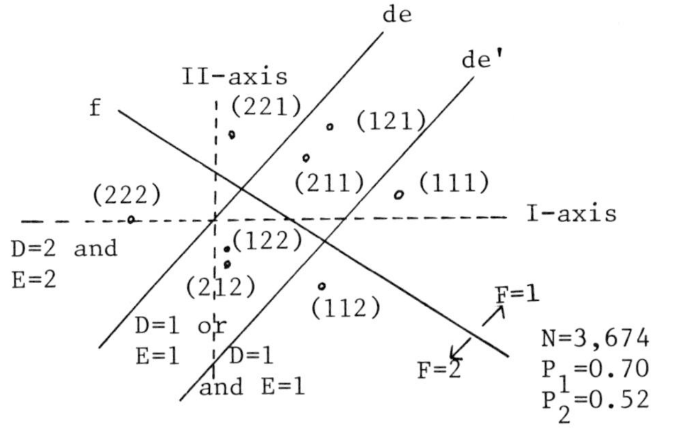

Fig.7 Distribution of response pattern (D,E,F)

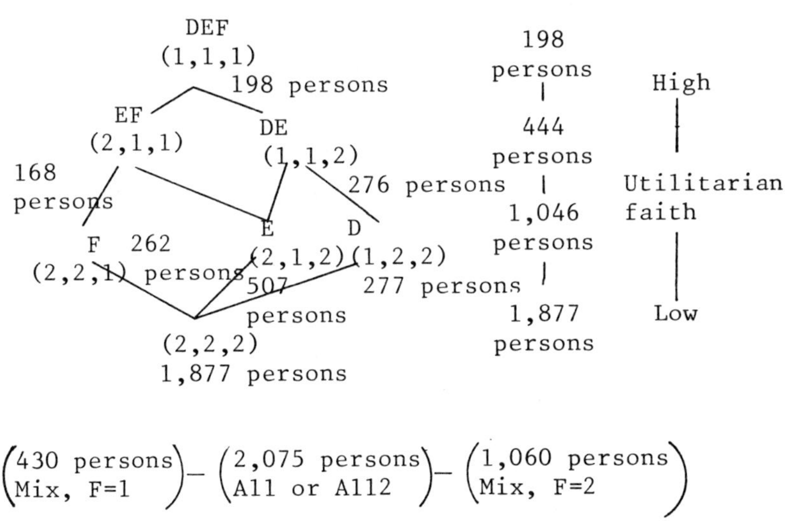

Fig.8 POSA on utilitarian faith (D,E,F)

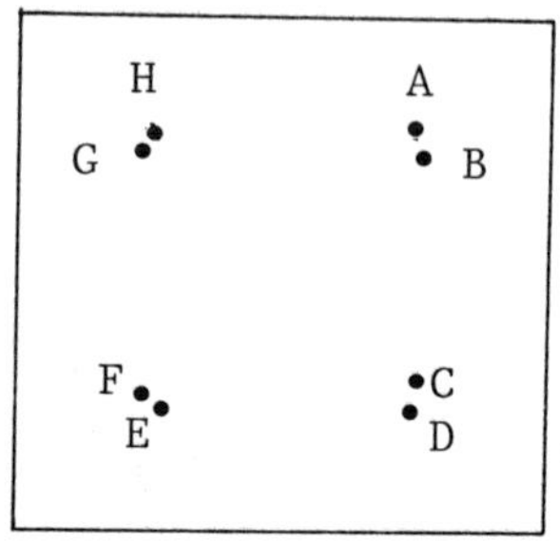

Fig.9 Original arrangement of 8 items

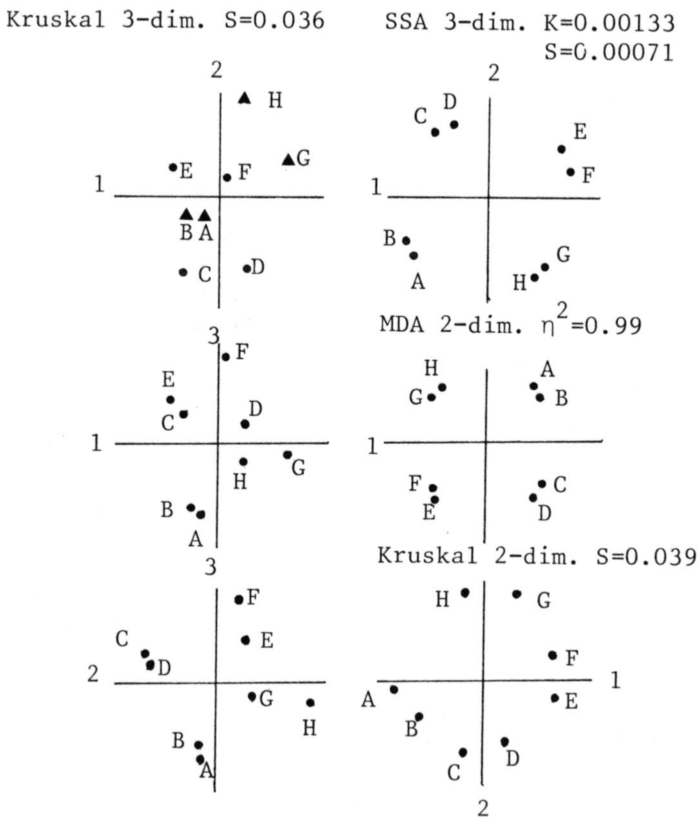

Fig. 10 Space arrangement of 8 items

frequencies. Based on the POSA conception, response patterns with only a single variation are linked together with a tine line.

There are several different flows of D, E and F, which range from all-No (2, 2, 2) to all-Yes (1, 1, 1). One is a flow of Yes being in the counterclockwise order of F-E-D. Another is a flow of Yes being in the clockwise order. In a flow of a large number of persons, E becomes Yes, then D or F becomes Yes, and finally all become Yes.

The reproducibility of this POSA is 97%, or 3565 persons in 3674 persons.

As described above, the POSA diagram could be drawn based on Quantification Theory No. Ⅲ comparatively easily for three utilitarian items, but it cannot be possible at all times. As the number of items is increased, the number of combinations of response patterns is increased, and so, by hand operation, 5 or 6 items will be the limit. Simplifying by eliminating some response patterns is a very delicate art and requires repeated experience.

F. The Problem of Dimensions of Kruskal, SSA and MDA[2]

Of the multidimensional scalogram techniques, three techniques, Kruskal's MDSCAL, Guttman's SSA and Hayashi's MDA-OR, have been studied using model data as to how the space arrangement is varied by limiting the number of dimensions in obtaining the solution.

The model data are 8 items, A through H, for which the original arrangement is shown in Fig.9. In these data, AB, CD, EF and GH are collected in pairs at 4 places, and it is assumed that the distance matrix between the 8 items based on Fig.9 is given.

In the case of Kruskal and SSA, the distance between

these items is not used directly, but is converted into the distance ranking number before use. In the case of MDA-OR, further it is converted into the rank group relation.

In the rank group relation, small differences are ignored in general, but in the distance ranking number, the difference of 0.01 and the difference of 0.99 become sometimes equally a difference of 1 rank.

On assumption of the peculiar properties of each input matrix mentioned above, a space arrangement as shown in Fig.10 is obtained.

Referring to Fig.10, in the case of Kruskal, the number of dimensions showing the minimum stress is taken as the solution. That is, 1-dimensional: $S = 0.383$, 2-dimensional: $S = 0.039$, and 3-dimensional: $S = 0.036$, and so, the 3-dimensional is the solution. The space arrangement in the case of 3 dimensions is resembling to the original arrangement in none of 1 x 2 axis, 1 x 3 axis and 2 x 3 axis, while the space arrangement in the case of 2 dimensions is rather near the original arrangement.

In the case of SSA, the space arrangement drawn for 3 dimensions with minimum coefficient of alientation K is slightly resembling to the original arrangement in the case of 1 x 2 axis (1-dimensional: $K = 0.33638$, 2-dimensional: $K = 0.00136$, and 3-dimensional: $K = 0.00133$).

In the case of MDA-OR, 1-dimensional; $\eta_1^2 = 0.415$ and 2-dimensional: $\eta_2^2 = 0.993$, that is, η^2 approaches 1.0 which is the maximum value in the case of 2 dimensions.

As examined above, the Kruskal's method has a drawback in determining the problem of dimensions. In general cases, unlike in the case of the model, analysis must be done without knowing the original arrangement. Therefore, it is a cause of anxiety that an example which leads to an erroneous conclusion exists.

IV. CONCLUSION

Along with the development of electronic computers, especially the development and consoldation of programs for such multidimensional analysis, the multidimensional analysis of survey data has been reported increasingly. Thus, it has become possible to grasp the entire aspect of the data, to know about the overall structure of the data or interrelation of questions, and improve the analysis, as a favorable trend.

However, it is regret to see frequently that these techniques are applied without adequate understanding. Any of the techniques assume certain conditions or requirements, and the results obtained require to be interpreted with understanding of these conditions or requirements.

First of all, it will be necessary to select the technique consistent to the properties of the survey data and the purpose of analysis. Calculations of this kind do not necessarily given good results at one trial. In many cases, items are added or replaced, and further, item classification criteria are modified, categories are combined together, and trial and error steps are repeated until good results are obtained. Handling of data requires many intricated hand operations.

REFERENCES

1. Akuto, H. (1964). Quantification Theory, Annual Report of Socio-psychology, 5, 73-103.
2. Hayashi, C., Akuto, H. et al. (1976). Multidimensional Scalogram Analysis, Science Publishing Co.
3. Takamiya, Y. and Sugiyama, M. (1978). Relation between Individual Audience Rating and Family Audience Rating, The NHK Report on Broadcast Research, 28(10)

4. Sugiyama, M. (1965). Study on Audience Prediction, The Annual NHK Report on Broadcast Research, 10

5. Sugiyama, M. Report on Studies of Organization Plans and on Surveys of Metropolitan Areas, The NHK Report on Broadcast Research, 17(9), (10), (11)

6. Study on the TV Evaluation, Phase I, May, 1978 TV News Research Group. (unpublished)

7. Sugiyama, M. (1984). A Woman's Place in Society--Comparative Attitudes in Japan, West Germany and The United States, Behaviormetrika, 15, 55-75.

DATA ANALYTIC APPROACHES TO HUMAN BEHAVIORAL RELATIONSHIPS IN A SURVEY OF ACCIDENTS

Koya Kishida*
Yoshio Sugasawa**
Tomio Ito**

(*) Takasaki City University of Economics, Professor
1300 Kaminamie, Takasaki, Gunma 370 JAPAN

(**) Nihon University, Lecturer
College of Industrial Technology
1-2-1 Izumicho, Narashino, Chiba 275 JAPAN

I. INTRODUCTION

To date most analyses of bicycle accidents among traffic accidents have been conducted from the viewpoint of the bicycle rider as the victim, or weaker participant, in traffic. However, when we examine actual bicycle accidents, cases in which the bicycle rider used a dangerous riding style are not few.

An objective of the present study is to re-analyze bicycle accidents from the viewpoint that the bicycle rider is the actual cause of the accident. The study attempts to clarify what constitutes a dangerous riding style. The study surveyed riding styles such as riding double, riding without lights, riding with traffic, riding bicycles with multiple gears, etc., and investigated both actual and near accidents.

In order to extract the cause of accident, the data obtained were analyzed using Hayashi's quantification method III.

ISBN 0-12-215485-1

II. METHOD

The survey was conducted in a relatively small area in Takasaki City, Japan, where bicycles are a major means of transportation. Two questionnaires were used in the survey. One was a survey of bicycle riding styles consisting of 26 items. The other was a survey of accidents and near accidents using bicycles and consisted of 20 items. Riders were questioned both on their riding style and their experience with accidents.

Subjects consisted of five groups: elementary school pupils junior and senior high school students, college students and adults. The survey was carried out by visiting schools and houses from March through April 1984. A total of 1784 subjects were given the questionnaires, but 20 subjects were excluded because of incomplete answers, giving a total of 1764. Deleting 567 persons who did not respond the items required for the present study, the number of subjects used for analysis was 1197, including 309 elementary school pupils (5th grade), 324 junior high school students (8th grade), 296 senior high school students (11th grade), 165 college students and 103 adults.

III. RESULTS AND DISCUSSION

The results of six questions (with 17 possible responses) are shown in Table 1. Subjects who had had near accidents during the last year were 35.2%. Of the subjects who had had a near accidents in the last year, 35.9% had also had near accidents associated with no lights in the past. Of those who has not had any near accidents, 11.9% had experienced near accidents in the past involving no lights. Subjects who

Table 1. Cross-Tabulation Between Riding Style And Accidents

Responses			Q_5: 14 (N = 421) Experienced a near accident during past year		15 (N = 776) Did not experience a near accident during past year		Total (N = 1197)	
			N	%	N	%	N	%
Q_1	1	Bicycle with 2~4 gears	61	14.5	82	10.6	143	11.9
	2	Bicycle with 5 or more gears	188	44.6	318	41.0	506	42.3
	3	Bicycle with no gears	172	40.9	376	48.4	548	45.8
Q_2	4	Always rode without lights after dark	236	56.0	448	57.7	684	57.1
	5	Occasionally rode without lights after dark	162	38.5	284	36.6	446	37.3
	6	Almost rode with lights after dark	23	5.5	44	5.7	67	5.6
Q_3	7	Frequently rode double	51	12.1	83	10.7	134	11.2
	8	Occasionally rode double	191	45.4	347	44.7	538	44.9
	9	Rarely rode double	179	42.5	346	44.6	525	43.9
Q_4	10	Almost always rode with traffic	65	15.4	142	18.3	207	17.3
	11	Always rode with traffic	198	47.0	331	42.7	529	44.2
	12	Occasionally rode with traffic	42	10.0	118	15.2	160	13.4
	13	Indiscriminately rode with or against traffic	116	27.6	185	23.8	301	25.1
Q_6	16	Experienced a near accident without lights in the past	151	35.9	92	11.9	243	20.3
	17	Did not experience a near accident without lights in the past	270	64.1	684	88.1	954	79.7

always rode without lights after dark were 56.1% in near accident group and 57.7% in no near accident group. Subjects who rode double frequently were 12.1% in the near accident group, 10.7% in the no near accident group. Subjects who always rode with traffic were 15.4% in the near accident group and 18.3% in the no near accident group. Subjects who rode bicycles with five gears or more were 44.7% in the near accident group; 41.0% in the no near accident group.

Based on this data, a cross-tabulation with sex and age group as columns, riding style as rows was made. This cross-tabulation was analyzed using Hayashi's quantification method III.

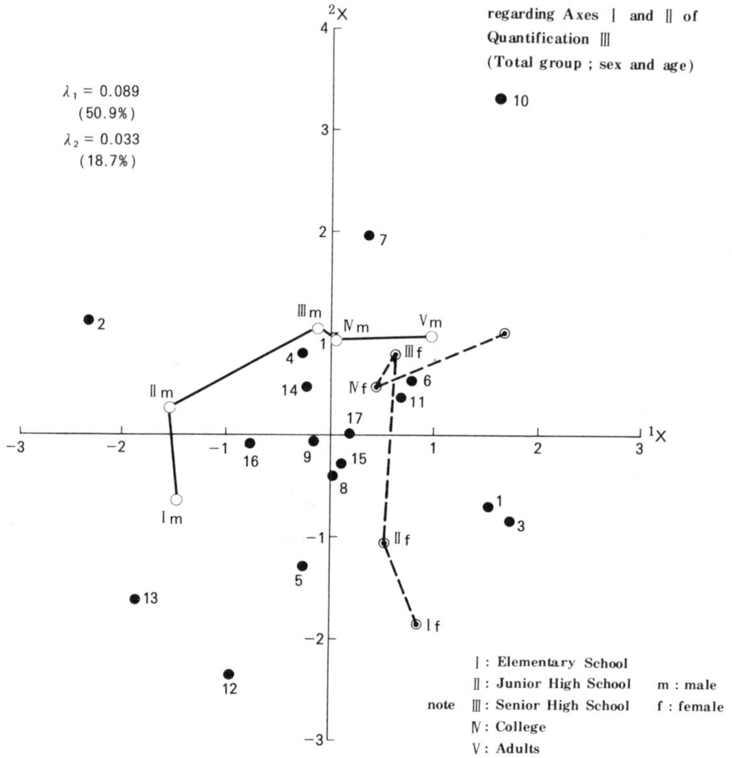

Fig. 1 Configuration of items regarding Axes Ⅰ and Ⅱ of Quantification Ⅲ (Total group ; sex and age)

The result is shown Figure 1. Eigenvalue of the first axis was 0.089, and of the second axis was 0.033. The meaning of the first axis was the relationship between riding with traffic and bicycles with gears. This relationship is between operating ability and observance of the law. The second axis is the relationship between riding double and riding with traffic.

The tendencies for sex and age suggest that in females, junior and senior high school students had bicycle with 2 - 4 gears, while elementary school pupils and adults had bicycles with no gear. Almost all senior high school and college students rode with traffic. Most junior and senior high school students also rode double occasionally. Many elementary pupils and junior high school students occasionally rode without lights after dark.

In males, many senior high school students always rode without lights. Junior high school students had bicycles with 5 or more gears. Most of the elementary pupils paid little attention to whether they rode with or against traffic. It seems that male had a slightly stronger tendency to experience near accidents. In both sexes, many senior high school students, college students and adults almost always rode double. On the other hand, most adults and college students observed the rule of riding with traffic. Most of these also had driving licenses; thus, they believed that riding with traffic is safer.

It was difficult to clarify the relationships among riding style, law observance and near accidents by the analysis above. Thus, a second analysis was carried out by applying Hayashi's quantification method III to a second cross-tabulation of near accidents.

Figure 2 shows one of the results of the second analysis. The first axis is the first axis of in Figure 1, but the second axis in Figure 1 was changed to the third axis in the second alaysis.

Figure 2 shows for males that elementary pupils who only occasionally rode with traffic or paid no attention to their direction experienced near accidents. Junior high school students who rode bicycle with five or more gears also experienced near accidents. Senior high school students who had bicycle with 2 - 4 gears, who frequently rode without lights after dark, and who rode double experienced near accidents. College students and adults who always rode with traffic and did not ride double had very few near accidents. This tendency is also suggested from the results of an analysis of the group with no near accidents. Among female

Fig. 2 Configuration of items regarding Axes I and III of Quantification III (Near accidents group ; sex and age)

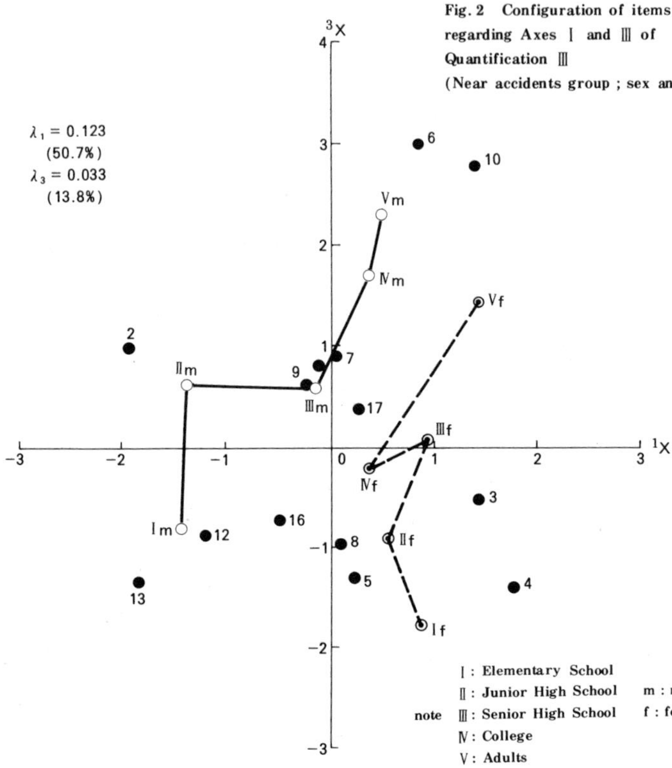

note
I : Elementary School
II : Junior High School
III : Senior High School
IV : College
V : Adults

m : male
f : female

The meaning of Responses see table 1.

junior high school students and elementary school pupils who occasionally rode double and without lights had near accidents.

Thus Figure 2 shows a clearer relationship between riding style and near accidents. However, in the present study, it seems that sexual differences riding style influenced the occurrence of bicycle accidents. The third analysis was done by applying Hayashi's quantification method III to a

cross-tabulation with changed columns and rows. This time the column was age group and the rows were male and female riding style.

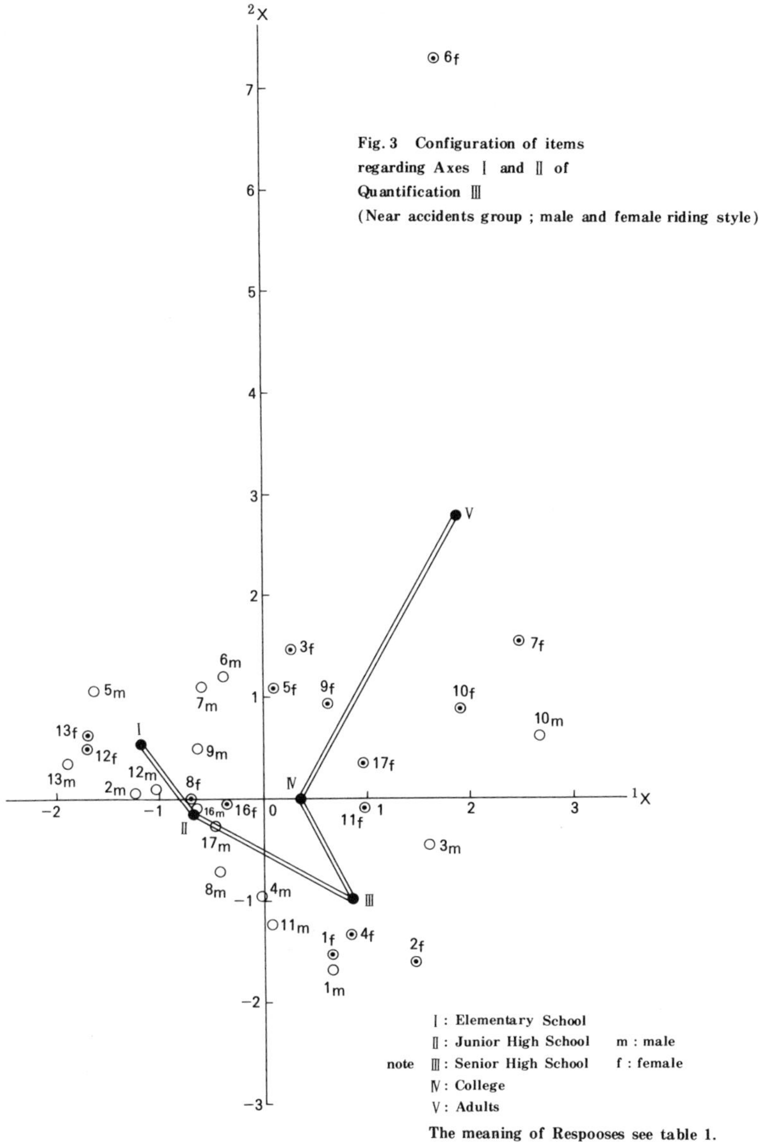

Fig. 3 Configuration of items regarding Axes I and II of Quantification III
(Near accidents group ; male and female riding style)

Figure 3 shows the results for subjects who had experienced near accidents. From Figure 3, male senior high school students who had bicycles with 2 - 4 gears, and rode without lights had near accident experiences. Female senior high school students who had bicycle with two or more gears, and always rode without lights experienced near accidents. Junior high school students who had bicycle with 2 - 4 gears, and only occasionally rode with traffic experienced near accidents. Some female junior high school students who occasionally rode double also had near accident experiences. It was revealed that male elementary school pupils who had bicycles with five or more gears, and rode freely with or against traffic or only occasionally with traffic experienced near accidents. Female adults with infants had very few near accidents. Though they rode double, they always rode with traffic and did not ride without lights.

In the present study, the data was analyzed by Hayashi's quantification method III to clarify the relationship between riding style and accidents. Using the results obtained, it was easy to explain the relationship between riding style and accidents.

ACKNOWLEDGEMENTS

The authors are grateful to Professor N. Ohsumi who introduced us to this seminar and made useful comments.

This research was originally supported by the TOYOTA foundation.

REFERENCE

Kishida, K. (1984). A Study on Causes of Bicycle Accidents (in Japanese), The Japanese Journal of Ergonomics, 20, Special Issue, 88 - 89.

ABOUT THE NUMERICAL AND STATISTICAL STABILITY OF FORECASTING ALGORITHMS

Guy der Megreditchian

Direction de la Météorologie Nationale
2 avenue Rapp 75007 Paris
France

I. THE STATISTICAL MODEL OF EVENTS FORECAST

Discriminant analysis is the adequate tool for the statistical forecast of some events. When the random experiment has two possible issues (the occurrence A_1 and the non-occurrence A_2 of the event A) :

$$\Omega = A_1 + A_2 \tag{1}$$

the forecast of the indicator variable y of the event A is realized by means of the discriminant function g(X) corresponding to the optimal decision rule. Such rule is obtained by means of the minimization of the quality criterion Q, defined as the mathematical expectation of the cost c associated with the consequences of the adopted decisions :

$$Q = E[c] \tag{2}$$

This optimal decision rule may be formulated in terms of the discriminant function g(X) defined by the expression [2,3] :

$$g[X] = \frac{p(A_1) \cdot f_{A_1}(X) \cdot c[\widetilde{A}_2 / A_1]}{p(A_2) \cdot f_{A_2}(X) \cdot c[\widetilde{A}_1 / A_2]}, \tag{3}$$

ISBN 0-12-215485-1

where $p_i = P[A_i]$ is the "a priori" probability of the event A_i, $f_{A_i}[X]$ the conditional probability density function of the vector X, if A_i occurs, and $c[\tilde{A}_i/A_j]$ the cost associated to the decision A_i, while A_j occurs.
The forecasted value $\tilde{y}$ of y (here y = 1 if A_1 occurs and y = 2 if A_2 occurs) is then given by the relation :

$$\tilde{y} = \frac{3 + \text{sign}\,[1 - g(X)]}{2} \qquad (3')$$

The concretization of the discriminant procedure requires the knowledge of the "a priori" probabilities p_i, the conditional probability density functions $f_{A_i}(X)$ and the costs $c[\tilde{A}_i/A_j]$.
The cost matrix has to be defined by the user on the basis of specific economical considerations involved by the concrete forecasting problem. From a general point of view, if A_1 is the dangerous phenomena, one considers that the cost $c[\tilde{A}_1/A_2]$ of the false alert is less than the cost of the non-detection :

$$c\,[\tilde{A}_1/A_2] < c\,[\tilde{A}_2/A_1]$$

The definition of the costs $c[\tilde{A}_i/A_j]$ is not easy, even for the specialist of the concerned field, but fortunately in discriminant analysis it's sufficient to know the proportion $c[\tilde{A}_1/A_2]/c[\tilde{A}_2/A_1]$, which is much more easy to determine. For such proportion one uses frequently the values 1/3 or 1/5. In the case of lack of any economical consideration it's convenient to choose a solution of simplicity, consisting to adopt the value :

$$\frac{c\,[\tilde{A}_1/A_2]}{c\,[\tilde{A}_2/A_1]} = \frac{p\,(A_1)}{p\,(A_2)} \qquad (4)$$

The theoretical probability density functions $f_{A_i}(X)$ are never known in practical problems and there exist two possible ways for their concretization.

First, it's possible to get some estimation $f_{A_i}(X)$ of such functions on the avalaible set of data by means of the application of one of the existing methods of estimation of probability density function. For example, we can use the Kernel method [10], or the Potential functions method [1], or the Fix and Hodges method [7], or a any other adequate method. Besides, another way leads to postulate a choice of the family of the distribution law of the vector X of predictors. More frequently one considers that X follows a normal distribution law $N[M_{(i)}, V_{(i)}]$ with conditional mathematical expectations

$$E_{A_i}[X] = M_{(i)} \quad ,$$

and covariance matrix :

$$E_{A_i}\left[\overset{o}{X} \cdot \overset{o}{X}'\right] = V_{(i)} \; .$$

Consequently, in the general case $V_{(1)} \neq V_{(2)}$ the discriminant function is quadratic, but if we suppose the equality of theoretical covariance matrix

$$V_{(1)} = V_{(2)} = V \tag{5}$$

then it becomes linear :

$$g_L(X) = M'_X(-)\, V^{-1}\left[X - M_X(+)\right], \tag{6}$$

where

$$M(-) = M_{(1)} - M_{(2)} \; ; \quad M(+) = \frac{1}{2}\left[M_{(1)} + M_{(2)}\right] \tag{7}$$

Nevertheless the imposed restrictions of normality and equality of covariance matrix, the linear discriminant function remains extremly popular and many programs are avalaible for its realization. In particular, the interest of such a classical approach is that it gives the possibility to associate to a geometrical index of separability of the two populations A_1 and A_2, the Mahalanobis distance

$$\Delta_n^2\left[A_1, A_2\right] = M'_X(-)\, V^{-1}\, M_X(-) \tag{8}$$

a corresponding index of informativity for the given number of concerned predictors. For that, we have a classical relation defining the decision error probability :

$$P(e) = \Phi\left[-\Delta/2\right], \tag{9}$$

where

$$\Phi(Z) = 1/\sqrt{2\pi} \int_{-\infty}^{Z} e^{-X^2/2} dx$$

is the distribution function of the (0,1) normal variable. The more the populations are separated, the higher is the Mahalanobis distance and lesser the error probability.

II. ATTENUATION COEFFICIENT OF THE MAHALANOBIS DISTANCE

A long practice of the meteorological forecast has shown that a "good" forecast corresponds to a percentage of error of 15 to 20 %, so a percentage of good forecasts of 80 to 85 %. For such values of P(e) corresponds through the formula (9) a Mahalanobis distance being between 2,8 and 4,2. Nevertheless, one can meet in the technical literature much higher unrealistic values of Δ_n^2, though it's well known that the quality of forecasts remains at the indicated very poor levels. Two reasons are the outcome of such overevaluation of the Δ_n^2.

The first is the bias arising due to the sampling fluctuations, that can be evaluated in the case of gaussian populations. So, even if we suppose that the predictors are totally uninformative, i.e. that the theoretical value of the Mahalanobis distance is zero :

$$\Delta_n^2 = 0$$

this does not prevent the empirical $\hat{\Delta}_n^2$ from having a value different from zero. As for every random variable, it's possible to characterize it by means of its conditional mathematical expectation $E[\hat{\Delta}_n^2 / \Delta_n^2 = 0]$, which we will study in the next paragraph.

So, the $\hat{\Delta}_n^2$ has an optimistic bias that can be eliminated by introducing the unbiased estimation $\tilde{\Delta}_n^2$ of Δ_n^2, which is of course more realistic than $\hat{\Delta}_n^2$.

But, it is not sufficient, because there is another reason for the artificial inflating of the empirical value of Δ_n^2. The problem is that, in fact, the formula (9) is valid rigorously speaking only for the balanced samples for which $p_1 = p_2$. In the opposite case, the correspondance between geometrical separability and informativity is no more avalaible, and it's necessary to introduce an appropriate correction for the Mahalanobis distance in order to make its utilisation for imbalanced samples possible. Such correction may be justified in the frame of the Discriminant Regression Model, which allows to build a bridge between the discriminant model and the associated regression model [4].

Thus we can establish the relation :

$$Q_D = \frac{Q_R}{p_1 p_2 [1 - Q_R]}, \qquad (11)$$

where Q_D is the quality criteria for the discrimination i.e. the Mahalanobis distance :

$$Q_D = \Delta_n^2 [A_1, A_2] = M'_X(-) V_{XX}^{-1} M_X(-) \qquad (12)$$

and Q_R is the quality criteria for the regression i.e. the square of the multiple correlation coefficient :

$$Q_R = r_n^2 [y, \tilde{y}] = R_{yX} R_{XX}^{-1} R_{Xy} \quad . \qquad (13)$$

Consequently we observe that when Q_R has a fixed value, but when $p_i \longrightarrow 0$, then $Q_D \longrightarrow \infty$, and the relation (9) becomes inapplicable, as Q_D gets more and more important values, while the intrinsic quality Q_R of the model remains unmodified.
In such a case and in connection with (11) the intrinsic quality of the forecasting model will be :

$$Q_R = 1 - 1 / \left[1 + p_1 \, p_2 \, Q_D \right] . \qquad (14)$$

Now we will evaluate a new value $\tilde{Q}_D$ for the criteria Q_D, which in the case of balanced samples (in that case the relation (9) will be valid) corresponds to the intrinsic quality Q_R : for that, it's sufficient to put $p_1 = p_2 = 1/2$ in the relation (11) :

$$\tilde{Q}_D = 4 \, \frac{Q_R}{1 - Q_R} . \qquad (15)$$

By shifting the value (14) into the relation (15), we obtain the corrected value $\tilde{Q}_D$ of Q_D :

$$\tilde{Q}_D = 4 \, p_1 \, p_2 \, \hat{Q}_D . \qquad (16)$$

We can write it , concretely :

$$\tilde{\Delta}^2_n = 4 \, p_1 \, p_2 \, \hat{\Delta}^2_n . \qquad (17)$$

For the explanation of the influence of correction, we introduce the attenuation coefficient γ for Δ^2_n, which is due to the imbalance of the sample :

$$\gamma = \tilde{\Delta}^2_n / \hat{\Delta}^2_n = 4 \, p_1 \, p_2 . \qquad (18)$$

The coefficient γ is equal to 1 for balanced sample ($p_1 = p_2 = 1/2$) and tends to 0, when $p_i \longrightarrow 0$.

III. THE DISTRIBUTION LAW OF THE MAHALANOBIS DISTANCE

The bridge we established between regression and discrimination, which is concretized by the relation (11), may also be utilized for the study of distibution law of the empirical Mahalanobis distance.

Indeed, instead of searching directly for such a law by means of the study of the distribution of the corresponding quadratic form of gaussian vectors (with a coefficient matrix, which is the inverse of the empirical covariance matrix), we can utilize the classical results of Fisher for the quality index Q_R for the case of regression. It gives us an easy way to solve a complex problem, which needs a difficult manipulation of the Wishart's law.

Let us first recall the classical result of Fisher for the conditional distribution law of the empirical multiple correlation coefficient : the case $r_n^2 = 0$.

The conditional probability density function of $\hat{r}_n^2$ (when $r_n^2 = 0$) is a Beta I law with parameters $p = n/2$ and $q = \frac{T-n-1}{2}$:

$$f_{\hat{r}_n^2}[x / r_n^2 = 0] = f_{\beta_I}[x/p,q] = \frac{1}{B[p,q]} x^{\frac{n}{2}-1} [1-x]^{\frac{T-n-1}{2}-1} \quad (19)$$

where

$$f_{\beta_I}[x/p,q] = \frac{1}{B[p,q]} x^{p-1} [1-x]^{q-1} \quad . \quad (20)$$

Introducing the Beta II law by means of the transformation :

$$\beta_{II} = \beta_I / [1-\beta_I]$$

the corresponding probability density function is then :

$$f_{\beta_{II}}[x/p,q] = \frac{1}{B[p,q]} x^{p-1} [1+x]^{-(p+q)} \quad (21)$$

The moments of such law is easy to compute :

$$m_\ell\left[\beta_I(p,q)\right] = \prod_{i=0}^{\ell-1} [p+i]/[p+q+i] \tag{22}$$

$$m_\ell\left[\beta_{II}(p,q)\right] = \prod_{i=0}^{\ell-1} [p+i]/[q-i-1] \tag{22'}$$

In particular, we have :

$$E\left[\beta_I(p,q)\right] = p/[p+q]\,;\ \mathcal{D}\left[\beta_I(p,q)\right] = \frac{pq}{(p+q)^2(p+q+1)} \tag{23}$$

$$E\left[\beta_{II}(p,q)\right] = p/[q-1]\,;\ \mathcal{D}\left[\beta_{II}(p,q)\right] = \frac{p(p+q-1)}{(q-1)^2(q-2)} \tag{23'}$$

As $\hat{r}^2_n \in \beta_I\left[\frac{n}{2}, \frac{T-n-1}{2}\right]$, we get immediately that $\hat{r}^2_n/[1-\hat{r}^2_n] \in \beta_{II}\left[\frac{n}{2}, \frac{T-n-1}{2}\right]$,

or in other terms : $f_{\frac{\hat{r}^2_n}{1-\hat{r}^2_n}}(x) = \dfrac{1}{B\left[\frac{n}{2}, \frac{T-n-1}{2}\right]}\, x^{\frac{n}{2}-1}\,[1+x]^{-\left(\frac{T-1}{2}\right)}$.

It follows from (22') and (25') the expression of the moments of this random variable :

$$m_\ell\left[\hat{r}^2_n/(1-\hat{r}^2_n)\right] = \prod_{i=0}^{\ell-1} (n+2i)/(T-n-2i-3) \quad ; \tag{24}$$

$$E\left[\hat{r}^2_n/(1-\hat{r}^2_n)\right] = \frac{n}{T-n-3}\,;\ \mathcal{D}\left[\hat{r}^2_n/(1-\hat{r}^2_n)\right] = \frac{2n(T-3)}{(T-n-3)^2(T-n-5)}\ . \tag{24'}$$

Let us now utilize the relation (11) to obtain the conditional distribution law of $\hat{\Delta}^2_n$, if $\Delta^2_n = 0$ putting $a = T^2/T_1T_2$

$$f_{\hat{\Delta}^2_n}\left[x/\Delta^2_n=0\right] = f_{a\frac{\hat{r}^2_n}{1-\hat{r}^2_n}}\left[x/r^2_n=0\right] = \frac{1}{a}\, f_{\beta_{II}}\left[\frac{x}{a}/\frac{n}{2}, \frac{T-n-1}{2}\right].$$

We obtain finally :

$$f_{\hat{\Delta}^2_n}\left[x/\Delta^2_n=0\right] = \frac{1}{B\left[\frac{n}{2}, \frac{T-n-1}{2}\right]}\left[\frac{T_1T_2}{T^2}\right]^{\frac{n}{2}} x^{\frac{n}{2}-1}\left[1+\frac{T_1T_2x}{T^2}\right]^{-\left(\frac{T-1}{2}\right)}. \tag{25}$$

So, the conditional moments of $\hat{\Delta}^2_n$ (if $\Delta^2_n = 0$) are obtained directly from (24) and (24') by taking into account the relation :

$$m_\ell[a\,\xi] = a^\ell\, m_\ell[\xi]\ . \tag{25'}$$

We obtain in this case under the condition $\Delta_n^2 = 0$, the expressions :

$$m_\ell \left[\hat{\Delta}_n^2 / \Delta_n^2 = 0\right] = \left[T^2/(T_1 T_2)\right]^\ell \cdot \prod_{i=0}^{\ell-1} [n+2i]/[T-n-2i-3]. \quad (26)$$

$$E\left[\hat{\Delta}_n^2 / \Delta_n^2 = 0\right] = \left[T^2/(T_1 T_2)\right] \cdot n/[T-n-3] \quad . \quad (27)$$

$$\mathfrak{D}\left[\hat{\Delta}_n^2 / \Delta_n^2 = 0\right] = \left[T^2/(T_1 T_2)\right]^2; \; 2n[T-3]/[T-n-3]^2 [T-n-5]. (27')$$

The obtained formulas reveal a pernicious (optimistic) effect of the sampling distribution law of $\hat{\Delta}_n^2$, expressing the fact that even when the n predictors don't bring any information for the separation of concerned populations ($\Delta_n^2 = 0$), the empirical value $\hat{\Delta}_n^2$ calculated on the avalaible sample seems to indicate some (illusory) possibility of separation of the two populations. Such illusory separability is as much important as the proportion n/T is high ; so, for the objectivisation of the forecasting model quality evaluation, we have a great interest to take the smaller number of predictors.

Such a "parcimony principle" will also find another justification from the point of view of numerical and statistical stability of the forecasting models.

IV. NUMERICAL STABILITY OF THE FORECASTING MODEL

Usually the statistician collects the maximum avalaible set of informations for the elaboration of its forecasting model. Then he finds himself confronted with the situation for which he has an overabundance of potential predictors.

Consequently he will have some links to overestimate the number of predictors. It may appear that one can only win by adding some complementary information, even if it is redundant. But that is not true. The famous temptation to take "always more and more" must be replaced here by the "parcimony principle" requiring the pityless elimination of redundant predictors. We shall illustrate such assertion on the example of the regression model [5].

Let X be an excellent vector of very informative predictors : the forecast given by the regression equation (for the centered and normalized variables)

$$\tilde{y} = R_{yX} R_{XX}^{-1} X$$

is a good forecast for y.

Let us suppose now, that we add a complementary predictor x to the vector X.

The new forecasted value $\tilde{\tilde{y}}[X,x]$ of y may be expressed by means of the previous expression of $\tilde{y}$ by applying an algebraical identity [6], very useful for the study of quadratic and bilinear forms

$$\tilde{\tilde{y}}\ (X,x) = \tilde{y}\ (X) + \frac{\left[r_{yx} - R_{xX} R_{XX}^{-1} R_{Xy}\right]\left[x - R_{xX} R_{XX}^{-1} X\right]}{1 - R_{xX} R_{XX}^{-1} R_{Xx}} \qquad (28)$$

which is possible to express in the form

$$\tilde{\tilde{y}}\ (X,x) = \tilde{y}\ (X) + \frac{\left[D \cdot \rho(y,x/X)\right]\left[x - \tilde{x}(X)\right]}{1 - r_n^2\left[x, \tilde{x}\ (X)\right]} \qquad (29)$$

What happens when the information brought by the new predictor x is redundant with the previous information given by the vector X ?

The term $D\rho(y,x/X)$ will be equal to 0 ; because $\rho[y,x/X]$ is the partial correlation coefficient between y and x, when X is fixed. So when x is well explained by X, then we have $\rho[y,x/X] = 0$.

Similarly in this case the multiple correlation coefficient is near to 1, so $1 - r_n^2(x, \tilde{x}(X)) \approx 0$.

Otherwise, if x is well explained by X its forecasted value $\tilde{x}(X)$, is very near to the real value x, so, we have once more $x \approx \tilde{x}(X)$ or $x-\tilde{x}(X) \approx 0$.

Taking into account these results, we can write for the forecasted value $\tilde{\tilde{y}}$ of y (introducing $\overset{+}{X} = \{X,x\}$) in the form :

$$\tilde{\tilde{y}}\,(\overset{+}{X}) = \tilde{y}\,(X) + \frac{0 \times 0}{0} \; . \tag{30}$$

The redundancy of the new predictor x with the previous vector X of predictors leads to the appearance of an indeterminated form of the type $\frac{0 \times 0}{0}$, that determines the numerical instability of the forecasting model.

In the same way, we obtain for the discriminant model a formula symetrical to (28).

V. STATISTICAL STABILITY OF THE FORECASTING MODEL

The usual procedure for the elaboration of a forecasting model begins with the definition of a decision rule :

$$G = \left\{ g_{\alpha}(X) \right\} \qquad , \text{ where } \quad \alpha \in I_G$$

and a quality criterion Q of the forecast. The set I_G characterized the values of the parameter α (which may be a vector) which concretizes the rule g_{α} (X).

For example, G may be the set of the linear decision rules and Q the mathematical expectation of the cost of the forecast.

After that we find the best decision rule optimizing the quality criterion.

For that, we have to distinguish between two values of the quality criterion : Q the value evaluated on a test sample.

$\hat{Q}$ the value evaluated on the learning sample ;
Usually we define on the learning sample "the best" decision rule $g_{\alpha_{opt}}(X)$ defined by the relation :

$$\alpha_{opt} = \arg \underset{\alpha \in I_G}{\mathrm{opt}} \hat{Q}\left[g_\alpha(X)\right] \qquad (32)$$

However, in pratical problems the objective quality of a decision rule will be defined by the value $Q[g_\alpha (X)]$, and not by $\hat{Q}[g_\alpha (X)]$; our problem is to give forecasts not for the past, but for the future data, not for the learning sample, but for the test sample. This is why it is of prime interest to give an evaluation of the discrepancy between Q and $\hat{Q}$, i.e. of the quality degradation $\delta Q = Q - \hat{Q}$.
Each practician knows well that such degradation depends on the complexity of the utilized rule. The more the decisional rule is complex, the more the degradation will be important. So, it's necessary to formalize the notion of "complexity" of a set of decision rules. The way in this field was shown in the works of Vapnik and Tchervonenkis [13]. We will be presenting a particular case for which the results are very clear.
Let us consider a finite set of decisional rules :

$$G = \{g_i(X)\} \ ; \ i = 1, N$$

In such a case the complexity of the set G may be assimilated to the richness of its elements, that is simply the number N of avalaible decision rules.
It's easy to understand, that it is necessary to evaluate the speed of uniform convergence of $\hat{Q}$ to Q on all α , and in this connection, some important theorems can be formulated in the following form [12].
<u>THEOREM</u> : If $G = \{ g_i \ (X)\}$ is a finite set of decision rules $(i = 1, N)$, then it's possible to affirm with a probability greater than $1 - \eta$, that the overall degradation $\sup |Q_i - \hat{Q}_i|$ of the quality will be less than ϵ . Here the confidence

level η will be a function of the size T of the sample, of the index N of the complexity of the set G, and of the level ϵ of the quality degradation.

$$P \left\{ \sup_i \mid Q(g_i(X)) - \hat{Q}(g_i(X)) \mid < \epsilon \right\} > 1 - \eta \ .$$

For the expression of η we have several results based on the precision of the evaluation of the discrepancy between the frequency and the probability of some event.

1) From the Tchebichev inequality :

$$\eta = N / 4 T \epsilon^2 \ ;$$

2) From the Bernstein inequality :

$$\eta = N . e^{-T\epsilon^2} \ ;$$

3) From the Hoeffding inequality :

$$\eta = 2 N . e^{-2T\epsilon^2}$$

Such results allow several interesting applications.

A) For a given size T of the sample, a given complexity N and confidence level $1-\eta$ we have an estimation of the potential level of degradation

$$\epsilon = \sqrt{[\ell n (2N) - \ell n \eta] / 2T} \ ;$$

B) For a given degradation $\epsilon > 0$, complexity N and confidence level $1 - \eta$, we have an estimation of the size of the data file, necessary to insure the validity of the assertion of the theorem.

$$T = [\ell n (2N) - \ell n \eta] / 2 \epsilon^2 \ ;$$

C) For a given value of degradation $\epsilon > 0$, of the sample size T and complexity N, we can define the confidence level $1-\eta$ corresponding to the assertion of the theorem :

$$\eta = 2N . \exp\left[-2T \epsilon^2\right] .$$

Such results are generalized in the case of an infinite set G of decision rules, and have an analoguous structure :

$$P \left\{ \sup_\alpha \left| Q(\alpha) - \hat{Q}(\alpha) \right| > \epsilon \right\} < \varphi_1 [G] \times \varphi_2 [\delta Q] ,$$

where $\varphi_1[G]$ is a measure of the complexity, of the richness of G, and $\varphi_2[\delta Q]$ is an estimation of the probability to obtain a given level of degradation for a unique decision rule.

In this connection different notions of complexity have been defined by Nilsson [8], Vapnik and Tchervonenkis [13], and others. They may be utilized in the problem of uniform convergence speed estimation of the degradation to zero.

Let us consider a sample

$X = (X(t))$, where $t \in \mathcal{T}$ and card $(\mathcal{T}) = T$,

and a family $B = (B_\alpha)$, $\alpha \in I_B$ of random events B_α associated with the partition of R^n :

$R^n = B_\alpha + \overline{B}_\alpha$; $B_\alpha \cap \overline{B}_\alpha = \emptyset$, and $\mathcal{T} = \mathcal{T}_\alpha + \overline{\mathcal{T}}_\alpha$, where $t \in \mathcal{T}_\alpha$, if $X(t) \in B_\alpha$.

So, we have associated to each event B_α a corresponding partition :

$$X = X(B_\alpha) + X(\overline{B}_\alpha)$$

of the sample X.

Now, we will give several definitions to precise the notion of complexity [12].

<u>DEFINITION 1</u> : The number $N_B(X)$ of all possible partitions $X = X(B_\alpha) + X(\overline{B}_\alpha)$ generated by the set B, is called the index of the set B relatively to the sample X.

<u>DEFINITION 2</u> : The maximum value $N_B(T)$ of $N_B(X_T)$ for all possible samples of size T is called the growing function of the set B :

$$C_B[T] = \max_{\forall X_T} N_B(X_T)$$

<u>DEFINITION 3</u> : The maximum value K of the size T of the sample X for which the set B allows to perform all the 2^T possible partitions of the sample $X \in R^n$ is called the capacity of the set B. In other terms :

$$C_B[K] = 2^K, \text{ but } C_B[K+1] < 2^{K+1}.$$

The definition of the capacity "K" is related to a remarkable property of the growing function $C_B(T)$, which is concretized by the following theorem [12].

THEOREM : For the growing function $C_B(T)$ there are two possibilities ; for any value "K" of the size T either :

1) $C_B [T] = 2^K$

or

2) $C_B [T] < 1,5 . T^K/_{K!}$

If $C_B[T] = 2^T$, then the capacity of the set B is called infinite.

If a value "K" exists, for which $C_B[T] < 1,5\, T^K/_{K!}$, then the capacity of the set B is said finite and equal to K : K(B) = K.

DEFINITION 4 : The entropy $H_B[T]$ of the set B is defined by the relation :

$H_B [T] = E \left\{ \ln [N_B(T)] \right\}.$

The entropy is in fact the most interesting notion of complexity, as derived from the following theorem.

THEOREM : The necessary and sufficient condition of uniform convergence of $\hat{Q}(\alpha)$ to $Q(\alpha)$ is :

$$\lim_{T \to \infty} H_B (T)/T = 0$$

Unfortunately, one cannot compute really the entropy in practical problems, as the probability density function $f[X,y]$ is usually unknown, and here appears the interest of the growing function.

THEOREM : The entropy and the growing function verify the relation : $\ln C_B(T) \geqslant H_B (T)$

Consequently, it gives a sufficient relation for the uniform convergence of $\hat{Q}$ to Q.

THEOREM : The relation $\left[\ln C_B(T) \right] / T \to 0 \ , (T \to \infty)$ is a sufficient condition for the convergence of $\hat{Q}$ to Q.

It is of great interest not only to establish the fact of the uniform convergence of $\hat{Q}$ to Q, but also to study the speed of such convergence which is expressed by the following theorem [12].

<u>THEOREM</u> : For an infinite set $G = \{g_\alpha(X)\}$ of decisional rules we have the relation :

$$P\left\{\sup_\alpha |Q(\alpha) - \hat{Q}(\alpha)| > \epsilon\right\} < 6\ C_G(2T) \,.\, e^{-T\epsilon^2/4}$$

Moreover if the set G has a finite capacity $K[G] = K$, then we have :

$$P\left\{\sup_\alpha |Q(\alpha) - \hat{Q}(\alpha)| > \epsilon\right\} < 9\,(2T)^K / K!\ e^{-T\epsilon^2/4}$$

For a practical application of such results, it is necessary to give the explicit form of the capacity for the most popular sets of decision rules : the linear set of hyperplans $G_L = \{a'X\}$, the quadratic set of second degree surfaces $G_Q = \{(X - a)'\, A(X - b)\}$ and generally for a polynomial set of degree "r" : $G_{P_r} = \{P_r(X)\}$. Such a combinatorial problem is not difficult to solve, and we obtain the following general result :

<u>THEOREM</u> : The capacity $K[G_{P_r}]$ of the set of polynomial decisional rules, associated to the vector X of predictors ($X \in R^n$), is given by the expression :

$$K\left[G_{P_r}\right] = C^r_{n+r}$$

In particular, we have :

$$K\left[G_L\right] = n+1 \ ; \quad K\left[G_Q\right] = (n+1)(n+2)/2$$

Now, it's possible to perform adequate computation to obtain for the most popular families G_L and G_Q corresponding useful tables. They allow us to answer various questions concerning the choice of ϵ, T, n, η according to the set of datas we dispose, as well as the choice between linear or quadratic discrimination.

VI. THE NATURAL LINK TO OVEREVALUATE THE NUMBER OF PREDICTORS

In regression analysis a popular method for the definition of the "optimal" number of predictors leads to compute the criteria :

$$Q_n = \frac{1}{T-n-1} \sum_{t=1}^{T} e^2(t) \quad ,$$

where

$$e(t) = y(t) - \tilde{y}(t) \quad .$$

We will suppose the predictors ranged by decreasing order of interest, and let n_0 be the unknown real number of predictors. The choice of n_{opt} is determined from the following procedure. It's the smaller number verifying the relations

$$Q_{n_{opt}} < Q_{n_{opt-1}} \quad , \quad Q_{n_{opt}} < Q_{n_{opt+1}}$$

What is possible to say about such strategy of choice ? In other terms what relation exists between n_0 and n_{opt} ?
To such question Orlov [9] has given a very interesting answer by means of asymptotic relations obtained under very large assumptions.
He has established that if $y(t) = \sum_{i=1}^{n_o} a_i x_i(t) + \epsilon(t)$,
then the following two relations arise :

1) $$\lim_{T \to \infty} P\{n_{opt} < n_o\} = 0 \; ;$$

2) $$\lim_{T \to \infty} P\{n_{opt} = n_o + k\} = \lambda(1-\lambda)^k \quad ,$$

where $$\lambda = \Phi(1) - \Phi(-1) = 1/\sqrt{2\pi} \int_{-1}^{+1} e^{-x^2/2} dx \approx 0,6827 \; .$$

Such result shows with evidence that even objective methods to optimize the number of predictors generally lead to an overevaluation of such number. The same result has been obtained for the number of predictors in discriminant analysis and the number of clusters in cluster analysis.

REFERENCES

1. AIZERMAN, BRAVERMAN, ROZENAUER. La méthode des fonctions potentielles (en russe). Nauka. 1970.
2. ANDERSON : An introduction to multivariate statistical analysis. Wiley. N.Y. 1958.
3. ANDREWS : Introduction to mathematical techniques in pattern recognition. Wiley Interscience. N.Y. 1972.
4. DER MEGREDITCHIAN : La prévision statistique des phénomènes météorologiques. Note EERM No 100. Paris. 1984.
5. DER MEGREDITCHIAN : Le traitement statistique des données multidimensionnelles. Tomes I, II. Séminaire sur les outils statistiques. Toulouse. 1983.
6. DER MEGREDITCHIAN : Une identité algébrique intéressante pour la statistique mathématique. La Météorologie. II Série N. 32 Mars 1983.
7. FUKANAGA : Introduction to statistical pattern recognition. Academic Press. N.Y. 1972.
8. NILSSON : Learning machines. McGraw Hill, N.Y. 1965.
9. ORLOV : Estimation de la dimension du modèle en régression. Algorithmes et programmes de l'analyse statistique. Moscou. Ed. Nauka. 1980 (en russe).

10. PARZEN : On the estimation of a probability function and mode. AMS V. 33. No 3. 1962.

11. SEBER : Multivariate observations. Wiley. N.Y. 1984.

12. VAPNIK : Estimation of dependences based on empirical data. Springer Verlag. 1982.

13. VAPNIK, TCHERVONENKIS : Théorie de la reconnaissance des formes (en russe). Nauka. 1974.

THE METHOD OF PATTERN CLASSIFICATION AND ITS APPLICATION TO PROGNOSIS OF DISEASE

Tsutomu Komazawa

Institute of Statistical Mathematics
4-6-7, Minami-Azabu, Minato-ku, Tokyo, Japan

I. INTRODUCTION

This is being done mainly with the use of every method of quantification theory and multi-dimensional analytical research, without outer standard on the data of collective medical examinations, as well as basic pathological data and clinical studies aimed at the prevention of arteriosclerosis related diseases. On the basis of this study, we will report the results, which are practical enough to be put into a concrete policy, to control the disease and to be applicable to the multiple test data obtained from collective medical examinations which are repeated every year.

II. QUANTIFICATION THEORY TYPE III AND ONE DIMENSIONAL SCALING STRUCTURE.

The artificial mode data of Tables 1 and 4, which were analyzed with the use of the Quantification Theory Type III, are typical data which afford an one dimensional scaling structure. In the data matrix Table 1, Guttman (1941), Okamoto and Associates (1974), and Iwatsubo (1979), made an

ISBN 0-12-215485-1

analytical examination or characteristic problem.

However, the development of applications in this field has not yet been made. Recently though, Iwatsubo and Associates made an examination of the Item Categorical Data Matrix of Table 4, taking a theoretical approach.

Table 1 : Guttman Type Data

CATEGORY	1	2	3	4	5	6	7	8	9	10
1	1	1	1	1	1	1	1	1	1	1
2		1	1	1	1	1	1	1	1	1
3			1	1	1	1	1	1	1	1
4				1	1	1	1	1	1	1
5					1	1	1	1	1	1
6						1	1	1	1	1
7							1	1	1	1
8								1	1	1
9									1	1
10										1

Then, we show the solution of quantification theory type III for item categorical data.

$D = \{\delta_{i(jk)}\}$ ... data matrix

where,

Individual ; $i = 1, 2, \ldots, n$

Item ; $j = 1, 2, \ldots, m$

Category ; $k = 1, 2, \ldots, 1_j$

$$\delta_{i(jk)} = \begin{cases} 1 \text{ ; respond to } C_{(jk)} \\ 0 \text{ ; non-response} \end{cases}$$

The condition $C_{(jk)}$ consists of item j and category k

$$f_{(jk)} = \sum_{i=1}^{n} \delta_{i(jk)}$$

$$f_{(jk)(uv)} = \sum_{i=1}^{n} \delta_{i(jk)} \delta_{i(uv)}$$

Quantify $X_{(jk)}$ given to $C_{(jk)}$ is,

$$AX = \lambda X$$

where,

$$A = \{a_{(jk)(uv)}\}, \; X = \{x_{(jk)}\}$$

$$j, u = 1, 2, \ldots, m$$
$$k = 1, 2, \ldots, 1_j$$
$$v = 1, 2, \ldots, 1_u$$

However,

$$a_{(jk)(uv)} = \alpha \, \{f_{(jk)(uv)} - f_{(jk)} f_{(uv)}/n\}$$

A parameter α is following criterion,

Case (1) Eigen-value λ is square of correlation coefficient in quantification theory.

$$\alpha = 1/(m \cdot f_{(jk)})$$

Table 2 : Individual Source Data Table of Item Category

Item	1	... j ...	m
individual \ category	$C_{(11)}$ ...	... $C_{(jk)}$ ...	... $C_{(ml)}$
1	$\delta_{2(11)}$ ...	... $\delta_{1(jk)}$ ...	... $\delta_{1(ml)}$
2	$\delta_{2(11)}$ ...	... $\delta_{2(jk)}$ ...	... $\delta_{2(ml)}$
3	$\delta_{3(11)}$ ...	... $\delta_{3(jk)}$ ...	... $\delta_{3(ml)}$
⋮	⋮	⋮	⋮
i	$\delta_{i(11)}$ ...	... $\delta_{i(jk)}$ ...	... $\delta_{i(ml)}$
⋮	⋮	⋮	⋮
n	$\delta_{n(11)}$ ...	... $\delta_{n(jk)}$ ...	... $\delta_{n(ml)}$

Table 3 : Contingency Table of Item Category

item		1 ---	u			--- m	
item	category	----	$C_{(u1)}$	$C_{(u2)}$	--- $C_{(uv)}$ ---	----	$mf_{(f1)}$
1 ⋮	⋮	⋮	⋮	⋮	--- ⋮ ---	⋮	⋮
	$C_{(j1)}$	----	$f_{(j1)(u1)}$	$f_{(j1)(u2)}$	---- $f_{(j1)(uv)}$ ---	----	$mf_{(j1)}$
	$C_{(j2)}$	----	$f_{(j2)(u1)}$	$f_{(j2)(u2)}$	---- $f_{(j2)(uv)}$ ---	----	$mf_{(j2)}$
j	⋮	⋮	⋮	⋮	⋮	⋮	⋮
	$C_{(jk)}$	----	$f_{(jk)(u1)}$	$f_{(jk)(u2)}$	---- $f_{(jk)(uv)}$ ---	----	$mf_{(jk)}$
	⋮	⋮	⋮	⋮	⋮	⋮	⋮
⋮ m	⋮	⋮	⋮	⋮	⋮	⋮	⋮
total		----	$mf_{(u1)}$	$mf_{(u2)}$	--- $mf_{(uv)}$ ---	----	$m^2 \times n$

Case (2) λ is variance in principal component analysis for variance and co-variance matrices.

$$\alpha = 1/n$$

Case (3) λ is variance in principal component analysis for correlation coefficient matrices.

$$\alpha = n/\sqrt{(n-n_{(jk)})(n-n_{(uv)})n_{(jk)}n_{(uv)}}$$

This solution $\{X^{(t)}{}_{(jk)}\}$ have eigen vectors $\{X^{(1)}{}_{(jk)}, X^{(t)}{}_{(jk)}\}$ $(j = 1, 2, \ldots, m, k = 1, 2, \ldots, l_j, t = 2, 3, \ldots)$ which correspond to eigen value $\lambda_1 \geqq \lambda_2 \geqq \ldots \geqq \lambda_t \geqq$ being two dimensional and three dimensional curve respectively as well as Figures when the data structure of an item has an order, that is, when one dimensional scaling structure is possible.

These Figures are shown characteristic of each case. Analysis of case (1) present to completed one dimensional scaling for data of example mode. But others case do not show from scatter diagram as well as case (1) whether data structure of an item category has an order.

Table 4 : Example Mode Data

Table 4.1: Source Data Table

ITEM	1	1	1	1	2	2	2	2	3	3	3	3	4	4	4	4
CATEGORY	1	2	3	4	1	2	3	4	1	2	3	4	1	2	3	4
No. 1	1				1				1				1			
2		1			1				1				1			
3		1				1			1				1			
4		1				1				1			1			
5		1				1				1				1		
6			1			1				1				1		
7			1				1			1				1		
8			1				1				1			1		
9			1				1				1				1	
10				1			1				1				1	
11				1				1			1				1	
12				1				1				1			1	
13				1				1				1				1

→

Table 4.2: Exchange Data Table

1	2	3	4	1	2	3	4	1	2	3	4	1	2	3	4
1	1	1	1	2	2	2	2	3	3	3	3	4	4	4	4
1	1	1	1												
	1	1	1	1											
		1	1	1	1										
			1	1	1	1									
				1	1	1	1								
					1	1	1	1							
						1	1	1	1						
							1	1	1	1					
								1	1	1	1				
									1	1	1	1			
										1	1	1	1		
											1	1	1	1	
												1	1	1	1

Figure 1 : Scatter Diagram of Each Analysis for Example Mode Data

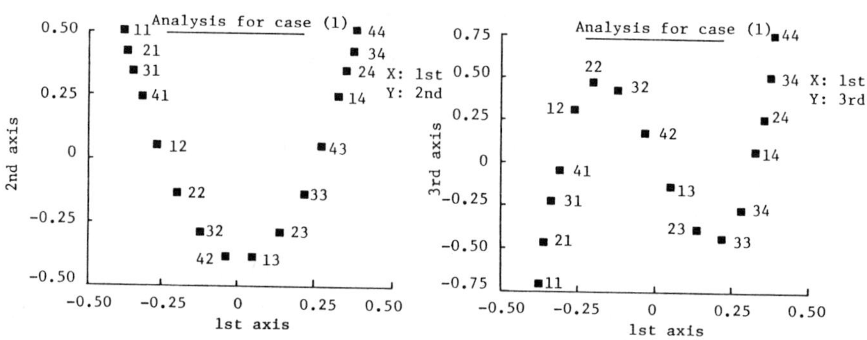

Figure 1.1 : Analysis of Case (1)

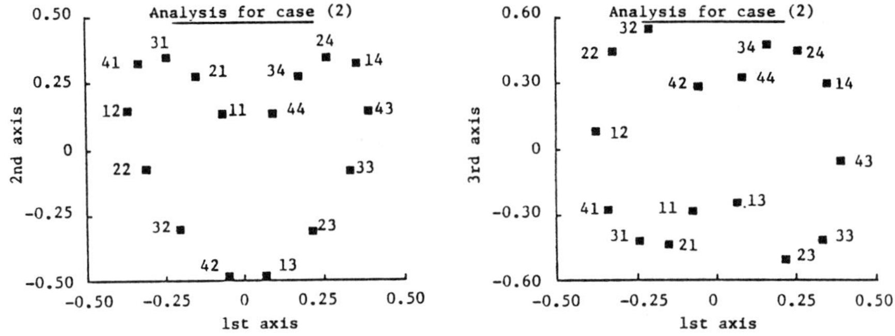

Figure 1.2 : Analysis of Case (2)

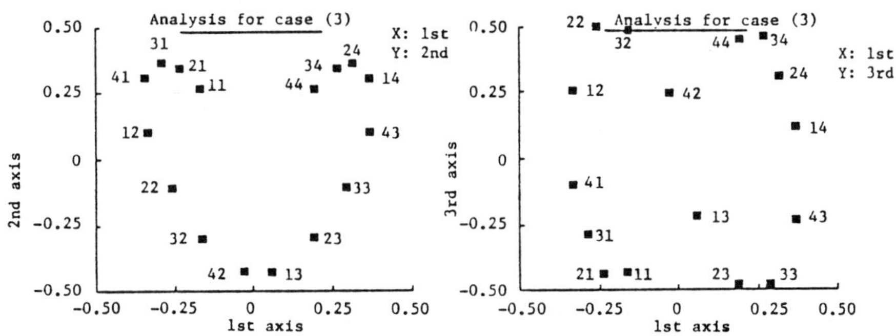

Figure 1.3 : Analysis of Case (3)

We made use of this property in order to put to practical use the vast amount of multiple data obtained at medical examinations of groups which are repeated annually.

The by making test items more important our study was of enough practical value to be able to be part of a concrete policy on health control.

III. APPLICATION

We have made an analysis of data obtained from medical examinations of groups. This analysis was performed for the prevention of diseases caused by arteriosclerosis. It also evaluated diseases which occurred one year after medical examinations, mainly by means of Quantification Theory Type III and in view of each component, as well as making an analysis of the prospects and distinctions, clarifying the parts which are likely to be infected with diseases caused by arteriosclerosis. The object of the analysis is following.

Among the 30,000 people who were chosen at random from over 150,000 people who were examined in groups during 1982 by the Institute of Arteriosclerosis attached to the Japan Labors, Health and Welfare Association (A Judicial Foundation), the 18,207 people who answered the inquiry card sent for the follow up survey are the objects. The patients are as many as 181 peoples among them. Their residences are in the Kansai, Chubu, Kanto and Tohoku districts.

The examination items included blood pressure, electro cardiogram, fuduscopy, serumlipid and velocity of pulse waves of the large artery (hardening index), which were considered to be important for the prediction of diseases caused by arteriosclerosis, we have made an analysis of Quantification Theory Type III.

The quantity given to each examination test items and the division showed a strong correlation between occurrence and non-occurrence as the first peculiarity.

The second peculiarity was hypertension and infarct constrict type. These were in an arrangement and of quantity to indicate a strong correlation. A dispersion drawing of the objective quantity regarding the first and second peculiarity showed 66.3% occupied by the non-occurrence cases in the minus area, and 33.7% occupied by the occurrence cases. Also, supporting an origin to be a dividing point of distinction, and judging the objective quantity of a minus value to be non-occurrence type, which that of plus value to be occurrence type, the hit rate is about 66.3%.

Figure 2 in case of examination is made up of blood pressure factors, Figure 3 in case consists of those factors with the addition of surum lipid factors. The former case appear simple factor that scatter diagram Figure 3.1 expands only from center of co-ordinates. But the latter case seem multiple factors that Figure 3.2 clear to finded arrangement of data structure.

IV. CONCLUSION

The analysis in distinction and prospect due to quantification theory of type III is the most adequate for the analysis of this kind of data. This is because it proper quantities, according to items and divisions, even for simultaneous treatment of analogous test items for medical examinations in groups, which were considered to be important from the standpoint of medical science in the judgement of occurrence and nonoccurrence type. Beside, the scatter diagram drawn using an objective quantity obtained from whole quantity, gives us a convenient, visual report for judgement for the decision of dispersion of analogous disease group which dispersing point indicate.

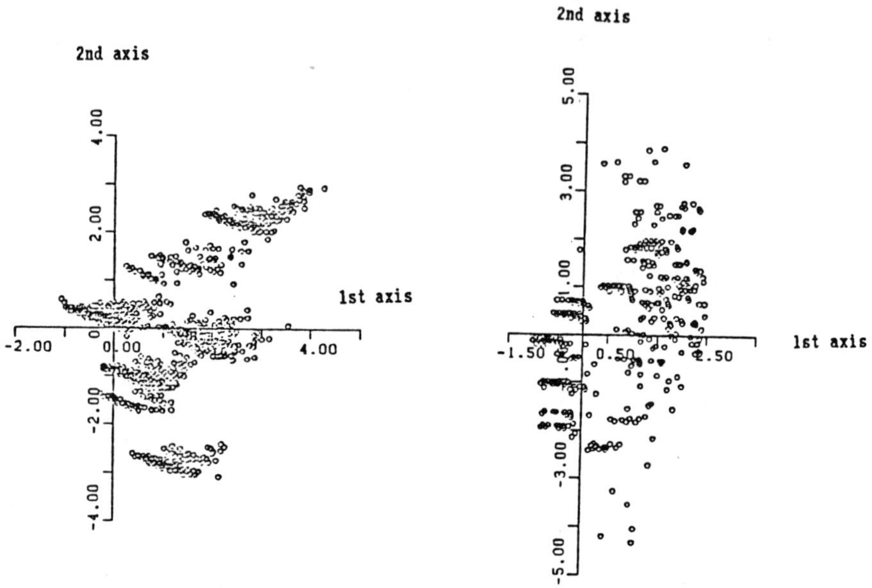

Figure 2.1: Analysis of Case(1) Figure 2.2: Analysis of Case(2)

Figure 2 : Scatter Diagram of Individual for Medical Examination

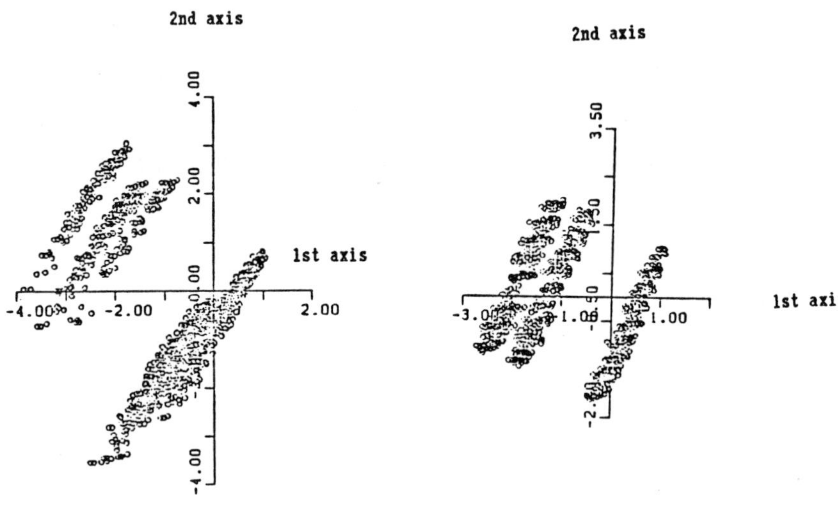

Figure 3.1: Analysis of Case(1) Figure 3.2: Analysis of Case(3)

Figure 3 : Scatter Diagram of Individual for Medical Examination (II)

REFERENCES

1. Greenacre, M.J. (1984) : Theory and Application of Correspondence Analysis, Academic Press.
2. Guttman, L. (1941) : The Quantification of a Class of Attributes, A theory and method of scale construction. The Prediction of Personal Adjustment (P.Horst, ed.), Soc. Sci.Res.Council.
3. Hayashi, C. (1956) : Theory and Examples of Quantification, II, Proc. Inst. Statist. Math., 4(in Japanese).
4. Iwatsubo, S. (1979) : The Analytical Solutions of Eigen Value Problem in the Case of Applying Optimal Scoring Method to Some Type of Data, The Second International Symposium on Data Analysis and Informatics, France.
5. Komazawa, T. (1982) : The Quantification Theory and Data Processing, Asakura-Shoten (in Japanese).
6. Okamoto, M. (1974) : An Asymptotic Theory of Categorical Canonical Correlation Analysis, Journal Japan Statist.

A NON PARAMETRIC DISCRIMINANT ANALYSIS BASED ON THE CONSTRUCTION OF A BINARY DECISION TREE

Jean-Pierre Nakache
Alice Gueguen

CNRS/INSERM U.88
91 Boulevard de l'Hôpital
75634 PARIS CEDEX 13, France

I. INTRODUCTION

The standard classical methods in classification problems are discriminant analysis and logistic regression. In these cases, the prediction rules are given in the form of algebraic expressions that are, sometimes, difficult to understand and interpret.

The discriminant analysis presented here takes a completely different approach, and the resulting prediction rules are given in the form of binary decision trees close to the physician reasoning, easy to understand, use, explain and interpret.

Use of binary decision trees in discriminant problems dates back to the Automatic Interaction Detection program (A.I.D.) proposed by Morgan and Sonquist (1) in the early's 1960. Recent developments are due to Breiman, Friedman, Olshen and Stone (2) of University of California.

The discrimination tree procedure is illustrated by a medical example concerning 382 patients with a liver nodule divided into 259 cases of malignant tumour and 123 cases of benign tumour. 13 qualitative measurements (Table 1) have been collected on these patients.

ISBN 0-12-215485-1

Table 1

Medical example : List of measurements and their code

LOSS OF WEIGHT	WLO	1- no	2- yes	
ASCITES	ASC	1- no	2- yes	
R.U. QUAD. PAIN	PAI	1- no	2- yes	
FEVER	FEV	1- no	2- yes	
JAUNDICE	JAU	1- no	2- yes	
HEPATOMEGALY	HEP	1- no	2- yes	
HEMOCHROMATOSIS	HEM	1- no	2- yes	
CIRRHOSIS	CIR	1- no	2- yes	
PRIM. CARCINOMA	PCA	1- no	2- yes	
PAST HIST. OF HEP.	PHH	1- no	2- yes	
SEDIM. RATE	SER	1- norm.	2- incr.	
ALK. PHOSPH.	ALP	1- norm.	2- incr.(+)	3- incr.(++)
G.G.TRANSFERASE	GGT	1- norm.	2- incr.(+)	3- incr.(++)

II. ILLUSTRATIVE BINARY DECISION TREE

A binary decision tree is constructed by repeated splits of subsets of patients into two descendant subsets. The fundamental idea is to select each split so that the data in each of the descendant subsets are "purer" than the data in the parent node. In the illustrative binary decision tree (Fig. 1) each subset is a node and we distinguish two types : intermediate or non terminal node (circled node) and terminal node (squared node).

III. CONSTRUCTION OF A BINARY DECISION TREE

The entire construction of a binary decision tree requires the following definitions :

A. Possible splits of a node t

A binary decision tree is obtained by means of the qualitative measurements of the data set. In a practical data set the qualitative variables may be of different types : binary, ordinal or nominal. A binary variable allows the splitting of a node into two descendant nodes. An ordinal variable with k categories provides (k-1) splits when a nominal variable with the same number of categories splits the node into 2^{k-1} -1 descendant nodes.
For example if, for a node t, we are dealing with 3 variables : one binary variable B (B_1, B_2), one ordinal variable N (N_1, N_2, N_3), then the 7 possible splits are :

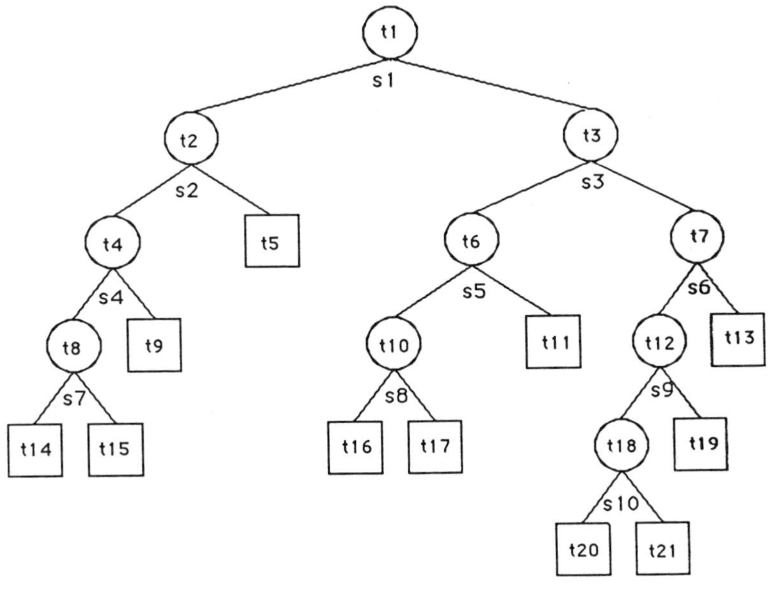

Figure 1
Illustrative binary tree diagram

$(B_1), (B_2)$
$(O_1), (O_2, O_3, O_4)$; $(O_1,O_2), (O_3,O_4)$; $(O_1,O_2, O_3), O_4$
$(N_1), (N_2, N_3)$; $(N_2), (N_1,N_3)$; $(N_3), (N_1,N_2)$

B. Impurity of node t

In the case of k groups submitted to discrimination, the impurity i(t) of node t is expressed as :

$$i(t) = \Sigma \{P(r/t)\, P(s/t)\ ;\ r=s=1,.,k\}$$

where P(r/t) is the proportion of cases of group G_r in the node t (r = 1, 2, .., k).

In the particular case of 2 groups, i(t) = 2 P (1/t) [1 - P(1/t)].

C. Relative decrease of impurity resulting from a split

For a node t, the relative decrease of impurity resulting from a split s is expressed as :

$$\Delta i(s/t) = i(t) - [P_L\, i(t_L) + P_R\, i(t_R)]$$

where t_L and t_R are the two descendant nodes and P_L (respectively P_R) is the proportion of the cases in t going into t_L (respectively t_R).

D. Best split s_0 of a node t

The best split s_0 of a node t is the split s of t leading to descendant nodes with a number of cases equal or greather than a fixed number n_0 and corresponding to the maximum decrease of impurity, that is : $i\,(s_0,t) = \max_s\{\Delta i(s,t)\}$.

Notice : the splitting criterion based on the decrease of impurity is more adequate than the one based on the decrease of the percentage of misclassified patients. In fact, it is possible to deal in practice with examples where the percentage of misclassification remains constant during the splitting when the impurity decreases.

The splitting rule allows us to grow a large tree having only a few cases (n_0 = 5) in each terminal node. Fig. 2 provides such a tree grown from a sample of 255 cases obtained from the total sample of 382 cases by random selection. This sample is the learning sample used later in one of the two methods of determination of the most reliable estimated true error rate.

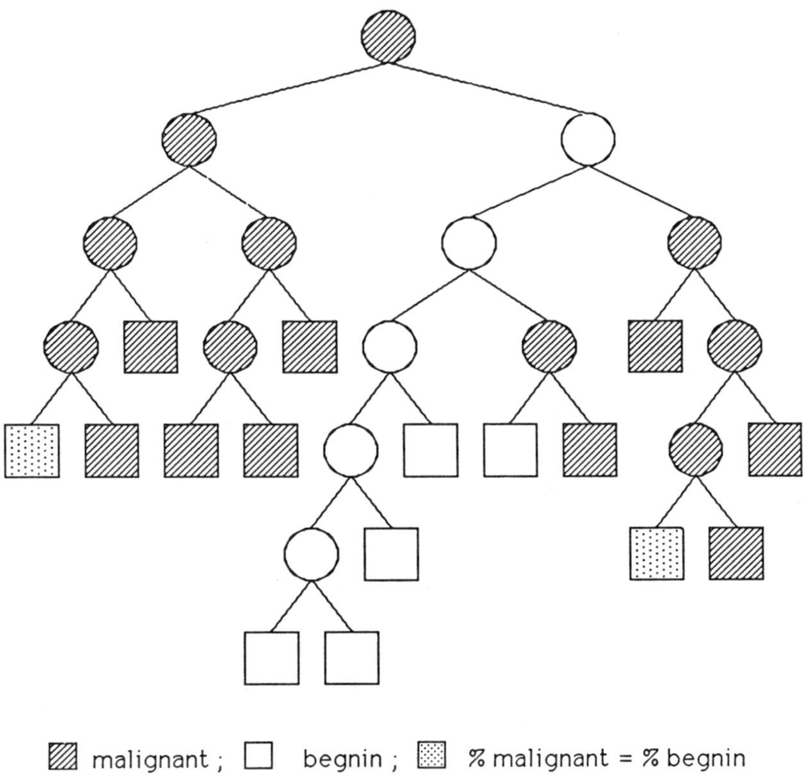

Figure 2

Tree Tmax obtained from the sample of 255 cases

The tree T_{max} of Fig. 2 contains 16 terminal nodes. A terminal node is assigned to the malignant group (respectively to the benign group) if it contains a major proportion of malignant tumours (respectively benign tumours). The apparent error rate (AER) assigned to T_{max} (8,63 %) represents the proportion of misclassified patients in all of the 16 terminal nodes.

As pointed in (2), one of the dangers in growing trees is that if the tree is too small, it may have an unnecessarily large error rate. But if it is too large, while the AER as measured in the learning set may be small, its true error rate may be large. Somewhere in between is the right-sized tree.

The determination of the right-sized tree from the large tree T_{max} requires the construction of an optimal sequence of smaller and smaller subtrees using a pruning algorithm.

IV. PRUNING PROCESS

Pruning a branch with root node t from a tree T consists of deleting from T all descendants of t, that is, cutting off all the branch except its root node t.

The pruning algorithm selectively prunes branches off of the large tree T_{max} providing an optimal sequence of smaller and smaller trees such that each tree in the sequence has a smaller AER than any other subtree of the same size. The right-sized tree is then selected out this sequence by means of a test sample method.

The pruning criterion is expressed as :

$$\beta(t) = [MC(t) - MCDT(t)] / NC(t).[NDT(t) - 1]$$

where MC(t) = number of misclassified cases in node t, NC(t) = number of cases in node t, MCDT(t) = the number of misclassified cases in the descendant terminal nodes of t and NDT(t) = number of descendant terminal nodes of t.

The pruning proceeds as follows :

- compute the value of ß(t) for each intermediate node of the tree T_{max}
- cut off the branch $B(t_0)$ with root node t_0 such that $ß(t_0)$ is minimum.

In so doing, $B(t_0)$ is the largest branch to cut off corresponding to the minimum increase of the AER. The first pruned tree T_1 of the optimal sequence is T_{max} where $B(t_0)$ is omitted. The second pruned tree T_2 is obtained from T_1 and so on...

Table 2 provides the pruned sequence of the learning sample (255 cases).

Table 2

Optimal Sequence { T1, T2, T3 } : Learning sample (255 cases)

Tmax	:	**16** terminal nodes		A.E.R. = **8.627 %**			
T1	:	**5** terminal nodes		A.E.R. = **8.627 %**			
Obtention of T2							
t	NC(t)	Begn.	Malign.	MC(t)	MCDT(t)	NDT(t)	ß(t)
1	255	82	173	82	22	5	.0588
3	131	77	54	54	17	4	.0941
6	**88**	**70**	**18**	**18**	**10**	**3**	**.0451**
13	14	4	10	4	2	2	.1428
T2	:	**3** terminal nodes		A.E.R. = **11.765 %**			
Obtention of T3							
1	**255**	**82**	**173**	**82**	**30**	**3**	**.1019**
3	131	77	54	54	25	2	.2213
T3	:	**1** terminal node		A.E.R. = **32.15 %**			

V. CHOOSING THE RIGHT-SIZED TREE : A PROBLEM OF DETERMINATION OF THE MOST RELIABLE ESTIMATED TRUE ERROR RATE

The selection of the right-sized tree out of this pruned sequence is done in two different ways according to the sample size.

A. Test sample procedure

If the data set has many cases in it, then the fasted way to estimate the true error rate (T.E.R.) is by means of a test sample. In this case, the total sample L is divided by random selection into a learning sample with 2/3 of the cases and a test sample with 1/3 of the cases.

Once the pruned sequence of the learning sample obtained, each case of the test sample is run down each subtree of the sequence, finally moving into a terminal node. Its predicted class is given by the class label attached to that terminal node. We so obtain an estimated T.E.R. for each subtree :

Trees	NTN	ß	AER	TER
Tmax	22		8.63	7.87
T1	5	.00	8.63	7.87
T2	3	.04	11.77	11.81
T3	1	.10	32.15	32.28

The estimated T.E.R. is usually large for the very large subtrees, decreases as the trees get smaller and then begins to rise again when the trees get too small. And the right-sized tree is the smallest tree giving the minimum estimated T.E.R.

In our example the right-sized tree is T_1 with 5 terminal nodes providing an estimated T.E.R. of 7,87 % with a 95 % confidence interval containing the values ranging between 3,40 % and 12,30 %.

B. Cross-validation procedure

The second procedure of estimation of the T.E.R. is the cross-validation procedure used for small sample sizes. In this procedure every case is used both in the tree construction and the estimation of the T.E.R.

The total sample L is divided by random selection into 10 subsets each containing the same number of cases :

- test samples (1/10 of the cases) : L_i (i = 1, ..,10)
- learning samples (9/10 of the cases) : $L^i = L - L_i$ (i = 1,..,10)

The cross-validation procedure involves the construction of 10 auxillary trees as well as the main tree. It requires much more running time that the test sample procedure since a pruned sequence (S_i ; i = 1, 2, .., 10) has to be constructed for each of the 10 auxillary trees and for the main tree as well.

The total sample L provides the following pruned sequence $S = \{T_k ; k=1,.., 5\}$

Trees T_k	T_1	T_2	T_3	T_4	T_5
nodes in T_k	10	7	5	3	1
$ß_k.10^3$	.0	1.8	16.8	46.4	102.1

The problem is then to compute for each subtree T_k of S an estimated T.E.R. using of the pruned sequences S_i. For this purpose, k being fixed, we choose for each i (i=1, 2,.., 10) the smallest subtree $T_{j,i}$ of S_i such that : $ß(T_{j,i}) \leq ß'_k$ where $ß'_k \leq \sqrt{ß_k\ ß_{k+1}}$ is the geometric midpoint of the interval $[ß_ß, ß_{k+1}[$. The estimated T.E.R. associated with T_k of S is the sum of misclassified cases in the 10 test samples L_i. The right-sized tree is then the subtree of the main sequence S providing the minimum estimated T.E.R. as shown in Table 3.

Table 3

Right-sized tree obtained by the cross-validation procedure

MICL = total number of misclassified patients

E.TER = estimated true error rate

N.T.N. = number of terminal nodes

	T1 ß'=0		**T2** ß'=5.53		**T3** ß'=27.94		**T4** ß'=68.85		**T5** ß'>68.85	
L_1	T1,1	1	T3,1	2	T4,1	2	T5,1	4	T6,1	13
L_2	T1,2	3	T2,2	2	T2,2	2	T3,2	2	T4,2	13
L_3	T1,3	6	T1,3	6	T1,3	6	T2,3	10	T3,3	12
L_4	T1,4	7	T2,4	7	T3,4	7	T4,4	9	T5,4	12
L_5	T1,5	4	T2,5	3	T4,5	0	T5,5	2	T6,5	12
L_6	T1,6	5	T1,6	5	T2,6	6	T3,6	6	T4,6	12
L_7	T1,7	4	T1,7	4	T3,7	2	T4,7	3	T5,7	12
L_8	T1,8	3	T1,8	3	T3,8	3	T4,8	3	T5,8	12
L_9	T1,9	1	T2,9	1	T3,9	1	T4,9	3	T5,9	12
L_{10}	T1,10	4	T1,10	4	T1,10	4	T2,10	6	T3,10	13
MISCL.	(38)		(37)		**(33)**		(48)		(123)	
E.TER	9.95%		9.69%		**8.64%**		12.57%		32.20%	
N.T.N.	10		7		**5**		3		1	

So, the cross-validation procedure leads to an estimated T.E.R. of 8,64 % with a 95 % confidence containing the values between 7,2 % and 10 %.

When comparing the two methods of estimation of the T.E.R., we notice that the test sample procedure provides the smaller estimated T.E.R. but the cross-validation procedure leads to the smaller standard deviation.

C. Diagram of the right-sized tree obtained by the cross-validation procedure.

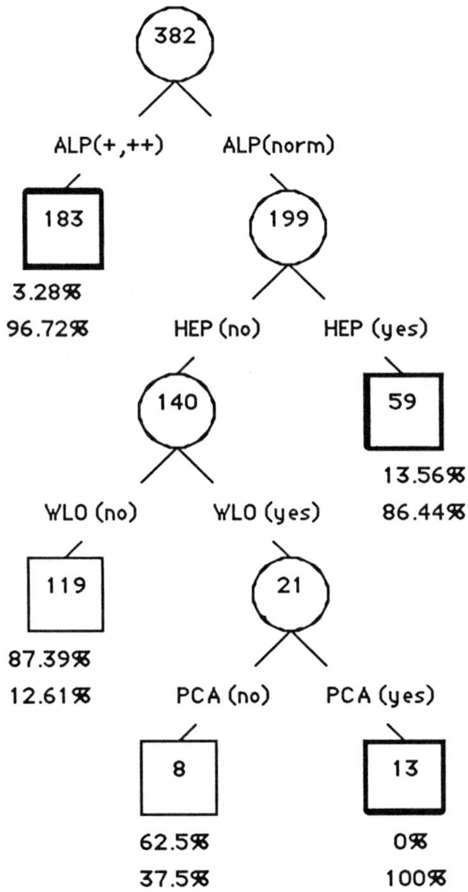

Figure 3

In the diagram (Fig.3) the prediction rule is given in the form of five paths with two of them in favour of benignity, and, the three bold ones in favour of malignancy. The percentages written under each terminal node represent the proportion of cases of each group in the node where the major proportion is underlined.

VI. CONCLUSION

The discriminant tree procedure is a non parametric method of discrimination on qualitative variables (binary, nominal or ordinal). Its prediction rule, given in the form of a binary decision tree, is easy to understand and close to the physician reasoning. This method takes into account interactions between variables, is able to handle missing data, provides the possibility of selecting splits to include boolean combinations of variables and is also able to deal with different costs of misclassification.

REFERENCES

1. Sonquist, J.A. and Morgan, J.N. (1964). The Detection of Interactions Effects, Rev. ed. Ann. Arbor : Institute for Social Research, University of Michigan.

2. Breiman, L. and Friedman, J.H. and Olshen, R.A. and Stone, C.J. (1984). Classification And Regression Trees, Wadsworth Intern. Group.

EXPERIMENTAL COMPARISON BETWEEN THE OPTIMAL DISCRIMINATE PLANE BASED ON SAMPLES AND GENERAL DISCRIMINANT ANALYSIS

Akihiko Miyake

Nihon Medical School, Division of Mathematics
2-297-2, Kosugi, Nakahara-ku, Kawasaki, JAPAN

I. INTRODUCTION

The Optimal Discriminant Function (O.D.F.) is obtained from the ratio between the population distribution functions of two groups. The boundary is determined differently by the likelihood, Bayes, risk methods.

When two population distributions are normal, O.D.F. is of quadratic form. When their variance-covariance matrixese are equal, it is of linear form. But in the general case about population distributions, the form of O.D.F. is not definite.

In this paper, the form of the discriminant function is confined to the linear function, and the optimal function among the linear discriminant functions is investigated.

When two population distributions are normal, it coincides with Auderson=Bahadour linear discriminant function. When the variance-covariance matrixese are equal, it coincides with O.D.F. above mentioned.

In this paper, the optimal linear discriminant functions about the general population distributions are investigated.

ISBN 0-12-215485-1

II. THEORETICAL CONSIDARATION

A. In the Population

There does not always exsist the optimal linear discriminant function (O.L.D.F.), even when two population distributions are given. Figure 1 shows one of the counter-example.

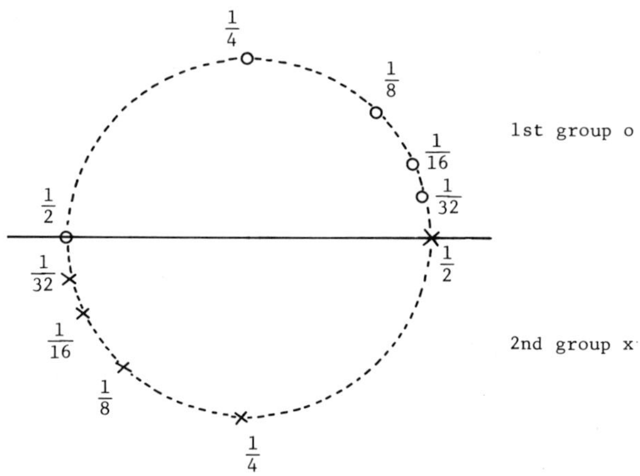

Figure 1

But there always exists the supremum value of hitting rate.

$P = \sup P_b$

(b: linear discriminant coefficient vectors, including constant term, limited to norm $|b|=1$)

and there always exists the coefficient sequence $\{a_n\}$, such that

$$\lim_{n\to\infty} P_{a_n} = \sup P_b \leqq 1 \text{ , exist } a = \lim_{n\to\infty} a_n \text{ .}$$

Let's call the sequence the optimal linear discriminant function sequence (O.L.D.F.S).[1)]

B. In the Samples

The two groups data such as

$$x_1, x_2, \ldots\ldots, x_{n1},$$
$$y_1, y_2, \ldots\ldots, y_{n2},$$

are given, then there always the sample optimal linear discriminant coefficients, which discriminate two group samples in the best way. The coefficients are obtained in the form of convex poly-hedron in the coefficient space.[2)]

Figure 2 shows the poly-hedron conceptionally.

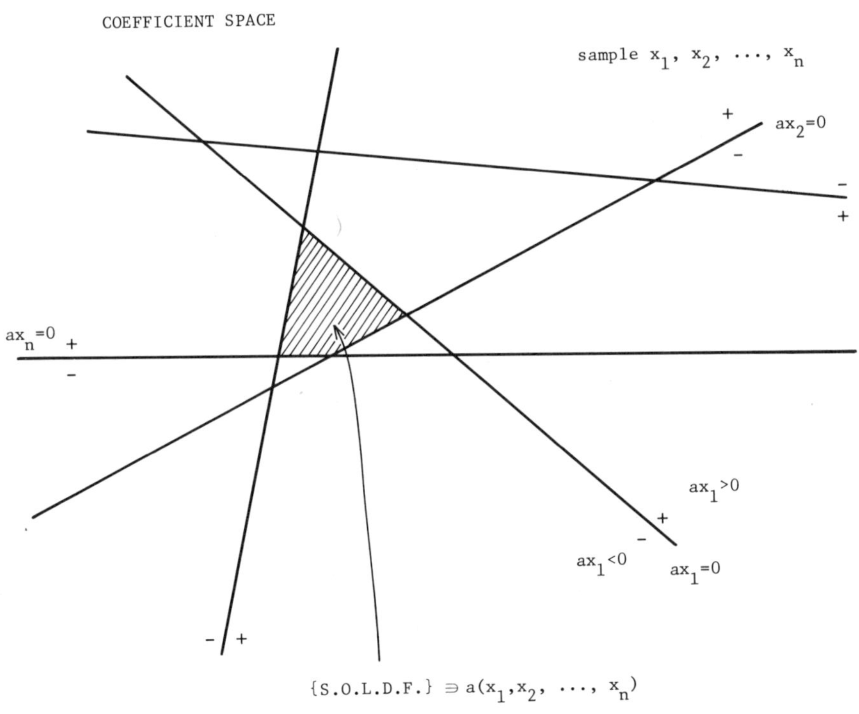

Figure 2

Table 1 gives an example which analyzes (100,100) samples from two 5-dim normal population distributions.

Table 1

Populations

$$N\left(\begin{pmatrix}0\\0\\0\\0\\0\end{pmatrix}, \begin{pmatrix}10000\\01000\\00100\\00010\\00001\end{pmatrix}\right) \qquad N\left(\begin{pmatrix}0.6\\0.6\\0.6\\0.6\\0.6\end{pmatrix}, \begin{pmatrix}10000\\01000\\00100\\00010\\00001\end{pmatrix}\right)$$

	Coefficients					Const.	Error Rate E.C.	Error Rate I.C.
O.L.D.F.	1.000	1.000	1.000	1.000	1.000	-1.500	0.250	0.250
F.L.D.F.	0.937	0.534	1.745	1.254	0.952	-1.500	0.265	0.205
S.O.L.D.F.1	3.405	1.350	7.800	3.030	2.655	-1.500	0.283	0.155
S.O.L.D.F.2	2.295	0.780	4.935	2.250	1.875	-1.500	0.280	0.155
S.O.L.D.F.3	2.340	0.810	4.995	2.175	1.905	-1.500	0.280	0.155
S.O.L.D.F.4	2.325	0.510	4.845	2.100	1.950	-1.500	0.282	0.155

Sample size (100,100)

Lets a_n be one of the sample optimal linear discriminant function (S.O.L.D.F.) when the sample size is n. Then, the function sequence $\{a_n\}$ becomes the optimal linear discriminant function sequence (O.L.D.F.S.) mentioned above.[1)]

C. Comparison with the Standard Discriminant Analysis

Fisher linear discriminant function (F.L.D.F.) is frequently used as the standard linear discriminant function, the linear coefficient of which maximizes the ratio of the between-class variance to the within-class variance.

When samples come from two normal population distributions which variance-covariance matrixese are equal, Fisher linear discriminant function (F.L.D.F.) sequence converages to the optimal linear discriminant function (O.L.D.F.).

Otherwise (non-normal or normal with non-equal matrixese), F.L.D.F. sequence does not converge to O.L.D.F., or strictly speaking, F.L.D.F. sequence is not O.L.D.F.S..

This paper will deal the Comparison among the Convergences of Fisher linear discriminant function (F.L.D.F.) sequence, Anderson=Bahadour linear discriminant function (A.B.L.D.F.) sequence and the sample optimal linear discriminant function (S.O.L.D.F.) sequence.

In Figure 3, the images of the converg. of F.L.D.F. seq. and S.O.L.D.F. seq. are drawn in the form of Internal Check and External Check.

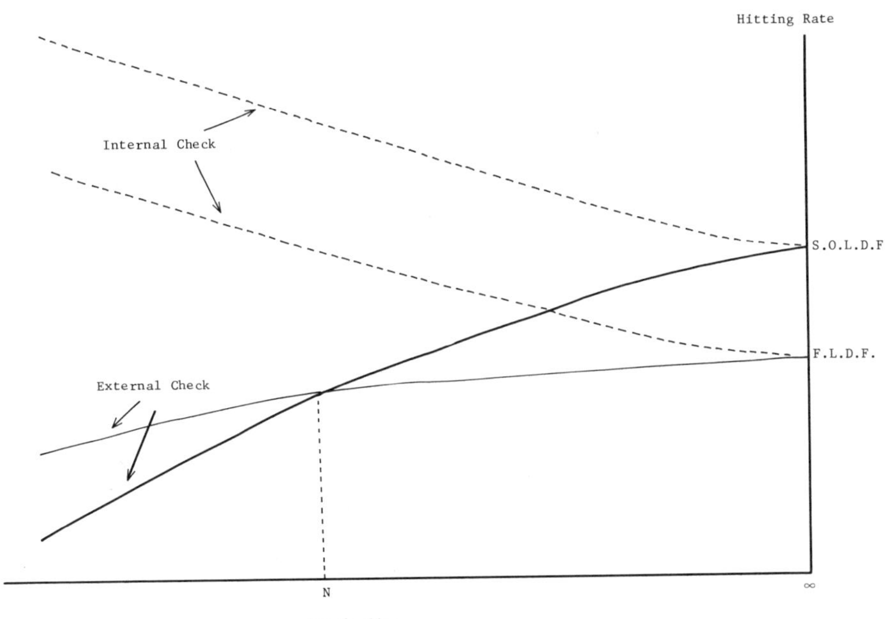

Figure 3

In Internal Check, properly S.O.L.D.F. surpass F.L.D.F..

In External Check, at small sample size, F.L.D.F. surpass S.O.L.D.F., at large sample size, properly S.O.L.D.F. surpass F.L.D.F..

F.L.D.F. is very excellent method, easy to calculate, less bias, efficient at the small sample size.

S.O.L.D.F. is difficult to calculate, almost impossible without electric computer, but it seems to be very clear or primitive logically and valuable at the large sample size.

In think it is not meaningless to investigate S.O.L.D.F..

III. EXPERIMENT

A. Simulation Experiment 1.

Figure 4.1: Let's consider two populations with 2-dim uniform distribution which are shown as one square and one rectangle, with Maharanobis distance 440.

12 lines are drawn, which are boundaries of S.O.L.D.F., A.B.L.D.F., F.L.D.F. derived from data with Sample Size 100, 400, 1000, ∞.

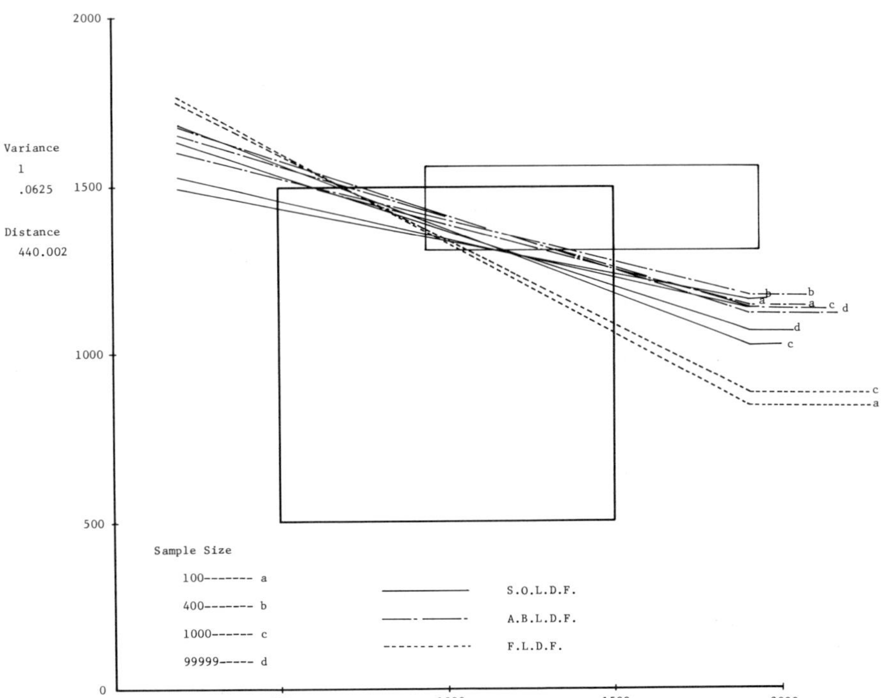

Figure 4.1

Figure 4.2: 24 values show the Error Rate on Internal Check, External Check of above 12 lines.

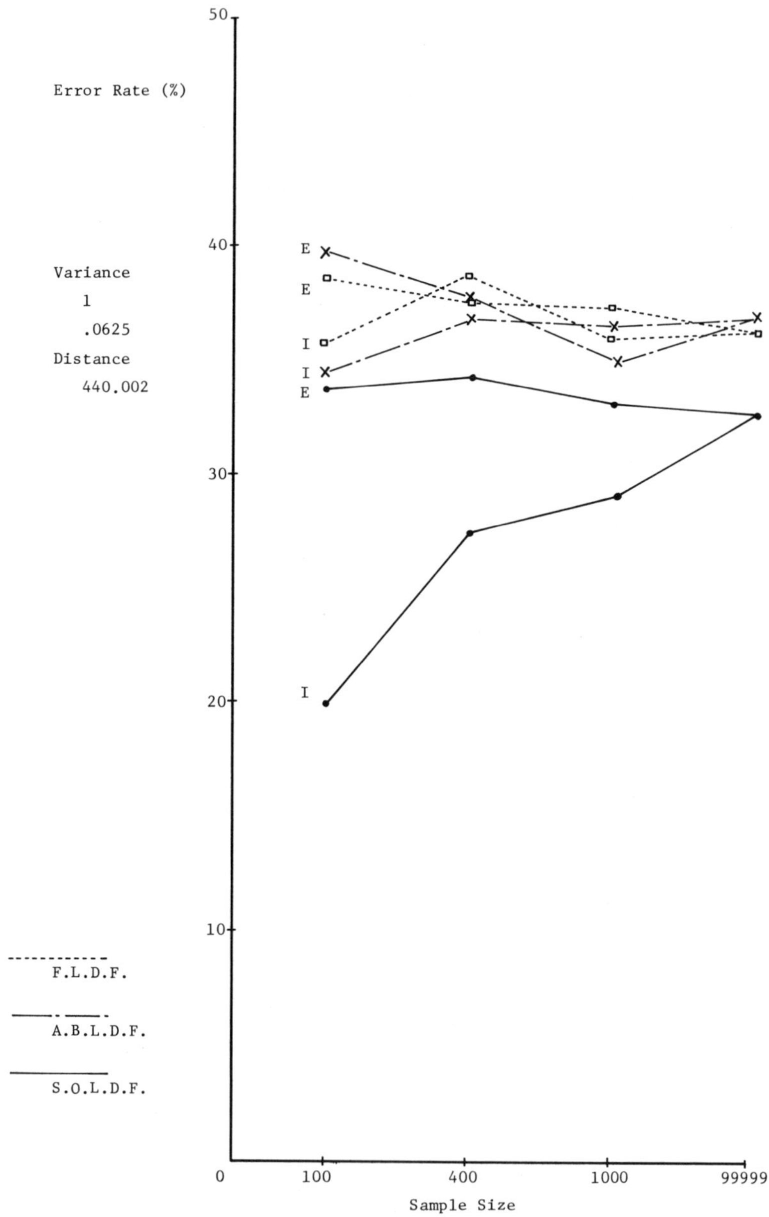

Figure 4.2

Figure 4.3: Two coefficients are shown in the coefficient space. Constant term is normalized to 1.

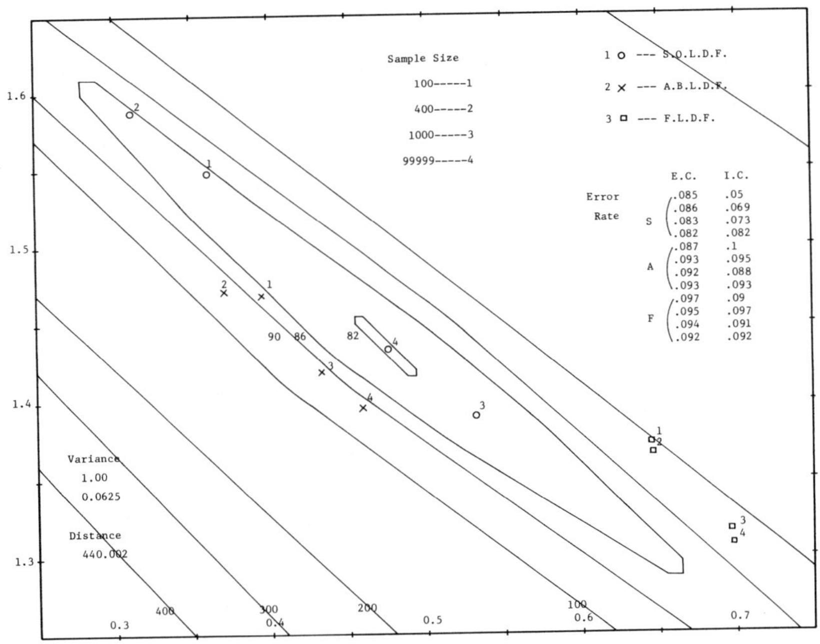

Figure 4.3

B. Simulation Experiment 2.

Figure 5: Let's consider two populations with 5-dim distribution which consist of independent uniform distribution

Table 2 shows the results derived from Data of Sample Size 100, 400, 1000, ∞.

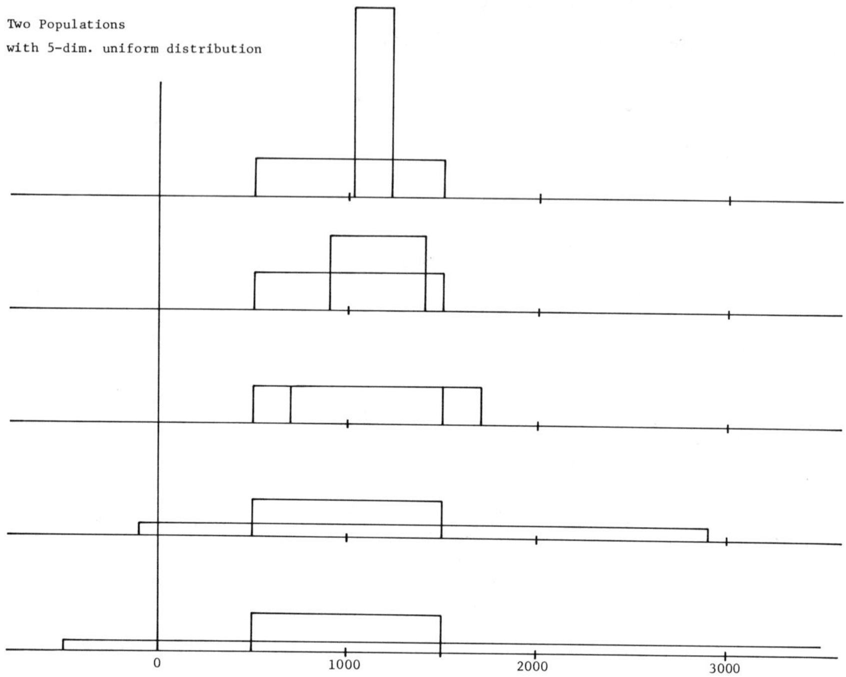

Figure 5

Table 2

S.O.L.D.F. Coefficients	Hitting Rate E	Hitting Rate I	Sample Size	Hitting Rate I	Hitting Rate E	F.L.D.F. Coefficients
0.3 ~ 0.6 0.1 ~ 0.9 1.2 ~ 2.4 0.3 ~ 1.4 1.1 ~ 2.0		.810	50x2 = 100	.720	.731	1.81 0.92 1.41 0.26 0.58
0.9 0.9 1.7 0.6 0.6	.750	.798	200x2 = 400	.753	.749	1.74 1.32 1.25 0.37 0.32
0.4 0.5 1.0 1.5 1.1	.770	.795	500x2 = 1000	.779	.749	1.65 1.24 1.29 0.42 0.37
0.5 0.5 0.75 1.5 1.25	.772	×	∞	×	.751	1.42 1.48 1.23 0.49 0.36

IV. CONCLUSION

It is doubtless that Fisher linear discriminant function is very usefull, especially efficient at small sample size.

Nevertheless, when population distributions are extremely different from normal distribution, Fisher linear discriminant function does not converge to the Optimal Linear Discriminant Function for the population distributions as sample size becomes infinitely large.

On the other hand, the Sample Optimal Linear Discriminant Functions needs vast calculation to obtain it, but its logic is very clear and always converges to the Optimal Linear Discriminant Function for any population distributions.

Occasionally, this non-parametric method may be very useful.

REFERENCES

1. Miyake, A (1979). Mathematical Aspects of Optimal Linear Discriminant Function, Proceeding of COMPSAC '79, 161-166.
2. Miyake, A (1982). On the haar condition in algorithms for the optimal solution of linear inequalities, NATO advanced Study Institute, D. REIDEL.

A METHOD OF DISCRIMINATION FOR ELECTRICAL BIOSIGNAL

Satoki P. Ninomiya

College of Science and Engineering,
Aoyama Gakuin University
6-16-1, Chitosedai, Setagaya-ku, Tokyo, 157 , JAPAN

I. INTRODUCTION

The patterns of electric displacements measured from the human body, (for examples, EEG, ECG, EMG), display various characteristic wave forms. And there are variations in the patterns of those electric displacements.

There exists no general tools to pick out a specific pattern of waveform in electric biosignals.

We intend to investigate the important things in developing a system which are able to detect a specific pattern of electric biosignals through studying spike detection method from electroencephalogram and fetal QRS detection method from fetal electrocardiogram.

II. SPIKE DETECTION

A. Introduction

One of the major neurological conditions in which EEG examination plays a vital role for diagnostic formulation is epilepsy. In this medical condition, the EEG pattern between

ISBN 0-12-215485-1

epileptic seizures often displays spike or spike and wave discharge, arising out of relative quiescent background activity. At times, such discharge appears only infrequently, and EEG examination for a prolonged period becomes necessary. In addition, the quantity of spike discharge present in the background activity provides some measure of theraputic effectiveness of anti-epileptic medication. Therefore, improved spike detection and quantification have a significant clinical implication. So We developed an automatic spike detection method.

B. Method

The EEG data were digitalized at 200 samples per second. Using the following procedure:

1) We computed the regression line

$$y = a(i) + b(i)t \tag{1}$$

with regard to digitalized EEG y(i) (i=1, 2, ... , n), where a(i) is an interception and b(i) is a slope. And we computed a variance S(i) for each regression line.

2) We constructed moment t(j) in which the absolute value of the slope is local maximum for "i" and larger than S_1 which is the boundary value of the slope. These moments are possibly located spikes.

3) We obtained

$$b(t(j))*b(t(j+1)) < 0. \tag{2}$$

Because the sign of the left side slope b(t(j)) of the extreme point of spikes is opposite that of the right side slope b(t(j+1)).

4) The duration of spikes must be smaller than I_1 and larger

than I_2. We tested

$$I_1 > t(j+1) - t(j) > I_2 \tag{3}$$

where I_1 is the overboundary value of spike duration and I_2 is the underboundary value.

5) The variance S(t(j)) of the regression line which is computed at the beginning interval of the spike must be small, because in this interval the change of the EEG is acute and nearly linear. But it depends on the value of the slope, so we must normalize this variance by the absolute value of the slope b(t(j)).

$$S(t(j))/|b(t(j))|^2 < S_2 \tag{4}$$

where S_2 is the overboundary value of normalized variance.

6) We checked the same value for the finishing interval of spikes.

7) The statistics computed in 5) and 6) must be balanced and the summation of those statistics must be small; it is smaller than S_3 ,where S_3 is the overboundary value.

8) The variance for the regression line in the interval which includes the extreme point of spikes must be large; it is larger than S_4 which is the boundary value.

$$S(t(j))+S(t(j)+1))+S(t(j)+L+1) > S_4 \tag{5}$$

9) Using auto-correlation function, we reject α-waves from possible spikes. Because non spikes are included in possible spikes.

C. Result

We analyzed the 40 seconds portion of 16 channels EEG data which were bipolarly recorded from seven epileptic patients who were resting with their eyes closed.

As a result, the accuracy for spike detection is high. Finally errors of the first kind are less than 3% and errors of the second are less than 1%.

III. FETAL QRS DETECTION

A. Introduction

Clinically, the fetal ECG is very important in determining the condition and function of the fetal heart in the mother's body, and records of the variation of the fetal heart rate baseline(FHR-baseline) are currently valued for the diagnosis of the fetal distress. In order, however, to obtain the most accurate records, any variation has to be detected through the study of the fetal electrocardiographic QRS waves. In such cases, it is often impossible to determine the presence of the fetal electrocardiographic QRS waves from leads on the maternal abdomen either because the amplitude is small or noise interferes. Not with standing this, it is vital that clinicians should be able to determine the presence of the fetal electrocardiographic QRS waves.

To this end,we set about to develop a computer-assisted fetal QRS complex detecting system for fetal ECGs with mixture noises.

B. Method

The method of detecting fetal electrocardiographic QRS complex using a computer is as follows.

Fetal ECG was digitalized at the rate of 400 samples per

second and the result obtained was expressed as X(i).
1) A first filter was prepared from X(i) using the method of moving averages for elimination of alternating current.

$$S(i) = \frac{1}{N_1} \sum_{j=0}^{N_1-1} X(i+j) \tag{6}$$

$$(N_1=8, \ i = 1, 2, \ldots)$$

where N_1 is 8.
2) We computed the regression lines

$$S(i) = a(i) + b(i)t \tag{7}$$

by the method of moving the least square with regard to moving averages S(i) obtained from digitalized fetal ECG X(i), where a(i) is an interception and b(i) is a slope. 3) We made B(i) based on a slope b(i).

$$B(i)=SIGN(b(i)) \tag{8}$$

$$\text{where} \quad SIGN(x)= \begin{cases} 1 : x>0 \\ 0 : x=0 \\ -1 : x<0 \end{cases}$$

4) we computed $A_1(i)$;

$$A_1(i) = \sum_{j=0}^{N_2-1} B(i+j) \tag{9}$$

$$(N_2=6, i= 1, \ 2, \ldots\ldots).$$

5) If we could obtain "i" which satisfied the following condition:

$$\begin{cases} A_1(i) \geqq P_1 \\ A_1(i+n) \leqq P_2 \end{cases} \qquad (10)$$

$(P_1 = 4 , P_2 = -4, n=6)$,

we decided that a fetal QRS complex was possibly present at "i".

6) We computed moving variance for S(i), and we recognized the maternal QRS waves. We took the location of maternal QRS wave away from possible QRS waves.

7) We estimated the fetal heart interval from possible QRS waves by using periodgram method which were weighted by slopes of possible QRS waves. And We detected the location of the fetal QRS waves by obtained the heart inteval.

C. Results

We analyzed about the 60 seconds portion of fetal ECG data which were bipolarly recorded from the same maternal wall of the 38th week of gestation. Almost fetal QRS waves could be detected from each fetal ECG except the fetal QRS waves on the maternal QRS waves and the fetal QRS waves with extremely minute amplitude. The fetal QRS waves which were detected by this method coincided with the results of a specialist's eyes.

We calculated fetal R-R intervals by using detected fetal QRS waves and the variation of R-R intervals was shown in Figure 1.

IV. DISCUSSION

In this paper, we proposed two wave detection algorithm. One is spike detection algorithm and the other is the fetal

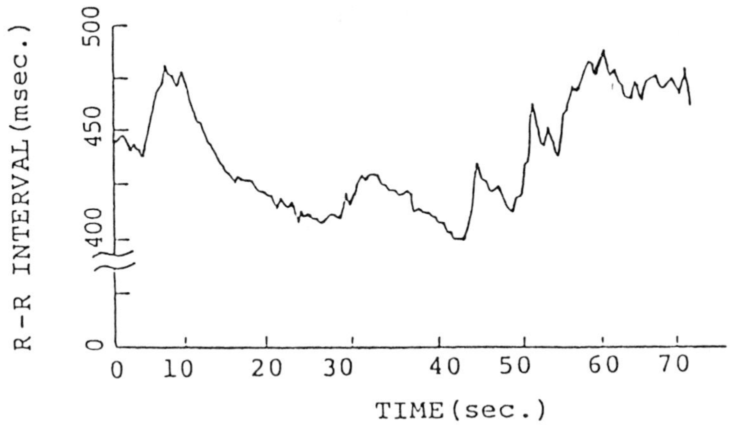

Figure 1: Variation of R-R intervals at 38th week.

QRS wave detection algorithm. The spike detection algorithm is a method to detect waves which appear at random and fetal QRS wave detectionalgorithm is a method to detect the waves which appear periodically. These two methods have something in common with each other in that two wave detection algorithms are based on the knowledge of the specialists. That is to say, the spike detection algorithm is based on the knowledge of an EEG specialist and the fetal QRS wave detection algorithm is also based on the knowledge of an fetal ECG specialist.

When we constructed our methods, we questioned each specialist about the characteristics of the object waves and we introduced above-mentioned statistics and constructed our algorithms based on the answers obtained from each specialist.

By the way, almost wave detection methods are constructed by the following procedure: one or some specialists pick out some samples of the object waves from the wave data and some statistics which can detect those sample of waves are introduced.

So our methods are very different from those methods.

After the investigation on our study, we believe that the method based on the knowledge of the specialists is more useful for the data in which there are variations.

V. CONCLUSION

We found that following problems were significant through this study:

1) Introducing statistics which include some parameters and constructing algorithm, we had better use some of the knowledge of specialists.
2) We should construct a detection algorithm using most suitable statistics.

ACKNOWLEDGEMENT

The EEG data used in this study was obtained from Dr. J. Gotman, Montreal Neurological Instisute and fetal ECGs from Dr. J. Nishimura, Toho University with the authors' acknowledgement.

REFERENCES

1. Yano, F., Ninomija, S. P. and Wada, J. A. (1983). Automatic Spike Detecting System with Global Data, The Proceedings of the 4th World Conference on Medical Informatics, 670-673.
2. Yano, F. and Ninomija, S. P. (1986). A Detecting Method for the Fetal QRS Complex with a Small Amplitude, Journal of Medical Systems, Vol.10, 307-320.

CLASSIFICATION OF BEHAVIORAL STATES OF THE MOUSE

Keiji Yajima

Institute of Japanese Union of Scientists and Engineers
4-3-30, Sendagaya, Shibuya-ku, Tokyo, 151 JAPAN

No established criterion is available for sleep depth in experimental animals. Such criterion exists only for human beings, and, therefore, identification and classification of the stages in animals tend to be subjective depending on the fact that animal has no language.

For the analysis of the free moving mouse, Dr. A. Higashi (1) has developed Dynamic Behavioral Stage Analysis Method/-Micro-sleep and Micro-wakefulness Processing (DBSAM/MMP) system. In his procedure, he focused attention on two kinds of time series, namely the integrated values of EEG (Electroncephalogram) based on the components of 2-4 Hz and 10 msec sampling system, and body movement data sequence transmitted by a compact accelerator fixed above the head of the mouse.

This report deals with a simpler treatment to classify the sleep state and aims at the same results as the DBSAM procedure. The interim results indicate applicability of the simpler method.

I. MICRO STAGE AGGREGATION PROCEDURE

DBSAM applies a micro stage aggregation technique. In consideration of the distribution in body movement measurement, the boundary value for the moving stage is set to, say, 270. A value higher than 270 corresponds to the MV (moving)

ISBN 0-12-215485-1

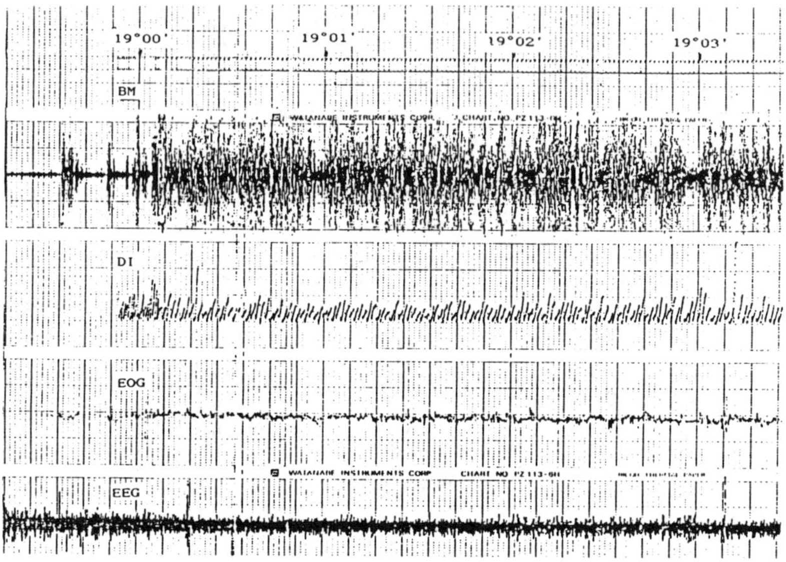

Figure 1 Direct measures of signals from moving mouse

Diagram essentially consists of following two measurements within approximately three minutes. BM: Body movement measures. DI: Integrated EEG at a frequency of 2-4 Hz in every second.

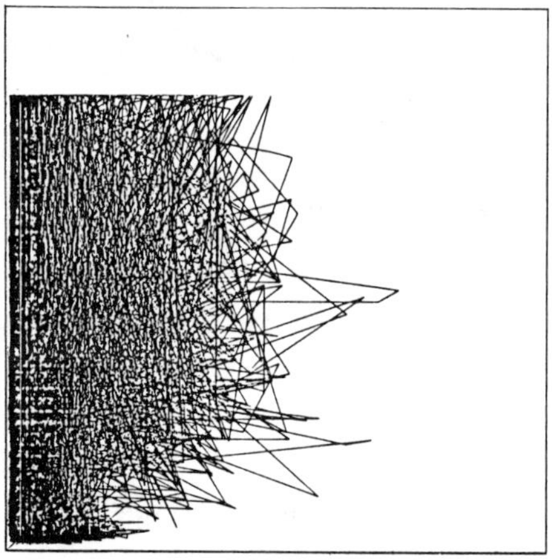

Figure 2 BM-DI Distribution

The x-axis corresponds to DI value and y-axis to BM value. A point representing DI and BM value is plotted continuously according to the time sequence order. Ranges of both axes are identical with Fig. 3 correspondingly.

stage with its domain covering 30% of the total area. The remaining domain where the values are less than 270 forms the non-MV stage area and needs further classification. (Fig. 3)

To classify 70% of the non-MV stages, EEG signals are integrated for each successive second (DI value). Distribution of the consecutive values is expected to lead to a definition of deep sleep. A boundary value of 30, for example, defines the extent of deep sleep. In some cases, a mouse is found to have had 8.04 hours of deep sleep depending on the threshold value.

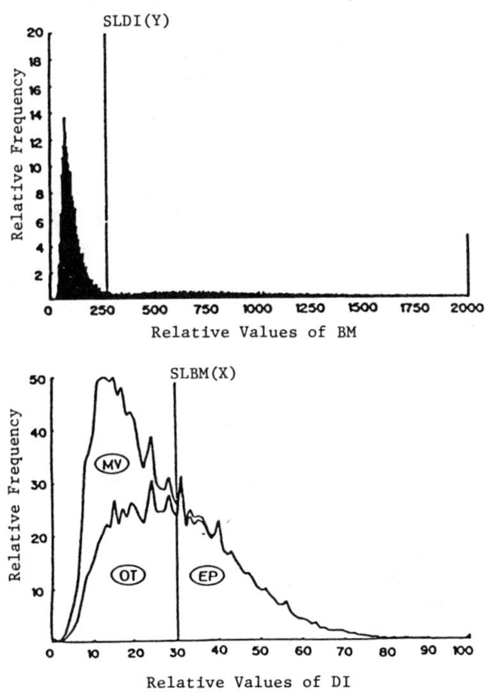

Figure 3 Distribution of BM and DI

Two boundary values are determined according to the characteristics of the respective mice. By assigning values to BM, we can classify the MV stage and non-MV state. Also, by assigning the boundary value, we can have data on deep sleep and transient states.

There are other remaining stage such as outliers, referred to as archfact phenomena, caused by violent body movements of the body and the REM (rapid eye movement) sleep stage which could be detected by EOG (electro ocular graph); thus, REM sleep stage identification could not be completed from the two time sequences.

The OT (Transient) state which belongs to non-MV condition could be classified on the basis of the preceding and succeeding stages with the stages (MV, OT, EP) indicated for each successive second (Fig. 4),

Time	1	2	3	4	5	6	7	8	9	10	11	12	13	14	15
MV (Moving) Stage	o				o			o							o
OT (Transient) Stage		o		o		o	o		o		o	o		o	
EP (Episode) Stage			o							o			o		

Figure 4 Successive stage

1. we define the OT stage as state 1 if this is preceded by the MV stage and succeeded by the EP stage,
2. state 2 is defined if OT is preceded by EP and succeeded by MV,
3. state 3 is defined if OT is preceded and succeeded by MV,
4. state 4 is defined if OT is preceded and succeeded by EP. We, therefore, identify four different OT stages.

It has been pointed out that the micro-sleep states frequently occur during long-term sleep.

DBSAM initiator emphasizes the following points:

(1) Among the various laboratory animals, the mouse has many experimental advantages including the cost advantage.

(2) Various dose effects, including, for example, the sleep-inducing dose effect, could be described by taking the relationship between natural sleep rhythm and the induced effect into consideration.

II. MACRO ANALYSIS

A. Detection of control states

The DBSAM procedure determines for each successive second the stage number adapting the classification algorithm. The control state concept was introduced by taking the period the mouse is awake as first step. As an illustration, Fig. 5 shows that a change of state occurs at 20°03', which agrees with the DBSAM procedure.

This control state method consists simply of elementary quality control:

(a) A mouse is supposed to be awake if the BM value (acceleration speed of the body) exceeds a value 500,

(b) If the BM value exceeds 2000 or the DI (Integrated EEG) value exceeds 200, then it is set as outlier,

(c) From the consecutive DI values, the mean and standard deviation are calculated and if the succeeding DI value exist in three sigma ranges, then the new DI value is pooled into the preceded interval. If it does not exist in the range, we accept it as indicating a change in state.

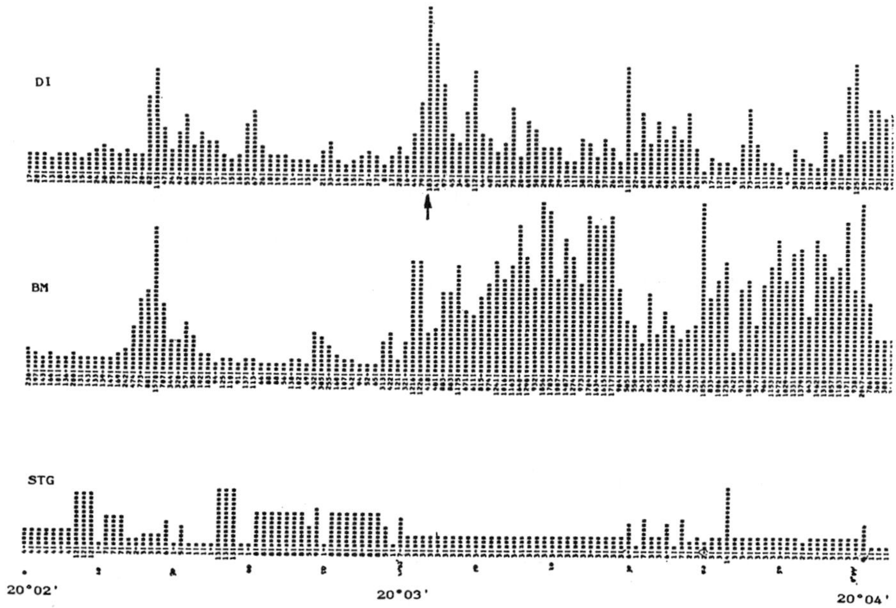

Figure 5 Details of EEG integral (DI) values and body moving (BM) measures

Interval average method introduces a border at approximately 20°03' time where the arrow is assigned. STG: Results of the former analysis (DBSAM method)

Applying this simple method, the entire body of data consisting of 84,600 seconds are divided into 344 stages, so the mean length of the introduced interval is 4 min 11 sec.

B. Unit aggregation and interval division

Test data consists of two values, or one DI value, each second, for a total of 14,400 DI values. Looking at the aggregation process has been tried by using the mean and standard deviation. Aggregation results in 210 stages.

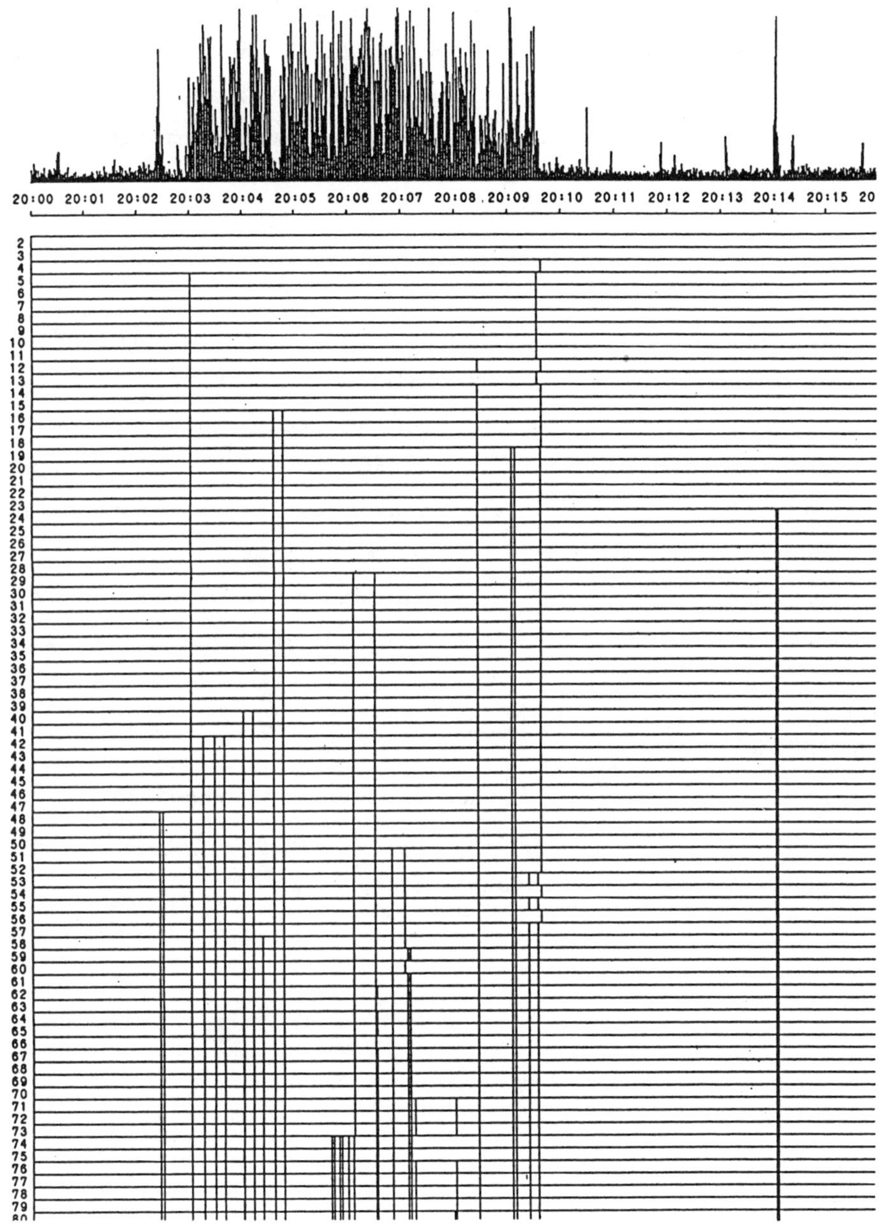

Figure 6 Interval division

The DI values are used to be devised by a method proposed by W. D. Fisher. DI values in this figure belong to similar time range as in Fig. 5.

The DI values are analyzed to be divided into intervals utilizing a method proposed by W. D. Fisher (1958) which relates with a total value of the interval variances. In Fig. 5 and Fig. 6 the major boundary value at 20:03 coincides each other.

C. Recognition of body movement

In the awake stage a mouse starts to move some and it happens that a movement diminishes in short time. A physiologist requires a detection of body movement to be utilized for an analysis of sleep stage and a variance oriented simple interval division process has been applied to the specific time interval. It seems that this method could be applied although the final procedure to determine the classification of body movement is still to be sophiscated.

REFERENCE

1. A. Higashi (1986), Method of Dynamic Behavioral Stage Analysis (DBSAM) in the Free Moving Mouse, Physiology and Behavior, 38, pp. 525-529, 1986.